ANIMAL CELL CULTURE AND PRODUCTION OF BIOLOGICALS

The Third Annual Meeting of the
Japanese Association for
Animal Cell Technology
(JAACT '90)

JAACT '90 Organizers (as of April 1990)

Meeting Chairman: Ryuzo Sasaki (Kyoto University)
Meeting Secretaries: Koji Ikura (Kyoto University)
                     Hiroshi Narita (Kyoto Women's University)
                     Yoshihito Shirai (Kyoto University)

JAACT Committee

Hiroki Murakami   Graduate School of Genetic Resources Technology, Kyushu University
Shuichi Kaminogawa   Department of Agricultural Chemistry, Faculty of Agriculture, University of Tokyo
Nobuo Fujiyoshi   Strategy & Planning Research & Development Division, Kyowa Hakko Co., Ltd.
Shuichi Hashizume   Morinaga Institute of Biological Science
Ryuzo Sasaki   Department of Food Science and Technology, Faculty of Agriculture, Kyoto University
Isao Karube   Research Center for Advanced Science and Technology, University of Tokyo
Yasuo Kitagawa   Institute for Biochemical Regulation, School of Agriculture, Nagoya University
Kazuki Shinohara   National Food Research Institute, Ministry of Agriculture, Forestry and Fisheries
Michiyuki Tokashiki Biotechnology Research Laboratory, Teijin Ltd.
Shun'ichi Dosako   Technical Research Institute, Snow Brand Milk Products Co., Ltd.
Takamoto Suzuki   Pharmaceutical Department, Kirin Brewery Co., Ltd.
Masato Shiraishi   Research Institute, Nichire Corporation
Kuniyo Inouye   Biotechnology Research Laboratories, Tosoh Corporation
Kazuhiro Nagaike   Research Laboratory Research Center, Mitsubishi Kasei Corporation
Takehiro Oshima   Bio pharmatech Center, Suntory Ltd.
Toshiaki Shimada   Pharmacia-LKB Biotechnology Group, Pharmacia K. K.
Kazuhiro Kitano   Applied Microbiology Laboratories, Central Research Division, Takeda Chemical Industries, Ltd.
Kin'ichi Kawamura   Komatsugawa Chemical Engineering Co., Ltd.

Animal Cell Culture
and
Production of Biologicals

Proceedings of the Third Annual Meeting of
the Japanese Association for Animal Cell Technology,
held in Kyoto, December 11–13, 1990

Edited by

RYUZO SASAKI

and

KOJI IKURA

*Department of Food Science and Technology,
Faculty of Agriculture,
Kyoto University, Kyoto, Japan*

SPRINGER SCIENCE+BUSINESS MEDIA, B.V.

Library of Congress Cataloging-in-Publication Data

Japanese Association for Animal Cell Technology. Meeting (3rd : 1990
: Kyoto, Japan)
    Animal cell culture and production of biologicals : proceedings of
the Third Annual Meeting of the Japanese Association for Animal Cell
Technology, held in Kyoto, December 11-13, 1990 / editors, Ryuzo
Sasaki, Koji Ikura.
        p.   cm.
    Includes index.
    ISBN 978-94-010-5572-7   ISBN 978-94-011-3550-4 (eBook)
    DOI 10.1007/978-94-011-3550-4
    1. Animal cell biotechnology--Congresses. 2. Cell culture-
-Congresses.   I. Sasaki, Ryūzō, 1938-   . II. Ikura, Kōji, 1947-
. III. Title.
TP248.27.A53J36   1990
660'.6--dc20   91-20694
                                                            CIP
ISBN 978-94-010-5572-7

Printed on acid-free paper

Contents

IV. PHYSICOCHEMICAL AND BIOCHEMICAL FACTORS FOR
CELL GROWTH AND PRODUCTION OF BIOLOGICALS

Preface

In the past two decades, the importance of animal cell technology has increased enormously. First, useful proteins can be produced by cultured animal cells, in which the desired product can be modified and organized so as to retain its biological function. Second, studies of cultured cells can provide information needed to understand molecular mechanisms that govern what happens in tissues, organs, and even entire organisms. For this second purpose, biochemists and molecular biologists may need a large number of such cells. Third, cultured cells can be used instead of tissues and organs clinically.

The Third Annual Meeting of the Japanese Association for Animal Cell Technology (JAACT), at which participants from abroad were warmly welcomed, was held in Kyoto on December 11–13, 1990. It was organized around the idea of providing a place for the review of much new data on such applications of cultured cells and for exchanges of the views of the participants about progress in the field. This volume, divided into seven sections, contains the proceedings of the meeting. The first section reviews the molecular basis of the control of animal cell growth. In the following sections, physicochemical and biochemical factors for cell growth and production of biologicals, cell culture systems including serum-free culture, new cell lines, specific products and their characteristics, and *in vitro* assays for toxic, carcinogenic, and pharmacological effects are taken up in their turn. The final section presents current views of regulations and guidelines for new substances to be produced by the use of animal cells. It is hoped that this volume will stimulate much thought about animal cell technology, helping to catalyze the rapid development of the field.

The editors express their gratitude to Drs. Hiroshi Narita of Kyoto Women's University and Yoshihito Shirai of Kyoto University for their help in organizing the meeting. We also thank Mr. Taketoshi Furudera of the Japan Bioindustry Association for his efforts in support of the special symposium on regulations and guidelines for new bioproducts, concerning which three papers are included in the final section.

The Editors

NOVEL SPECIFIC INHIBITORS FOR ANALYSIS OF EUKARYOTIC CELL CYCLE CONTROL

Teruhiko Beppu and Minoru Yoshida
Agricultural Chemistry, Faculty of Agriculture, The
University of Tokyo
1-1-1 Yayoi, Bunkyo-ku, Tokyo 113, Japan

ABSTRACT. Trichostatin A (TSA), isolated as a potent inducer of differentiation in murine erythroleukemia cells, caused reversible arrest of rat 3Y1 fibroblast proliferation in G1 and G2 phases without any inhibition of macromolecular synthesis. Leptomycin B (LMB), isolated as an agent inducing morphological abnormalities in fungi, also blocks the cell cycle of 3Y1 and fission yeast cells in both G1 and G2. Cells released from the G2-arrest by both drugs entered a new S phase without passage through M phase, resulting in the formation of proliferative tetraploid cells. On the other hand, staurosporine, which we rediscovered as a new inhibitor of the cell cycle blocking G1 and G2, induced synchronous G2/M transition after release from the G2 arrest. These results suggest the presence of two distinct stage in G2, which is divided by a putative commitment point for mitosis. Biochemical and genetic analyses using their resistant mutants suggest that the molecular target of TSA is the histone deacetylase while that of LMB is involved in the cdc2 regulatory network.

Introduction

The major events in the eukaryotic cell cycle, DNA replication in S phase and mitotic cell division in M phase, are separated by two gap phases, G1 and G2. Two general regulatory points have been identified in the cell cycle, one in G1 controlling entry into S phase and another in late G2 controlling entry into mitosis. These controls have been described in a wide range of cells, from yeast to human, suggesting that they are fundamental features of the eukaryotic cell cycle [1-4]. For instance the cdc2[+] gene that encodes the p34 protein kinase has been established as a control gene operating in both G1 and G2 phases in the fission yeast Schizosaccharomyces pombe [5-7]. Structurally similar and functionally interchangeable genes were also identified in the budding yeast Saccharomyces cerevisiae [8] and human cells [9], and confirmed as encoding similar protein kinases [10, 11]. These results together with a recent finding that the p34 protein kinase is a component of the maturation-promoting factor in frog [12],

1

R. Sasaki and K. Ikura (eds.), Animal Cell Culture and Production of Biologicals, 1–10.
© 1991 Kluwer Academic Publishers.

starfish [13], and clam oocytes [14] led to the proposal that p34 acts as a universal regulator of G2/M in all eukaryotic cells [15]. In addition, there are many indications that regulatory mechanisms in G1 and G2 phases might play an essential role in controlling cellular differentiation and carcinogenesis.

For further analysis of the mechanism of the cell cycle regulation, inhibitors of the functions of these general control genes might be useful. Such types of inhibitors are also expected to become selective antitumor agents, since abnormal regulation of the cell cycle is suggested to be responsible for the transformed phenotype in many tumor cell lines.

Trichostatins and leptomycins were the products of Streptomyces strains isolated as potent inducers of Friend erythroleukemia cells [16, 17] and morphological abnormalities in the lower eukaryotic microorganisms [18, 19], respectively. Both of these agents were found to possess new activities of causing the reversible arrest of the eukaryotic cell cycle in both G1 and G2 phases and inducing the formation of proliferative tetraploid cell after removal of the drugs from the G2-arrested culture [20, 21]. During the course of our new screening for cell cycle inhibitors using flow cytometry, staurosporine, which had been reported as a potent inhibitor of protein kinase C [22] and other kinases [23], was found to cause G1 and G2 blocks similar to that by TSA or LMB. However, the cells arrested at G2 phase by staurosporine proceeded to mitosis synchronously upon removal of the drug in contrast to TSA or LMB [24]. Herein we describe the characteristic modes of action of these agents and discuss their usefulness for the analysis of cell cycle control.

Trichostatins

The murine erythroleukemia cell line established by Friend et

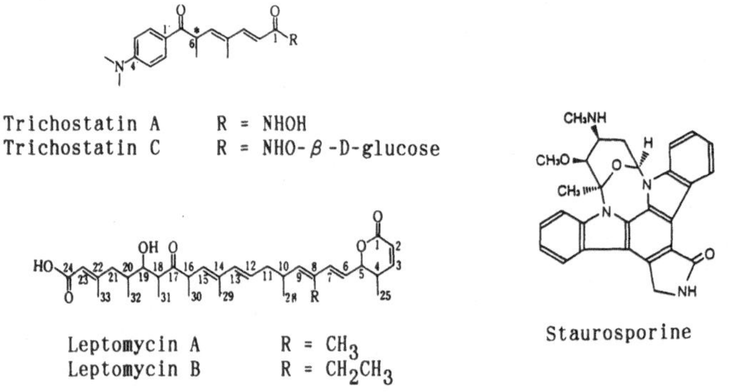

Fig. 1. structures of trichostatins, leptomycins, and staurosporine.

al. [25] consists of transformed cells in which erythroid differentiation is blocked by infection with Friend virus. Differentiation is reinitiated effectively by various polar compounds such as dimethyl sulfoxide (DMSO) [25]. The cell line has been used as a model system for eukaryotic cell differentiation as well as an assay system for screening new agents with differentiation-inducing activities.

As a result of such screening with microbial metabolites, we detected the potent activity in a broth of a strain which was identified to be Streptomyces platensis. The active agents were found to be trichostatin A (TSA) and trichostatin C (TSC) (Fig. 1) [16, 17], which had been reported to be antifungal antibiotics. TSA and TSC induced efficient differentiation of Friend cells with the maximum induction of about 80% benzidine-positive cells at 0.015 μM and 0.5 μM, respectively. These compounds have a chiral center and recently Mori and Koseki succeeded in the chemical synthesis of both enantiomers of TSA [26]. We showed that an unnatural (S)-enantiomer of TSA was totally inactive, suggesting that a target molecule with strict stereospecificity is involved in the biological effects [27].

During the course of these studies, we observed that TSA caused inhibition of cell growth without any inhibitory effect on macromolecular synthesis. Detailed analysis of the inhibitory effect was performed with proliferating rat 3Y1 fibroblast cells [20].

When the cells arrested in G0 phase by serum starvation were transferred to fresh minimum essential medium containing 12% fetal calf serum, proliferation of the cells started synchronously from G0 and proceeded through G1 to S phase. Addition of TSA to the synchronous culture within 6 h after the serum stimulation caused complete inhibition of entering S phase, indicating that the arrest occurred in early G1 phase.

On the other hand, TSA was found to cause arrest in G2 phase in

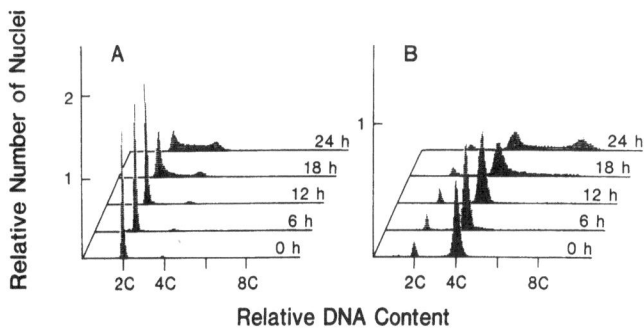

Fig. 2.   Distribution of DNA contents in cultures released from TSA-arrest.   At time 0, TSA was removed from the G1-arrested culture (A) and the G2-arrested culture (B).

the synchronous culture starting from early S phase. Proliferating
3Y1 cells were accumulated in early S phase by incubation with 1 mM
hydroxyurea. Then the cells were released from the hydroxyurea-arrest
and challenged with TSA. Distribution of DNA contents in nuclei in
the cell population was analyzed during the following cultivation by
flow cytometry. TSA did not inhibit DNA synthesis in S phase as
suggested by transition of the 2C peak toward 4C, but completely
blocked the following process to produce the 2C peak in the control
culture. Thus it is evident that TSA causes arrest of the cell cycle
in both gap phases.

Another striking effect of TSA was found when the cells arrested
in G2 phase by TSA were released from the inhibition. In such a
culture, almost all the cells with 4C DNA skipped M phase and entered
S phase after an 18-h lag, resulting in the formation of proliferative
tetraploid cells (Fig. 2B). This characteristic mode of action of TSA
on the cell cycle can be illustrated as in Fig. 3.

Recently, we found that TSA caused an accumulation of acetylated
histone species in a variety of mammalian cell lines. Pulse-labeling
experiments indicated that the release of acetyl groups in histones
was blocked by TSA, and in vitro experiments using the partially
purified histone deacetylase clearly showed that TSA was a potent and
specific inhibitor of the enzyme. Furthermore, newly isolated mutant
cells resistant to TSA did not show the accumulation of the acetylated
histones in the presence of TSA. The histone deacetylase from the
mutant showed decreased sensitivity to TSA. These results indicate
that the inhibition of the enzyme is the primary reason for all of the
in vivo effects of TSA [28].

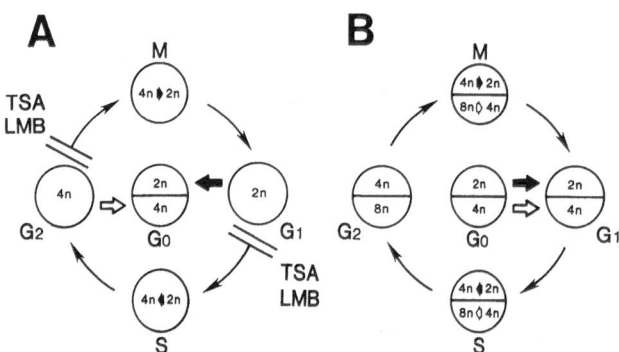

Fig. 3. Possible pathways of TSA- or LMB- arrested cells. TSA or LMB
blocks the cell cycle at both G1 and G2, and then cells with 2 n and 4
n DNA are introduced into G0 (A). After release, both of them
reinitiate their growth through G1 (B).

Leptomycins

In order to find a new type of antifungal antibiotics, we had
conducted a screening project to detect activity of inducing
morphological abnormalities in various fungi and yeast cells [29].
Leptomycins A and B (LMA, LMB) (Fig. 1) were obtained from the broth
of a Streptomyces strain as the agents inducing cell elongation of
fission yeast Schizosaccharomyces pombe [18, 19].

LMB caused growth inhibition of S. pombe following elongation of
cells at 0.02 μM, while only slight inhibition of DNA and RNA
synthesis was observed at this concentration. Addition of LMB to a
synchronous culture in G2 phase blocked subsequent events in the cell
cycle. Analysis of the effect of LMB on cdc2 and cdc7 mutants also
supported the G2-arrest by LMB in S. pombe [30]. In addition, recent
results revealed that LMB caused arrest of the cell cycle not only in
G2 phase but also in G1 phase in S. pombe [21]. The results with S.
pombe described above prompted us to examine whether LMB affects the
mammalian cell cycle. Flow cytometric analyses using both G0- and
early S-synchronous cultures of rat 3Y1 fibroblasts revealed that LMB
blocked the cell cycle in both G1 and G2 at the concentration of 0.001
μM. Removal of LMB from the G2-arrested culture also induced the
formation of proliferative tetraploid cells [21]. Although the
apparent mode of action of LMB closely resembles that of TSA (Fig. 3),
several results indicated that their targets are different.

A recent important observation is that the cdc mutants of S.
pombe carrying mutations in the genes of the cdc2 regulatory network
show abnormal sensitivity to LMB. We have newly isolated LMB-
resistant mutants which showed the cdc phenotype at the nonpermissive
temperature. These mutants belonged to the complementation groups
different from known cdc mutants. These results suggest that the
molecular target of LMB may be the product of an unidentified gene
involved in the cdc2 regulatory network. Cloning and sequencing of
the genes that confer LMB-resistance to wild-type cells are in
progress.

Staurosporine

By monitoring the effects of various microbial metabolites on the cell
cycle of normal rat fibroblasts (3Y1) with flow cytometry, we found a
metabolite of a Streptomyces strain, which cause a characteristic
arrest of the cell cycle without any inhibitory effect on DNA
synthesis. The agent was soon identified as staurosporine, which had
originally been described as a microbial alkaloid by Omura et al. [31]
and later revealed to be a potent inhibitor of protein kinases [22,
23]. Analyses of the effect of staurosporine on the cell cycle
progression with G0- and early S-synchronous cultures revealed that a
low concentration (0.02 μM) of the drug caused the early G1 arrest
whereas a higher concentration (0.2 μM) blocked the cell cycle at late
G2 [24]. An important observation was that the agent induced

synchronous G2/M transition from the G2 block after removal. When the cells arrested at G2 were released by washing, round mitotic cells appeared 30 min and the mitotic index reached a maximum at 60 min. Finally, more than 90% of the cells completed cell division and the cell number almost doubled within 3 h (Fig. 4). These released cells continued to grow until reaching confluency with a normal growth rate and a significant decrease in the plating efficiencies was not observed before or after staurosporine removal. These results suggest that the reversible arrest of G2 by staurosporine can provide a useful system to obtain a large population of cells that progress synchronously through G2/M without extensive cell death.

   The thus obtained synchronous culture from G2 enable us to analyze the effects of other cell cycle inhibitors such as TSA and LMB. These agents along with Colcemid, an inhibitor of the microtubule assembly, were added to the G2-synchronous culture at 0 h after release from staurosporine-arrest and the patterns of DNA contents at 6 h were analyzed. Colcemid blocked the progression at 4C DNA state. However, no inhibitory effect on the transition from G2 through M to G1 was observed with TSA and LMB. These results indicated that the arrest point of staurosporine exists at late G2 phase after those of TSA or LMB [24].

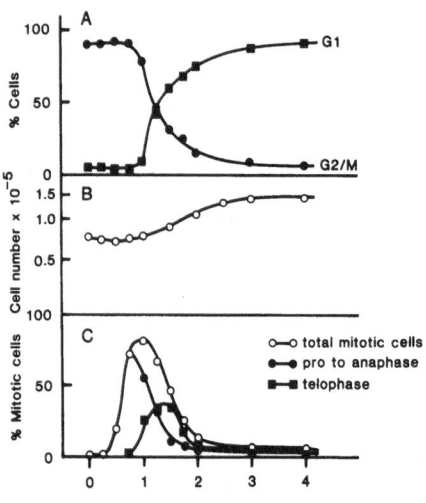

Fig. 4. Synchronous progression of G2/M phase after removal of staurosporine. The G2-arrested culture was released and then cell populations in the cell cycle (A), cell numbers (B), and mitotic indices (C) were determined.

Discussion

TSA and LMB were found to block the cell cycle progression specifically in both G1 and G2 phases without any inhibition of macromolecular synthesis. It seems probable that TSA and LMB attack different target molecules, both of which are commonly involved in the progression of G1 and G2. Both agents have the characteristic activity of inducing the formation of proliferative tetraploid cells from normal diploid fibroblasts after release from the G2-arrest. Several results suggest that these drugs introduce the growing cells into G0 phase even from G2 and reinitiate the cell cycle from G0 through G1 upon removal of the drugs (Fig. 3). Such a direct pathway from G2 through G0 to G1 implies the existence of an identical cellular state in both G1 and G2.

On the other hand, a high concentration of staurosporine, a potent inhibitor of protein kinases, caused a reversible G2 arrest without introducing the cells into G0 phase. The results obtained by the combination use of these drugs clearly showed that TSA or LMB could no longer inhibit the G2 progression after staurosporine removal. Different modes of action between these drugs suggest the presence of at least two stages divided by a putative commitment point in G2, the former of which can be converted to G1 of the tetraploid cell cycle during the arrest, whereas the latter is committed to enter mitosis.

The molecular target of TSA was identified to be the mammalian histone deacetylase. Potent inhibition of the enzyme by TSA causes histone hyperacetylation in nuclei. At present, it is still unclear how histone hyperacetylation leads to specific inhibition of the cell cycle. However, it seems reasonable that hyperacetylation of histones causes relaxation of the chromatin structure to make a variety of factors accessible to DNA, as suggested by many studies on chromatin structure-function relationship [32]. Therefore, the enzymes controlling the equilibrium of histone acetylation may play an important role in the regulation of cell cycle-dependent gene expression. In the case of LMB, the exact target molecule has not yet been determined. However, several observations with cdc mutants and LMB-resistant mutants strongly suggest its involvement in the cdc2 regulatory network. Structural analysis of genes for the LMB-resistance will help to elucidate the molecular action of LMB. We expect that a new class of growth-associated genes will be identified.

The most likely target of staurosporine in G2 may be p34 protein kinase encoded by the general cell cycle regulatory gene $cdc2^+$, whose activity increases in G2 and M. Another candidate may be protein kinase C, which is shown to be essential for G2/M transition in yeast [33]. Further analyses of the effects on protein phosphorylation are in progress.

Although the effects of TSA and LMB on the cell cycle of normal fibroblasts were reversible, they caused irreversible killing of SV40-transformed fibroblasts. These results suggest the possibility that

G1- and/or G2-specific inhibitors might be selective antitumor agents. Actually, LMB showed distinct antitumor activity against several transplantable murine tumors in vivo [34]. It seems probable that the past screening project to find potent killing agents for tumor cells has facilitated the detection of the DNA-attacking agents but missed another type of cell cycle inhibitors. Complicated regulatory machineries operating in G1 and G2 phases might be targets of a variety of new agents, whose possibilities in cell biology and tumor chemotherapy have now turn clear and will become fruitful in the future.

References

1. Prescott, D. M. (1976) Reproduction of Eukaryotic Cells, Academic Press, New York.
2. Gelfant, S (1977) 'A new concept of tissue and tumor cell proliferation', Cancer Res. 37, 3845-3862.
3. Pardee, A. B., Dubrow, R., Hamlin, J. L., and Kletzien, R. F. (1978) 'Animal cell cycle' Annu. Rev. Biochem. 47, 715-750.
4. Lee, M. and Nurse, P. (1988) ' Cell cycle control genes in fission yeast and mammalian cells', Trends Genet. 4, 287-290.
5. Nurse, P. and Bissett, T. (1981) 'Gene required in G1 for commitment to cell cycle and in G2 for control of mitosis in fission yeast', Nature 292, 558-560.
6. Costello, G., Rodgers, L., and Beach, D. (1986) 'Fission yeast enters the stationary phase G0 state from either mitotic G1 or G2', Curr. Genet. 11, 119-125.
7. Moreno, S., Hayles, J., and Nurse, P. (1989) 'Regulation of p34^{cdc2} protein kinase during mitosis', Cell 58, 361-372.
8. Beach, D., Durkacz, B., and Nurse, P. (1982) 'Functionally homologous cell cycle control genes in budding and fission yeast', Nature 300, 706-709.
9. Lee, M. G. and Nurse, P. (1987) 'Complementation used to clone a human homologue of the fission yeast cell cycle control gene cdc2', Nature 327, 31-35.
10. Lorincz, A. T. and Reed, S. I. (1984) 'Primary structure homology between the product of yeast cell division control gene CDC28 and vertebrate oncogenes', Nature 307, 183-185.
11. Draetta, G. and Beach, D. (1988) 'Activation of cdc2 protein kinase during mitosis in human cells: cell cycle-dependent phosphorylation and subunit rearrangement', Cell 54, 17-26.
12. Dunphy, W. G., Brizuela, L., Beach, D., and Newport, J. (1988) 'The Xenopus cdc2 protein is a component of MPF, a cytoplasmic regulator of mitosis', Cell 54, 423-431.
13. Labbe, J. C., Lee, M. G., Nurse, P., Picard, A., and Doree, M. (1988) 'Activation at M-phase of a protein kinase encoded by a starfish homologue of the cell cycle control gene cdc2$^+$', Nature 335, 251-254.
14. Draetta, G., Luca, F., Westendorf, J., Brizuela, L., Ruderman,

J., and Beach, D. (1989) 'cdc2 protein kinase is complexed with both cyclin A and B: evidence for proteolytic inactivation of MPF', Cell 56, 829-838.

15. Nurse, P. (1990) 'Universal control mechanism regulating onset of M-phase', Nature 344, 503-508.

16. Yoshida, M., Iwamoto, Y., Uozumi, T., and Beppu, T. (1985) 'Trichostatin C, a new inducer of differentiation of Friend leukemic cells', Agric. Biol. Chem. 49, 563-565.

17. Yoshida, M., Nomura, S., and Beppu, T. (1987) 'Effects of trichostatins on differentiation of murine erythroleukemia cells', Cancer Res. 47, 3688-3691.

18. Hamamoto, T., Gunji, S., Tsuji, H., and Beppu, T. (1983) 'Leptomycins A and B, new antifungal antibiotics I. Taxonomy of the producing strain and their fermentation, purification and characterization', J. Antibiot. 36, 639-645.

19. Hamamoto, T., Seto, H., and Beppu, T. (1983) 'Leptomycins A and B, New antifungal antibiotics II. structure elucidation', J. Antibiot. 36, 646-650.

20. Yoshida, M. and Beppu, T. (1988) 'Reversible arrest of proliferation of rat 3Y1 fibroblasts in both the G1 and G2 phases by trichostatin A', Exp. Cell Res. 177, 122-131.

21. Yoshida, M., Nishikawa, M., Nishi, K., Abe, K., Horinouchi, S., and Beppu, T. (1990) 'Effects of leptomycin B on the cell cycle of fibroblasts and fission yeast cells', Exp. Cell Res. 187, 150-156.

22. Tamaoki, T., Nomoto, H., Takahashi, I. Kato, Y., Morimoto, M., and Tomita, F. (1986) 'Staurosporine, a potent inhibitor of phospholipid/Ca^{2+} dependent protein kinase', Biochem. Biophys. Res. Commun. 135, 397-402.

23. Davis, P. D., Hill, C. H., Keech, E., Lawton, G., Nixon, J. S., Sedwick, A. D., Wadsworth, J., Westmacott, D., and Wilkinson, S. E. (1989) 'Potent selective inhibitor of protein kinase C', FEBS Lett. 259, 61-63.

24. Abe, K., Yoshida, M., Usui, T., Horinouchi, S., and Beppu, T. (1991) 'Highly synchronous culture of fibroblasts from G2 block caused by staurosporine, a potent inhibitor of protein kinases', Exp. Cell Res. 192, 122-127.

25. Friend, C., Patuleia, M. C., and deHarven, E. (1966) 'Erythrocytic maturation in vitro of murine (Friend) virus-induced leukemia cells', Natl. Cancer Inst. Monogr., 228, 505-520.

26. Mori, K. and Koseki, K. (1988) 'Synthesis of trichostatin A, a potent differentiation inducer of Friend leukemic cells, and its antipode', Tetrahedron 44, 6013-6020.

27. Yoshida, M., Hoshikawa, Y., Koseki, K., Mori, K., and Beppu, T. (1990) 'Structural specificity for biological activity of trichostatin A, a specific inhibitor of mammalian cell cycle with potent differentiation-inducing activity in Friend leukemia cells', J. Antibiot. 43, 1101-1106.

10

28. Yoshida, M., Kijima, M., Akita, M., and Beppu, T. (1990) 'Potent and specific inhibition of mammalian histone deacetylase both in vivo and in vitro by trichostatin A', J. Biol. Chem. 256, 17174-17179.
29. Gunji, S., Arima, K., and Beppu, T. (1983) 'Screening of antifungal antibiotics according to activities inducing morphological abnormalities', Agric, Biol. Chem. 47, 2061-2069.
30. Hamamoto, T., Uozumi, T., and Beppu, T. (1985) 'Leptomycins A and B, new antifungal antibiotics III. mode of action of leptomycin B on Schizosaccharomyces pombe', J. Antibiot. 38, 1573-1580.
31. Omura, S., Iwai, Y., Hirano, A., Nakagawa, A., Awaya, J., Tsuchiya, H., Takahashi, Y., and Masuma, R. (1977) 'A new alkaloid AM-2282 of Streptomyces origin. Taxonomy, fermentation, isolation and preliminary characterization', J. Antibiot. 30, 275-282.
32. Csordas, A. (1990) 'On the biological role of histone acetylation', Biochem. J. 265, 23-38.
33. Levin, D. E., Fields, F. O., Kunisawa, R., Bishop, J. M., and Thorner, J. (1990) 'A candidate protein kinase C gene PKC1, is required for the S. cerevisiae cell cycle', Cell 62, 213-224.
34. Komiyama, K., Okada, K., Tomisaka, S., Umezawa, I., Hamamoto, T., and Beppu, T. (1985) 'Antitumor activity of leptomycin B', J. Antibiot. 38, 427-429.

TRANSGENIC ANIMALS AS A SOURCE OF GENETICALLY-ENGINEERED TRANS-IMMORTALISED CELL LINES.

S. JALLAT, F. PERRAUD, W. DALEMANS, A. BALLAND, T. FAURE, P. MEULIEN[1] and A. PAVIRANI.
Department of Animal Systems and Department of Analytical Biochemistry, TRANSGENE, 11 rue de Molsheim, 67082 STRASBOURG CEDEX, FRANCE.
[1]*Present address : TRANSGENE-ICGM, 22 rue Méchain, 75014 PARIS, FRANCE.*

ABSTRACT. We have evaluated whether transgenic mouse technology could be used to derive immortalised recombinant cell lines stably expressing heterologous proteins, such as human α1-antitrypsin and human factor IX. In this system, the expression of an onc gene and the gene of interest is targetted to a specific cell type in transgenic mice using tissue-specific regulatory DNA sequences. This leads to the generation of specific tumours : onc gene-induced lymphomas and hepatomas. The tumour cells are then cultured and permanent producer cell lines are established. Under appropriate culture conditions, several cell lines derived from trans-hepatomas maintain a differentiated phenotype, and are capable of expressing biologically active and correctly processed factor IX.

1. Introduction

At present, several recombinant proteins destined for therapeutic use in humans are produced in mammalian cells, the only host system capable of ensuring full activity of the recombinant product. In certain cases however, expression of human proteins in heterologous cells can result in incorrect posttranslational modifications of the recombinant molecule. This is particularly true in the case of proteins requiring complex processing. For example recombinant coagulation Factor IX (FIX), which undergoes at least four posttranslational modifications in vivo, all of which being essential for biological activity (Jallat et al., 1990a), has been expressed by many groups using a variety of vectors and host cell types (Anson et al., 1985; Busby et al., 1985; De La Salle et al., 1985 ; Kaufman et al., 1986 ; Choo et al., 1987 ; Balland et al., 1988; Rees et al., 1988; Derian et al., 1989). High expression levels of human FIX have been obtained in CHO cells, but only partial biological activity has been demonstrated (Kaufman et al., 1986). This is partly due to the inefficient posttranslation processing performed by the heterologous host cell. It would be therefore advisable to express complex recombinant proteins in a more specialised cell. Unfortunately, such cell types are not always available as established cell lines.

In this paper, we have investigated the potential of transgenic mice as a source of novel cell lines engineered for the expression of complex proteins. This approach, which we have termed trans-immortalisation (Pavirani et al., 1989) consists in the co-expression of an onc gene and the gene of interest in a defined target cell type of the transgenic mice. This can lead to the generation

R. Sasaki and K. Ikura (eds.), Animal Cell Culture and Production of Biologicals, 11–19.
© 1991 *Kluwer Academic Publishers.*

of tumours that secrete the recombinant human (rh) protein. Cells derived from such tumours are then cultured <u>in vitro</u> resulting into the establishment of permanent cell lines secreting the protein of interest. At the present time, we have been able to obtain cell lines from transgenic lymphomas (trans-lymphomas) and hepatomas (trans-hepatomas) that have been studied in terms of differentiation state and expression of rh proteins such as the anti-protease α1-antitrypsin (α1AT) and FIX.

2. Material and Methods

2.1. CONSTRUCTION OF THE TRANSGENES

2.1.1. *Lymphoid Cell Targetting*. The cDNA for the human α1AT Pittsburgh variant containing a Met to Arg substitution at amino acid position 358 [α1AT (Arg)] (Courtney et al., 1985) was linked to a DNA fragment containing the mouse immunoglobin (Ig)k light chain promoter (Pk) and the Igk signal coding region which contains the first intron of the Igk gene. DNA fusion between α1AT (Arg) and the Igk signal sequence was made at the signal cleavage site so that precise processing of the primary translated product could release the mature α1AT sequence. The murine c-<u>myc</u> containing only the coding exons 2 and 3, or the SV40T-antigen <u>onc</u> genes were inserted downstream from the mouse Igμ heavy chain promoter (Pμ). The two expression units were placed in divergent organisation under the control of the mouse Ig heavy chain enhancer (Eμ).

2.1.2. *Liver Cell Targetting*. pTG2984 (Dalemans et al., 1990) was constructed by first subcloning into ppolyIII-I* (Lathe et al., 1987) a 1.8 kilobase (kb) KpnI fragment corresponding to the 5' portion of the α1AT gene. This fragment included 1.5 kb of 5' upstream sequence from the cap site, the entire non-coding first exon and 0.22 kb of the first intron. A XbaI fragment bearing the second and third exons of the murine c-<u>myc</u> gene (Bernard et al., 1983) was inserted into the XbaI site of the previous construct using a BamHI-XbaI adapter. A synthetic acceptor site derived from the sequence of the 1st intron of the murine c-<u>myc</u> was inserted upstream of the c-<u>myc</u> translation initiation codon located in the second exon. Such fusion creates a functional hybrid α1AT-c-<u>myc</u> intron. A 16 kb SalI fragment containing the entire hα1AT gene with 1.5 kb and 4 kb of 5' and 3' flanking sequences, respectively, was then introduced downstream of the c-<u>myc</u> expression block.

In pTG4912 (Perraud et al., submitted) SV40T-antigen transcription is driven by the hα1AT promoter. This vector was generated by replacing the SV40 early promoter region of pTG173 (Jallat et al., 1990b) with that of the hα1AT gene (from nucleotide (nt) - 1542 relative to the cap site).

pTG3960 (Jallat et al., 1990a) consisted of the FIX genomic sequences including 5 kb and 0.3 kb of 5' and 3' flanking sequences, respectively. The human FIX gene was cloned from a λEMBL3 library made from a 4XY chromosome lymphoblastoid cell line (Pavirani et al., 1987). This construct contains the majority of the FIX gene with the exception of 4.8 kb and 7.1 kb of internal sequences in intron A and intron F, respectively.

2.2. GENERATION AND ANALYSIS OF TRANSGENIC MICE

Purified linear fragments devoid of vector DNA were microinjected into the male pronucleus of fertilized eggs from F1 C57B1/6xSJL hybrid crosses as described previously (Hogan et al., 1986). Detection of transgenic mice among the offspring was made by dot blot hybridization of tail DNA with DNA probes specific for each transgene (Palmiter et al., 1982).
Quantification of the secreted rh proteins (FIX and α1AT) and of mouse albumin was made by enzyme-linked immunoadsorbent assay (ELISA) using antisera against hα1AT (Cappel, Malvern, PA, USA), hFIX (kit Diagnostica Stago, Asnière, France) or mouse albumin (Nordic Immunology, Tilburg, Netherlands). FIX procoagulant activity was determined in transgenic plasma samples using a Cephalin kaolin one stage assay as described by Austen and Rhymes (1975). Immunopurification and N-terminal amino acid sequencing of rhFIX was performed as described in Meulien et al., (1990).

2.3. CULTURE CONDITIONS FOR TRANS-HEPATIC CELLS

Tumoral liver nodules were collected from transgenic mice and treated with collagenase. The cells were cultured in William's E medium containing 5 % of dialysed FCS, 200 mM of glutamine, 25 ng/ml of EGF, 10 μg/ml of insuline, 10 μg/ml of transferrin and 300 ng/ml of hydrocortisone. Vitamin K1 (20 μg/ml) was added to the culture medium before measuring clotting activity of rhFIX.

2.4. RNA ANALYSIS

Expression of the mRNA for major hepatic markers was studied by dot blot analysis. Sequences of the specific oligonucleotides used as hybridization probes are given in Dalemans et al., (1990) and Jallat et al., (1990a).

3. Results

3.1. TRANS-LYMPHOMA-DERIVED CELL LINES

Immunoglobulin regulatory DNA sequences were chosen to target co-expression of an exogeneous human protein [the Pittsburgh variant of α1AT (α1AT Arg)] and of an onc gene (murine c-myc or SV40T-antigen) to the lymphocytes of transgenic mice. For this purpose Pμ and Pk were used to drive onc gene expression and α1AT (Arg) expression, respectively. The two promoters were placed adjacent to a centrally located Eμ element for maximal enhancer effect.
Transgenic mice expressing either c-myc or SV40T-antigen developed tumours at high

frequency. The mice displayed multiple abnormalities of lymphoid tissues (spleen, lymph nodes, mesenteric ganglia and occasionally thymus) which were histologically characterised as lymphomas. When tumours were injected into syngenic mice, a tumour pathology identical to that found in the transgenic mice was observed. Malignancy was confirmed by tumour propagation. c-myc mice displayed a more agressive malignancy compared to SV40T-antigen mice : tumours developed more rapidly, were more invasive and were more readily propagated.

Founder transgenic mice had plasma α1AT (Arg) levels of up to 20 μg/ml. Plasma α1AT (Arg) concentration paralleled the increase in tumour mass size. Furthermore, similar levels of circulating α1AT were measured in syngenic mice in which tumours had been propagated. These observations confirmed that tumour cells secreted α1AT (Arg).

High expression of the mRNA for SV40T-antigen as well as for the endogeneous c-myc was observed in the SV40T-antigen-induced tumours (primary and transplanted). In the splenic tumours from the c-myc transgenic line, expression of the exogeneous c-myc was abundant and accompanied by almost complete loss of endogeneous c-myc mRNA expression, possibly reflecting a regulatory feed back mechanism.

Tumour cells were examined by flow cytometry analysis of cell surface antigens. Surprisingly both SV40T-antigen and c-myc-induced transplanted tumours were negative for surface Ig, but positive for Thy-1, a T-cell surface antigen marker, indicating that the immortalised cells were of the T-cell lineage. Furthermore, rearrangement of the T-cell receptor β-chain gene was detected in almost all tumours thus confirming the T-cell origin.

Splenic malignant cells of a c-myc-derived transplanted tumour were cultured. The cells failed to grow in conditions optimized for B-cell lines as described by Adams et al., (1985) and only proliferated by addition of concanavalin A and mouse thymocyte conditioned medium. Cells were cloned and stability of growth was monitored for at least 3 months. α1AT (Arg)-secreting clones were isolated, the highest antigen level being 60 ng/10^6 cells/24 hours.

3.2. GENERATION OF TRANS-HEPATOMAS

Transgenic hepatomas can be derived by expression in transgenic mice of an onc gene under the control of a liver specific promoter element. Two DNA constructions were used to generate transgenic mice prone to develop liver tumours. In pTG2984 the murine c-myc expression is under the control of the hα1AT promoter. This transgene contains in addition to the onc gene unit, a genomic copy of the hα1AT gene. On the other hand, pTG4912 carries the SV40T-antigen under the control of the hα1AT promoter.

Six stable transgenic lines were generated, of which two were derived from the c-myc onc gene and four from the SV40T-antigen. All were prone to hepatoma development, a phenotype which was transmitted to the progeny. However, the time of the tumour onset and tumour morphology were directly related to the nature of the onc gene.

The two TMTG2984 lines develop hepatic tumours after an average delay of 12 months. Circulating levels of more than 1 mg/ml of hα1AT were detected in plasma of these mice, with a correlation being observed between the plasma expression level of hα1AT in animals (up to 8 mg/ml) and the size of the liver tumour mass. Phenotypically these tumours consisted of a few large separated nodules, no morphological abnormalities in other tissues being observed. In TMTG4912 mice, whereas tumorigenesis occured rapidly, tumour onset was subjected to fluctuations (from 2.5 to 7 months) probably depending among other possibilities on the chromosomal integration site of the transgene. Histological examination proved that both kind of malignant cells were of hepatocyte origin and characterized as hepatocellular carcinoma cells. The expression of several hepatic marker genes indicative of an adult differentiated phenotype was studied at the RNA level. Thus, mRNA for albumin, murine and recombinant human α1AT, transferrin, acute phase protein, tyrosine amino transferase and glutathion S transferase were shown to be present in tumour tissue of the transgenic livers, this suggesting that these liver tumours induced by both onc genes had preserved a hepatic differentiation status.

3.3. ESTABLISHMENT AND CHARACTERIZATION OF TRANS-HEPATOMAS-DERIVED CELL LINES

Protocols were optimized in order to maintain a hepatocyte differentiation phenotype during cell culturing. The use of the medium described in Material and Methods, resulted in maintenance of a hepatocyte-like morphology. However, a comparative study between cell lines derived from the two different onc genes revealed differences in responsiveness to EGF (Perraud et al., submitted).

RNA from different cultured hepatomas was isolated and steady state levels of several hepatic marker mRNA determined by RNA dot blot analysis. Levels of mRNAs were comparable in the cells derived from the two types of onc genes (TMhepTG2984 and TMhepTG4912 cell lines), thus confirming their differentiated status.

Cell lines secreted mouse albumin at similar levels (around 5 μg/ml/24h), a concentration which was stable for at least 40 passages. In the case of cell lines from hepatomas of TMTG2984 transgenic mice, the secretion of hα1AT was also measured and was estimated at up to 5μg/ml/24h.

3.4. EXPRESSION OF ACTIVE HUMAN FIX IN TRANS-HEPATOMAS-DERIVED CELL LINES

Trangenic mice with rhFIX sequences integrated in their genomes were generated by microinjection of the pTG3960 construct. This transgene carries the entire hFIX gene with 5 kb and 0.3 kb of 5' and 3' flanking sequence, respectively. One transgenic mouse was identified with rhFIX plasma levels of 6.9 μg/ml. The transgene was transmitted to the progeny. Higher rhFIX antigen plasma levels were present in the offspring (25 to 40 μg/ml) probably due to mosaicism of the founder. Northern blot analysis revealed that the transgene was expressed only in liver tissue demonstrating that the 5 kb of 5' flanking sequence of the hFIX gene is able to confer the relative tissue-specificity.

The rhFIX clotting activity was measured in plasma of transgenic mice. Full FIX activity was demonstrated (ratio of activity/antigen around 80 %), being similar to that of plasma-derived FIX. Furthermore, rhFIX was capable of being adsorbed to barium salt in a similar fashion as the natural molecule, thus confirming γ-carboxylation of the molecule. On Western blotting, a specific band with a mobility equivalent to the natural molecule was revealed by anti-hFIX antiserum. Human FIX was then immunopurified from pooled transgenic mouse plasma in order for biochemical characterisation to be performed. SDS PAGE and Western blot displayed a profile similar to that of plasma-derived hFIX and N-terminal amino acid sequence confirmed the presence of a unique sequence corresponding to that of mature hFIX.

FIX transgenic mice were crossed with hepatoma-prone mice in order to generate animals bearing FIX expressing tumours. A series of double transgenics that originated from crosses between the TMTG2984 and TMTG3960 mouse lines were thus obtained. rhα1AT and rhFIX levels in the plasma of these animals correlated with the expression levels detected in parent mice and all animals were prone to develop hepatic tumours after a period of 12 months.

In order to accelerate oncogenesis, triple transgenic mice (TMTG3960 x 2984 x 4912) were generated. Hepatomas were observed after a delay of three months. Tumour cells were cultivated using the conditions described previously and hepatic cell lines could be established (Fig. 1). Expression of the hepatic differentiation markers (albumin, mouse α1AT, glutathion S tranferase and transferrin) and of transgene-specific mRNA (c-myc, SV40T-antigen, hα1AT and hFIX) was studied by RNA dot blot analysis. In the case of the TMhepTG48 cell line mouse albumin and rhα1AT were present in the supernatant of concentration of 17 to 21 μg/10^6 cells/24h and 4.6 to 8 μg/10^6 cells/24h, respectively. This cell line stably secreted rhFIX at average levels of 0.7 μg/ml/24h. Productivity was determined to be within a range of between 0.29 and 0.55 μg/10^6 cells/24h and clotting activity close to 100 %. Analysis by Western blotting showed that rhFIX migrated with a relative mass similar to that of the natural molecule. After purification by immunoaffinity chromatography, the N-terminal amino acid sequencing demonstrated correct processing of the recombinant molecule.

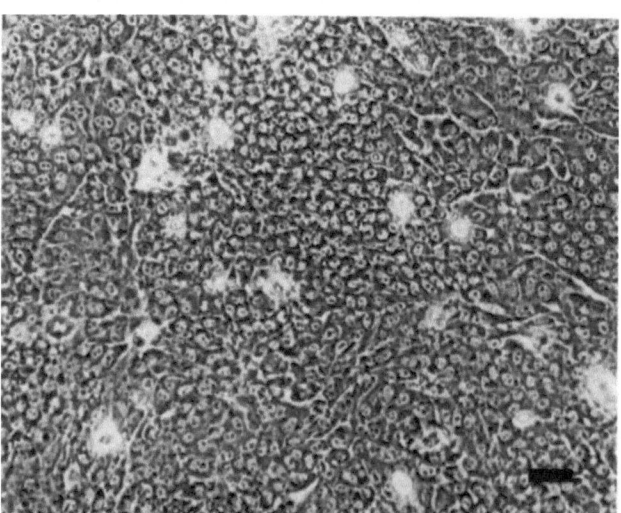

Figure 1. Phase contrast micrograph of cultured hepatoma cells TMhepTG48 derived from a transgenic mouse, after 8 weeks in culture. Cells were grown in 75 cm² tissue culture flask according to the conditions described in Material and Methods. Bar = 100 μm.

4. Discussion

In this paper, we have investigated the feasibility of establishing novel permanent cell lines secreting recombinant human proteins from onc gene-induced tumours generated in transgenic mice. The constructions were designed to drive co-expression of both onc genes and heterologous genes coding for human proteins in a (lymphoid or hepatic) tissue-specific manner. As a preliminary evaluation of this strategy, we have focused on the expression of a model protein which does not require specialised host cell functions for biological activity : α1AT (the Pittsburgh variant or the wild type). Subsequently the technology was applied to the production of a more complex protein such as coagulation FIX, which is subjected to at least 4 posttranslational modifications.

Ig regulatory sequences (Eμ, Pk and Pμ) were used to promote expression of the human α1AT (Arg) variant cDNA and onc-genes (SV40T-antigen or murine c-myc, respectively). This has led to the generation of lymphoid tumours in transgenic mice. Despite differences (e.g. onset of oncogenesis and degree of malignancy), both onc genes induced tumours of the T-cell lineage. This result is in contrast with the findings by Adams et al., (1985) who reported that transgenic mice carrying the c-myc onc gene under the control of its own promoter and Eμ frequently develop tumours of pre B or B cell lineage. Such a discrepancy can be due to the use in our construct of Pμ to drive onc gene expression resulting in disturbance of normal lymphoid development. The observed T-cell origin of tumours could explain the growth requirement of the c-myc-derived lymphoma cell lines for concanavalin A and mouse thymocyte conditioned medium in the culture. Nevertheless, circulating α1AT (Arg) levels were found in all examined transgenic mice (up to 20 μg/ml) and in several cultured clones (up to 60 ng/10^6 cells/24h), thus demonstrating the validity of the technology.

We then explored the possibility of isolating differentiated hepatic cell lines capable of specific posttranslational modifications as in the case of FIX expression. Firstly we generated hepatic tumours in transgenic mice by targeting expression of c-myc or SV40T-antigen to the liver using the hα1AT promoter. The transgene bearing the c-myc expression block also contained a genomic copy of the hα1AT gene to evaluate the potential of expressing the rh protein in transgenic mice and derived cell lines.

In both cases, transgenic mouse lines prone to hepatoma development could be established. High levels of α1AT were present in the plasma of mice bearing the hα1AT gene similar to those found in humans. The tumour cells could be cultured in vitro in order to produce immortalised hepatic cell lines which retained a differentiated status. Furthermore, stable secretion of rhα1AT was obtained. Similar results were obtained when a genomic FIX construction was used to generate transgenic mice. Thus several novel trans-hepatic cell lines were established. At present, the potential of these transgenic cell lines is being exploited for the production of fully active and correctly processed rhFIX.

In summary, the tissue-specific trans-immortalisation could be of general application for the derivation of specialised cell lines required for the expression of a complex recombinant human protein.

5. Acknowledgements

We thank J.P. Lecocq for his encouragement and interest in this work ; P. Chambon, P. Kourilsky for their support and critical suggestions ; M. Courtney, T. Skern, C. Roitsch, M. Lemeur, P. Gerlinger, R. Lathe and H. De La salle for scientific advice ; D. Fair for providing the neutralizing anti-FIX antiserum ; P. Lepage for N-terminal amino acid analysis ; D. Ali-Hadji, A. Dieterlé, D. Dreyer, D. Villeval, Y. Cordier, J.F. Spetz, and P. André for excellent technical assistance ; S. Périnel and N. Monfrini for typing the manuscript. The work on FIX was supported by Institut Mérieux, Lyon, France.

6. References

Adams, J.M., Harris, A.W., Pinkert, C.A., Corcoran, L.M., Alexander, W.S., Cory, S., Palmiter, R.D. and Brinster, R.L. (1985) 'The c-*myc* oncogene driven by immunoglobulin enhancers induces lymphoid malignancy in transgenic mice', Nature 318, 533-538.

Anson, D.S., Austen, D.E.G. and Brownlee, G.G. (1985) 'Expression of active human clotting factor IX from recombinant DNA clones in mammalian cells', Nature 315, 683-685.

Austen, D.E.G. and Rhymes, I.L. (1975) A Laboratory Manual of Blood Coagulation (Blackwell Scientific)

Balland, A., Faure, T., Carvallo, D., Cordier, P., Ulrich, P., Fournet, B., De La Salle, H. and Lecocq, J.P. (1988) 'Characterisation of two differently processed forms of human recombinant factor IX synthesised in CHO cells transformed with a polycistronic vector', Eur. J. Biochem. 172, 565-572.

Bernard, O., Cory, S., Gerondakis, S., Webb, E. and Adams, J.M. (1983) 'Sequence of the murine and human cellular *myc* oncogenes and two modes of *myc* transcription resulting from chromosome translocation in B lymphoid tumours', EMBO J. 2, 2375-2383.

Busby, S., Kumar, A., Joseph, M., Halfpap, L., Insley, M., Berkner, K., Kurachi, K. and Woodbury, R. (1985) 'Expression of active human factor IX in transfected cells', Nature 316, 271-273.

Choo, K.H., Raphael, K., McAdam, W. and Peterson, M.G. (1987) 'Expression of active human blood clotting factor IX in transgenic mice: use of cDNA with complete mRNA sequence', Nucleic Acids Res. 15, 871-884.

Courtney, M., Jallat, S., Tessier, L.H., Benavente, A., Crystal, R.G. and Lecocq, J.P. (1985) 'Synthesis in *E. coli* of α_1-antitrypsin variants of therapeutic potential for emphysema and thrombosis', Nature 313, 149-151.

Dalemans, W., Perraud, F., Le Meur, M., Gerlinger, P., Courtney, M. and Pavirani, A. (1990) 'Heterologous protein expression by trans-immortalised differentiated liver cell lines derived from transgenic mice', Biologicals 18, 191-198.

De La Salle, H., Altenburger, W., Elkaim, R., Dott, K., Dieterle, A., Drillien, R., Cazenave, J.P., Tolstoshev, P. and Lecocq, J.P. (1985) 'Active γ-carboxylated human factor IX expressed using recombinant DNA techniques', Nature 316, 268-270.

De La Salle, H., Altenburger, W., Elkaim, R., Dott, K., Dieterle, A., Drillien, R., Cazenave, J.P., Tolstoshev, P. and Lecocq, J.P. (1985) 'Active γ-carboxylated human factor IX expressed using recombinant DNA techniques', Nature 316, 268-270.

Derian, C.K., VanDusen, W., Przysiecki, C.T., Walsh, P.N., Berkner, K.L., Kaufman, R.J. and Friedman, P.A. (1989) 'Inhibitors of 2-ketoglutarate-dependent dioxygenases block aspartyl β-hydroxylation of recombinant human factor IX in several mammalian expression systems', J. Biol. Chem. 264, 6615-6618.

Hogan, B., Costantini, F. and Lacy, E. (1986) 'Manipulating the mouse embryo', Cold Spring Harbor Laboratory Press, Cold Spring Harbor, NY, pp. 332.

Jallat, S., Perraud, F., Dalemans, W., Balland, A., Dieterle, A., Faure, T., Meulien, P. and Pavirani, A. (1990a) 'Characterization of recombinant human Factor IX expressed in transgenic mice and in derived trans-immortalized hepatic cell lines', EMBO J. 9, 3295-3301.

Jallat, S., Meulien, P., Pavirani, A. and Perraud, F. (1990b) Patent Cooperation Treaty (PCT)/FR90/00606.

Kaufman, R.J., Wasley, L.C., Furie, B.C., Furie, B. and Shoemaker, C.B. (1986) 'Expression, purification and characterization of recombinant γ-carboxylated Factor IX synthesized in chinese hamster ovary cells', J. Biol. Chem. 261, 9622-9628.

Lathe, R., Vilotte, J.L. and Clark, A.J. (1987) 'Plasmid and bacteriophage vectors for excision of intact inserts', Gene 57, 193-201.

Meulien, P., Balland, A., Lepage, P., Mischler, F., Dott, K., Hauss, C., Grandgeorge, M. and Lecocq, J.P. (1990) 'Increased biological activity of a recombinant Factor IX variant carrying alanine at position +1', Protein Engineering 3, 629-633.

Palmiter, R.D., Chen, H.Y. and Brinster, R.L. (1982) 'Differential regulation of metallothionein-thymidine kinase fusion genes in transgenic mice and their offspring', Cell 29, 701-710.

Pavirani, A., Meulien, P., Harrer, H., Schamber, F., Dott, K., Villeval, D., Cordier, Y., Wiesel, M.L., Mazurier, C., Van de Pol, H., Piquet, Y., Cazenave, J.P. and Lecocq, J.P. (1987) 'Choosing a host cell for active recombinant factor VIII production using vaccinia virus', Bio/Technology 5, 389-392.

Pavirani, A., Skern, T., Le Meur, M., Lutz, Y., Lathe, R., Crystal, R.G., Fuchs, J.P., Gerlinger, P. and Courtney, M. (1989) 'Recombinant proteins of therapeutic interest expressed by lymphoid cell lines derived from transgenic mice', Bio/Technology 7, 1049-1054.

Perraud, F., Dalemans, W., Gendrault, J.L., Dreyer, D., Ali-Hadji, D., Faure, T. and Pavirani A. 'Characterisation of trans-immortalised hepatic cell lines established from transgenic mice' (submitted).

Rees, D.J.G., Jones, I.M., Handford, P.A., Walter, S.J., Esnouf, M.P., Smith, K.J. and Brownlee, G.G. (1988) 'The role of β-hydroxyaspartate and adjacent carboxylate residues in the first EGF domain of human Factor IX', EMBO J. 7, 2053-2061.

CD8$^+$ SUPPRESSOR T CELL CLONE 13G2 SECRETES A SUPPRESSIVE LYMPHOKINE, IMMUNE SUPPRESSIVE FACTOR-T (ISF-T)

TATSUHIRO HISATSUNE, KEN-ICHI NISHIJIMA, YUJI MINAI, ATSUSHI ENOMOTO, and SHUICHI KAMINOGAWA

Department of Agricultural Chemistry, The University of Tokyo, Bunkyo-ku, Tokyo 113, Japan

1. INTRODUCTION

Suppressor T cells (Ts) suppress the immune responses without any detectable cytotoxicity for responding cells, and their suppressor functions are partly mediated by soluble suppressive lymphokines (1). In our previous study, we established the CD8$^+$ Ts clone 13G2 from αs1-casein-primed lymph node cells (2). The clone suppressed the antigen (Ag)-induced proliferation of helper T cell (Th) clones, and the suppression was not antigen-specific. The clone was, however, unable to suppress Th proliferation induced either by IL-2 or immobilized anti-CD3 antibody when APC was absent. In this report, we show that Ts clone 13G2 secreted a soluble factor, immune suppressive factor T (ISF-T), which suppressed the proliferative response of Th clones induced by Ag plus APC but not by IL-2.

2. MATERIALS AND METHODS

2.1 T cell clones

A long-term cultured CD8$^+$ Ts clone 13G2 was established from lymph node cells of C57BL/6 and maintained by 50 U/ml of IL-2 for more than 3 years. The clone suppressed Ag-induced proliferation of Th cells in a dose-dependent manner (2). αs1-casein-specific CD4$^+$ Th clones, 3D20 and 11D4, derived from C57BL/6 were maintained by stimulation with irradiated spleen cells plus Ag (αs1-casein) as described previously (2). The determinants of 3D20 and 11D4 were mapped to the regions 136-155 and 101-122 of αs1-casein, respectively. KLH-specific CD4$^+$ Th clones, 28-4 and 9-16 (3), were the kind gift of Dr. T. Tada, The University of Tokyo, Tokyo, Japan.

R. Sasaki and K. Ikura (eds.), Animal Cell Culture and Production of Biologicals, 21–26.

2.2 Suppressor activity of 13G2 supernatant

To study the suppressor activity of supernatant fluid for Ag-induced T cell proliferation, 2×10^4 of CD4$^+$ Th clones were stimulated with Ag in the presence of 2×10^5 irradiated spleen cells in 200 μl of medium. The cells were cultured for 4 days, and serial dilutions of supernatant fluid of 13G2 cells were added to the mixture at the start of the culture. The cells were pulsed with 37 KBq of ^{3}H-thymidine during the last 20 h. The incorporation of ^{3}H-thymidine was measured by standard liquid scintillation counting. Percent suppression was calculated by the formula: 100 x (1-(cpm of the culture with the supernatant)/(cpm of the culture without the supernatant)).

3. RESULTS and DISCUSSION

3.1 Suppression of Ag-induced proliferation of Th clones by the culture supernatant of 13G2 cells

We have reported that the CD8$^+$ suppressor T cell clone 13G2 suppressed Ag-induced proliferation of Th clones in a dose-dependent manner (9). In this study, we first tested whether the culture supernatant of 13G2 cells could suppress the proliferative response of Th cells. Serial dilutions of the culture supernatant of 13G2 cells maintained by 50 U/ml of rIL-2 were added to the culture of Th clone 3D20 cells, which were stimulated with Ag plus APC. As shown in Figure 1, the supernatant of Ts clone 13G2 cells induced a dose-dependent suppression of the proliferative response of 3D20 cells, while the supernatant of Th clones 11D4 and 3D20, maintained under the same conditions, had no effect. The results indi-

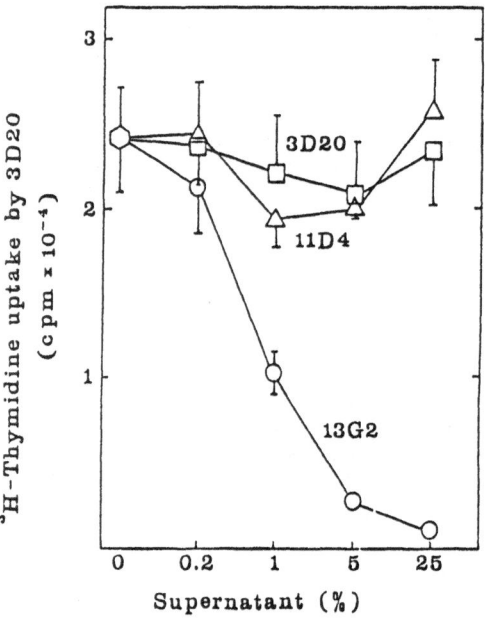

Figure 1. Suppression on T cell proliferation by supernatant of 13G2 cells

cated that 13G2 cells produced a soluble suppressor factor, termed ISF-T, which suppressed the proliferative
response of Th cells in response to Ag plus APC. However,
ISF-T was unable to inhibit the proliferative responses
of T cells induced either by IL-2 or immobilized antibodies when APC was absent (data not shown).

3.2 Lack of Ag specificity of suppressor activity by ISF-T

To assess the Ag specificity of the suppression by ISF-T,
α s1-casein-specific Th clones (3D20 and 11D4) and KLH-
specific Th clones (28-4 and 9-16) were used. All Th
clones were cultured in the presence of irradiated spleen
cells plus homologous Ag for 4 days, and ISF-T was added
to this mixture at the initiation of cultivation. As shown
in Table 1, ISF-T suppressed the proliferative responses
of 3D20 and 11D4. It also suppressed the proliferative
responses of Th clone 28-4 (specific for I-A^k plus KLH)
and 9-16 (specific for I-E^k plus KLH) equally well. Hence,
the suppressor function by ISF-T was not antigen-specific
and was not restricted to MHC.

3.3 ISF-T first acts on APC

Whether the first target cells of suppressor function of
ISF-T were APC or responding T cells was examined. Spleen
cells (APC) were precultured with the ISF-T sample (5%
(v/v) of 13G2 supernatant) for 24 h before the assay,

Table 1. Lack of Ag specificity of the suppression

Responding Th clones				^{3}H-TdR uptake by Th cells (k cpm)	
Name	Specificity		ISF-T added		
	MHC	Ag		Exp. 1	Exp. 2
3D20	A^b	α s1-casein (136-155)	− +	33.2±5.1 3.5±0.3	46.8±4.3 7.5±1.2
11D4	A^b	α s1-casein (101-122)	− +	24.5±3.9 0.5±0.0	20.1±3.1 1.2±0.5
28-4	A^k	KLH	− +	23.1±3.5 3.2±0.0	23.0±1.2 5.1±0.5
9-16	E^k	KLH	− +	8.1±1.9 1.1±0.3	N.T. N.T.

and 3D20 cells were stimulated with Ag plus the ISF-T-treated APC, or untreated APC. As shown in Table 2, the proliferative response of 3D20 cells stimulated with ISF-T-treated APC was significantly lower than that with untreated APC, indicating that pretreatment of APC with ISF-T diminished the response of Th cells. On the other hand, pretreatment of Th cells with ISF-T did not affect the response (Table 2). These results strongly indicate that the first target cells for the suppressor function were APC.

Table 2. First target cells for suppressor function

Th	APC	Cells preincubated with ISF-T	^{3}H-TdR uptake (k cpm)		Suppression
			Exp. 1	Exp. 2	
3D20	Spleen cells	-	52.2±8.2	54.5±4.1	
		Spleen cells	18.2±0.4	19.7±4.2	Yes
		-	52.2±8.2	92.4±3.7	
		3D20	50.4±4.5	98.7±6.5	No

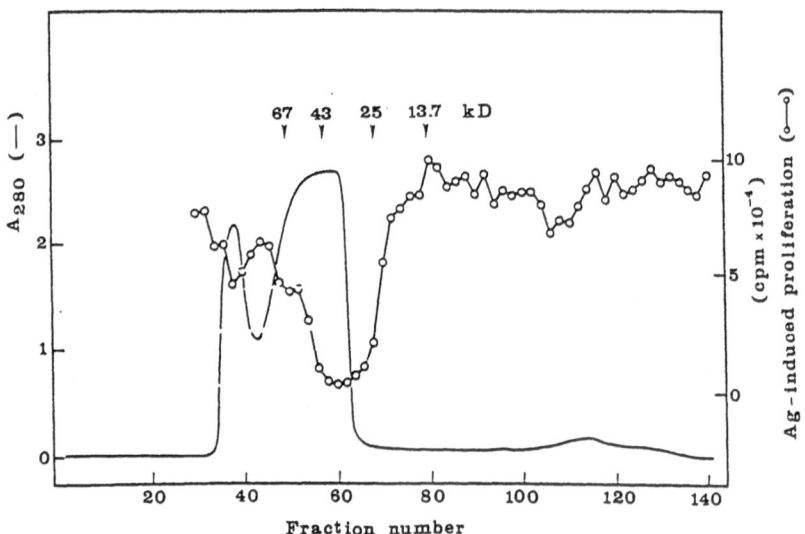

Figure 2. Gel filtration of ISF-T

Table 3. Effects of physicochemical treatment on ISF-T

Responding cells	ISF-T added	ISF-T treated with	^{3}H-TdR uptake (k cpm)	Percent suppression
3D20 plus APC+Ag	-	-	46.8±4.3	
	+	None	7.5±1.2	84
	+	56°C, 30 min	50.3±3.5	-7
	+	DTT, 0.1 M	52.1±5.4	-11
	+	pH 4.5	53.1±7.7	-13
	+	pH 6.0	8.3±1.2	82
	+	pH 10.0	7.2±0.8	85
	+	pH 11.5	43.7±0.3	7

3.4 Biochemical characterization of ISF-T

The supernatant of 13G2 concentrated with ultrafiltration
was put on a Sephadex G-100 column. Each fraction was
examined for its suppressor activity in the Ag-induced Th
proliferation. Figure 2 shows that the suppressor activity
in the Ag-induced response was eluted at a place corresponding to apparent molecular masses of 30-40 kD. We also
investigated the pH stability of ISF-T, and the sensitivity of ISF-T to DTT and heating. As shown in Table 3, ISF-T
was unstable in 0.1 M DTT and at 56°C, and was stable at
pH 6.0 to 10.0.

3.5 Functional properties of ISF-T were compared with
other known lymphokines

Other lymphokines that have been reported to have an
inhibitory effect on immune responses are as follows (4-
9): tumor necrosis factor (TNF)-α, TNF-β, interferon
(IFN)-γ, transforming growth factor (TGF)-β , soluble
immune response suppressor (SIRS), and IL-10. To assess
whether ISF-T differed from these lymphokines, biological
and biochemical characteristics of ISF-T were compared
with these lymphokines. As shown in Table 4, TGF-β significantly suppressed Ag-induced Th proliferation, while
TNF-α, TNF-β, and IFN-γ did not at all. TGF-β also
inhibited the Th response induced by anti-CD3 or IL-2, but
ISF-T was not effective on these responses, indicating
that ISF-T was not TGF-β. SIRS inhibited a wide variety
of immune responses including IL-2-dependent T cell proliferation when APC was present, and was stable at 56°C.
On the contrary, ISF-T never suppressed IL-2-dependent
proliferation of Th clones even in the presence of splenic
macrophages and was unstable at 56°C, suggesting that
ISF-T was different from SIRS. IL-10 inhibited IFN-γ

synthesis and was found to be a 17-21 kD glycoprotein. On the other hand, the m.w. of the factor was 30-40 kD. Parental 13G2 cells also secreted IFN-γ upon stimulation, suggesting that the Ts clone did not belong to the cell type which secreted IL-10. Besides, there is no distinct evidence that IL-10 inhibits Ag-induced proliferation of T cells. Therefore, we think that ISF-T probably differs from IL-10.

Table 4. suppressor activity of other lymphokines

Stimulation Factor	APC/Ag	Anti-CD3	IL-2
ISF-T	○	×	×
TGF-$\beta_{1/2}$	○	○	△
TNF-α	×	×	×
TNF-β	×	×	×
γ-IFN	×	×	×

○ : suppression
△ : partial suppression
× : no suppression

In conclusion, it is probable that ISF-T is a novel suppressive lymphokine different from other known lymphokines. Now we are purifying ISF-T to homogeneity and isolating a cDNA clone of the molecule. We believe that biological and molecular genetic studies of ISF-T can provide an important clue to the understanding of immune suppression by Ts cells.

4 REFERENCES

1 Tada,T., Asano,Y. and Sano,K., *Res.Immunol.* 1989.*140:291*
2 Hisatsune,T., Enomoto,A., Nishijima,K., Asano,Y., Tada, T., Kaminogawa,S., *J.Immunol.* 1990.*145:2421*
3 Nakayama,T., Kubo,R.T., Kubo,M., Fujisawa,H., Kishimoto, H., Asano,Y. and Tada,T., *Eur.J.Immunol.* 1988.*18:761*
4 Beutler,B. and Cerami,A., *Annu.Rev.Biochem.* 1988.*57:505*
5 Paul,N.L. and Ruddle,N.H., *Annu.Rev.Immunol.* 1988.*6:407*
6 Fernandez-Botran,R., Sanders,V.M., Mosmann,T.R. and Vitetta,E., *J.Exp.Med.* 1988.*168:543*
7 Wrann,M., Bodmer,S., de Martin,R., Siepl,C., Hofer-Warbinek,R., Frei,K., Hofer,E. and Fontana,A., *EMBO J.* 1987.*6:1633*
8 Aune,T.M. and Pierce,C.W., *J.Immunol.* 1981.*127:368*
9 Moore,K.W., Vieira,P., Fiorentino,D.F., Troustine,M.L., Khan,T.A. and Mosmann,T.R., *Science* 1990.*248:1230*

DEVELOPMENT, AVAILABILITY AND CHARACTERIZATION OF ATCC HUMAN AND ANIMAL CELL LINES

R.J. HAY
American Type Culture Collection
12301 Parklawn Drive
Rockville, Maryland 20852 USA

ABSTRACT. The development, acquisition otherwise, cryopreservation, characterization and distribution of cell lines represent the major activities of our department. For example, cell lines of fibroblastic, epithelial and lymphoblastic origin plus hybridomas have been developed. Standard and serum-free systems have been utilized. A wide variety of human and animal lines have been acquired otherwise and accessioned. The characteristics of representative cultures are discussed. Cell lines are tested routinely for microbial contaminants including mycoplasma and species is verified by isoenzymology. Karyology, the DNA fingerprint for human lines, functional traits, tumorigenicity and other characteristics are confirmed as appropriate. Computerized databases on cell lines are now available internationally for ATCC clients via CODATA and TYMNET or Telenet networks. By using the databases one can quickly retrieve information about specific lines or create lists of lines with specific characteristics.

1. Introduction

The American Type Culture Collection (ATCC) is best known for activities related to banking and distribution of microorganisms, cell lines and recombinant DNA biologicals. However, research and development programs which contribute to the collections form a part of day-to-day laboratory efforts. Staff and visiting scientists in our cell culture department, for example, participate both in extramurally-funded research and in the acquisition, preservation and characterization of cell lines for distribution. The following is an outline of some of these mutually complimentary functions.

R. Sasaki and K. Ikura (eds.), Animal Cell Culture and Production of Biologicals, 27–39.
© 1991 *Kluwer Academic Publishers.*

2. Development of Cell Lines

Grant and contract supported programs for the isolation
and development of cell lines supplement the more general
cell banking efforts at the ATCC and have been described
previously (Ruoff and Hay, 1979, Hay, et al., 1982, Hay,
1984). The established methods for accessioning and
certifying lines, discussed below also are applied to cell
lines developed at the ATCC.

2.1 FIBROBLASTS

Fibroblastic lines from skin, bronchus, lung and colon of
human and other animal origins (Table 1) have been
produced at the ATCC using a standard explant culture
method described in detail elsewhere (Hay, et al. 1982).
These are useful in a very wide variety of metabolic
studies including work relating to matrix synthesis and as
controls in experiments involving other cell types.

2.2 EPITHELIAL-LIKE CELL STRAINS

Epithelial strains have been isolated at the ATCC using
normal and tumor tissues from humans and animals (Table
1). For example, normal bronchial tissues from rhesus and
african green monkeys were used to establish epithelial
strains ATCC.CCL 208 (4MBr-5) and ATCC.CRL 1576 (12MBr6)
respectively. Glass chips or penicylinder cloning were
utilized to select for the epithelial component in cell
outgrowths. Cells from these cultures have PAS-positive
inclusions and require epidermal growth factor for optimal
propagation (Caputo, et al. 1979).
  Thompson, et al (1985) devised and utilized a technique
for serum-free cultivation of colonic cells to isolate an
epithelial-like strain from female fetal tissue. The
cells of ATCC.CRL 1790 (CCD841CoN) are propagated in ACL-4
medium in vessels coated with bovine serum albumin,
fibronectin and collagen. The morphology is similar to
that of other colonic mucosal lines and like those,
CCD841CoN cells do not contain keratin. Thus definitive
evidence for epithelial character is lacking. The
cultured cells may have lost ability to synthesize keratin
or may represent a primordial stem cell component. The
cells are diploid and no consistent marker chromosomes
have been observed.

TABLE 1. Representative cell lines developed at the ATCC

ATCC No.	Designation	Source/Comment
FIBROBLAST		
CCL 202	CCD-11Lu	Normal human lung/control
CRL 1485	CCD-32Lu	Lung/Congenital heart disease
CRL 1498	CCD-39Lu	Lung/Hyaline membrane disease
CCL 192	RFL6	Rat germ free lung
CCL 195	AHL1	Armenian hamster lung
EPITHELIA		
CCL 208	4MBr-5	Normal rhesus bronchus
CRL 1790	CCD841CoN	Human colon/lacks keratin
CRL 1492	AR42J	Rat exocrine pancreatic tumor/functional
HYBRIDOMAS		
HB 155	Ep-16	Monoclonal to human keratinocyte surface antigen
HB 172	10B9	Monoclonal to human endothelia and macrophages

See Hay et al (1988), review the cell line database available through CODATA TYMNET or contact the ATCC for further information.

An extremely interesting, and useful epithelial-like
cell strain ATCC.CRL 1492 (AR42J) was isolated at the ATCC
from an exocrine rat pancreatic tumor induced by azaserine
(Jessop and Hay, 1980). When grown in the presence of
dexamethasone or other steroids AR42J cells contain
pleomorphic zymogen droplets, significant amounts of
amylase and its mRNA, and secrete the enzyme in response
to cholecystokinin (Logsdon, et al. 1985). Under serum-
free culture conditions AR42J cells exhibit a
proliferative response to the peptide hormone gastrin as
measured by ^{3}H-thymidine incorporation (Seva et al. 1990).

2.3 HYBRIDOMAS

A number of hybridomas producing monoclonal antibodies to
cell surface antigens have been developed and
characterized at the ATCC. Epithelial-specific antigens
on human foreskin keratinocytes can be located, for
example, using monoclonal Ep-16 from ATCC.HB 155. This
hybridoma was developed by A. Hamburger, et al. (1985a)
after immunization of BALB/c mice with membrane-enriched
preparations from primary cultures of human keratinocytes.
NS-1 cells (ATCC.TIB 18) were used as fusion partners, and
a modified ELISA screening method was used to identify
positive clones. The antibody reacts very strongly with
stratified squamous epithelia. The cell surface antigen
detected is distinct from other epithelial-specific
antigens as shown by differences in tissue distribution
and cellular localization.

Using similar technique Hamburger, et al (1985b) were
also able to develop hybridomas secreting monoclonal
antibodies to human umbilical vein endothelia. In this
case intact primary passage, cultured endothelia were used
as immunogens and P3X63Ag8.653, a non-secreting mouse
myeloma line (ATCC.CRL 1580) was the fusion partner.

Of the various hybridoma lines developed, one designated
14E5 (from ATCC.HB 174) yields an antibody which reacts
with macrophages, but not fibroblasts or bovine endothelia
while another, 12C6 (ATCC.HB 173) reacts with human and
primate fibroblasts and endothelia from bovine arteries
but not with mature macrophages. A third clone, 10B9
(ATCC.HB 172) reacts only with human endothelia and
immature macrophages. These antibodies should prove
useful to detect differentiation markers of endothelia and
in studies of endothelial cell function.

2.4   HUMAN B LYMPHOBLASTIC LINES

Human B lymphocytic cells, once successfully transformed
by Epstein-Barr virus (EBV) are continuous and may be used
to provide a uniform supply of donor DNA for genetic and
other analyses. The method for EBV transformation of fresh
or cryopreserved human B lymphocytes as modified and
applied at The ATCC is extremely efficient giving a
success rate greater that 95% (Caputo, et al. 1991).
Lymphocytes are separated from whole blood by
centrifugation through Ficoll.   They are transformed by
exposure to Epstein-Barr Virus (EBV) obtained as a
supernatant from ATCC.CRL 1612 (B95-8) cells.   Virus
production is verified in advance by immunoperoxidase
staining after application of monoclonal antibody to EBV
capsid antigen (ATCC.HB 168) (72A1).   The addition of
irradiated feeder cells (ATCC.CCL 171) (MRC-5) is
important to enhance efficiency in lymphoblast culture
initiation.
    Literally hundreds of human lymphoblastic lines have
been developed at the ATCC using this procedure over the
past two years.

3.   Recent Acquisitions

The majority of cell lines added to The Collection have
been developed elsewhere.   They are made available through
the generosity of numerous originators, most of whom
supply their material without charge and with minimal
restrictions on distribution.   A listing of selected cell
lines added recently is provided in Table 2.   Included are
one of the very few human medullary thyroid carcinoma
lines available, TT, (CRL 1803); a transformed epithelial-
like strain from normal fetal colon (CRL 1807); a
pancreatic carcinoma of ductal origin, SU.86.86. (CRL
1837); and a mucoepidermoid lung carcinoma, NCI-H292, (CRL
1848) used for the isolation of all 6 human myxoviruses
and for hepatitis B virus cultivation.   Additional lines
in the human tumor cell bank include SW1353 (HTB 94) a
chondrosarcoma trisomic for chromosome 7 but otherwise
free of chromosomal markers; SW579 (HTB 107) from a
thyroid squamous cell carcinoma which produces malignant
spindle and giant cell tumors upon inoculation to nude
mice; plus a series of lung carcinoma lines developed by
A. Gazdar and associates (Carney et al., 1985; Gazdar, et

TABLE 2. Recent additions to the ATCC cell repository*

ATCC No.	Designation	Type	Comment	Contributor
		HUMAN		
CRL 1803	TT	Thyroid carcinoma	Calcitonin & CEA	Leong
CRL 1807	CCD841CoTr	Colon	ts SV40-epithelial	Thompson
CRL 1837	SU.86.86	Pancreatic carcinoma	CEA positive, LAK assays	Holder
CRL 1848	NCI-H292	Lung carcinoma	Supports HBV replication	Gazdar
CRL 1864	RF-1	Gastric carcinoma	CEA & Mucin positive	Shaver
CRL 1872	A375.S2	Melanoma	Assay for IL-1	Newman
HTB 93	SW982	Axillar sarcoma	Hyperdiploid	Leibovitz
HTB 94	SW1353	Chondro-sarcoma	Hyperdiploid	Leibovitz
HTB 107	SW579	Thyroid	Grade III giant cell tumor	Leibovitz
HTB 118	SW962	Vulva	Lymph node metastasis	Leibovitz
HTB 171	NCI-H446	Small cell carcinoma	Positive for enolase and creatine kinase	Gazdar
HTB 181	NCI-H820	Papillary lung adeno-carcinoma	Lamellar bodies Surfactant proteins AB & C	Gazdar
HTB 182	NCI-H520	Squamous lung carcinoma	Keratin and vimentin positive	Gazdar
HTB 184	NCI-510A	Small cell carcinoma	Elevated levels enolase, creatine kinase, dopa decarboxylase and bombesin	Gazdar
HTB 187	D341	Medullo-blastoma	UJ13A and enolase positive	Friedman

TABLE 2. (continued)

ATCC No.	Designation	Type	Comment	Contributor

PRIMATE AND OTHERS

ATCC No.	Designation	Type	Comment	Contributor
CRL 1835	IB,RS-2 D10	Porcine kidney	Virus propagation	House
CRL 1840	OK	Opossum kidney	Hormone receptor positive	Miller
CRL 1850	PUTI	Orangutan	EBV transformed	Lawlor
CRL 1851	7.TD1	Mouse hybridoma	IL-6 dependent	Nordan
CRL 1854	ROK	Gorilla	EBV transformed	Lawlor
CRL 1857	CARL	Chimp	EBV transformed	Lawlor
CRL 1858	CRE BAG2	NIH 3T3 derivative	Packaging line	Cepko
CRL 1861	EM9	Chinese hamster ovary	DNA repair mutant	Thompson
CRL 1875	8A3B.6	Hybridoma	Anti Blue tongue virus	House
CRL 1895	RK3A	Rat kidney	E1A oncogene transformed	Kinzler

*See Hay et al (1990), review the cell line database through
 CODATA TYMNET contact The ATCC for more information.

al., 1985). Of the latter, NCI-H446 (HTB 171), although
from a small cell lung cancer (SCLC), is atypical in both
morphology and biochemistry. It expresses neuron specific
enolase and the brain isoenzyme form of creatine kinase
but has no detectable L-dopa decarboxylase or bombesin-
like immunoreactivity. In contrast NCI-510A (HTB 184) has
elevated levels of all of these four biochemical markers
characteristic of SCLC.

The NCI-H820 (HTB-181) may be an extremely important
line for studies on type 2 pneumocyte physiology. It
reportedly has osmiophilic lamellar bodies and is positive
for the three surfactant associated proteins; SP-A
constitutively, and SP-B and SP-C after dexamethasone
induction (Gazdar, et al., 1990). The D341 (HTB 187) is
one of three medulloblastoma cell lines developed by H.S.
Friedman and associates (1988). This line and the tumors
produced from it after inoculation to nude mice, are
positive for neurofibrillary proteins, glutamine
synthetase and neuron-specific enolase but negative for
glial fibrillary acidic and S-100 proteins. The
neuroectodermal antigen is present as demonstrated by
monoclonal UJ13A.

New animal cell lines include a porcine kidney line, IB,
RS-2 B10 (CRL 1835) useful for propagating numerous swine
viruses as well as foot and mouth disease virus and
epizootic hemorrhagic disease virus (House, 1988); a NIH
3T3 murine line (CRE BAG 2, CRL 1858); an IL-6 dependent
murine line, 7TD1, (CRL 1851); an IL-3 dependent murine
line M-NSF-60, (CRL 1838) plus a series of EBV transformed
primate B- cell lines (chimpanzee, gorilla and orangutan)
developed at Stanford University (Lawlor, 1988).

4. Characterizations

General descriptions of the seed stock concept and
specific, detailed protocols for cell line
characterization have been presented elsewhere (Hay, 1986,
1988; Hay et al. 1989). In brief, key characterizations
are performed on cell line progeny from a seed stock of
ampules. These include tests to assure freedom from
microbial contamination, to confirm specific synthetic
function (eg. immunoglobulins, hormones, structural
proteins etc.), to verify cell line identity and so forth.
For the latter, DNA fingerprinting with probes such as the
33.6, pYNH24 and others represents a new rapid and
efficient technique for identification of human cell lines
(Gilbert, et al. 1990; Reid, et al. 1990).

By returning to seed stock material to produce each new distribution or replenishment freeze, the cell bank manager can assure recipients that the cultures supplied are similar both to those provided previously and to those which will be supplied in the future. Research comparability as a function of time and geographical location in thus sustained.

5. Distribution

Cell cultures are distributed either frozen in 1-ml sealed ampules or as growing populations in T25 flasks. Most frozen shipments are sent early in the week to ensure arrival before the weekend. Frozen ampules are removed from the distribution stocks and left in a liquid nitrogen refrigerator until ready for packaging. All shipments going by one carrier are then packaged within a short period of time to minimize the exposure to dry ice temperatures. Cell cultures are shipped in an insulated container with a capacity of 15 lb of dry ice and a holding time of 5 days. The shipping of frozen material eliminates quality control problems associated with the preparation of actively growing cultures. The latter involves greater risk of contamination, both from microorganisms and in the cell culture laboratory, and it puts greater pressure on recipients to coordinate their own laboratory work with the suppliers shipping schedule.

In Japan ATCC cultures are distributed by Sumisho Pharma Corporation, 4-7-8 Kitahama chuo-ku, Osaka 541 and through The Japanese Cancer Research Resources Bank, Division of Genetics and Mutagenesis, National Institute of Hygienic Sciences, 1-18-1 Kami Yooga, Setagaya-ku, Tokyo 158.

The demand for reference cell cultures has increased steadily and dramatically over the years (Table 3). Without doubt this is a reflection not only on the growth in utility of cell culture techniques for studies on a wide range of problems, but also an increased awareness by the scientific community of the importance of standardized cultures in research.

6. Information and availability

The availability and characteristics of cell lines are documented not only in ATCC catalogues but also via electronic networks. Online databases reside on a Dialcom Inc. computer. These can be accessed **worldwide** by personal computer and telecommunications software through Tymnet or Telenet and local communications networks which connect to these public networks. The cell line database, like databases from other ATCC departments on other ATCC offerings, is updated periodically to provide timely information on new accessions.

TABLE 3   Annual distribution of ATCC cell cultures 1975-1990

Year	Cell cultures provided per year	Change from 1975
1975	4,359	0
1980	6,883	158%
1985	24,457	561%
1990	49,703	1140%

7. Acknowledgements

This work was supported in part by USPHS Contracts NO1-CB-71014 from the NCI and NO1-RR-9-2105 from The NIH Center for Research Resources plus Grant 1-R26-CA25635 from The National Large Bowel Cancer Project of The NCI.

8. References

Caputo, J.L., Hay, R.J. and Williams, C.D. (1979)
'The isolation and properties of an epithelial cell
strain from rhesus monkey bronchus'. In Vitro 15,
222-223.

Caputo, J.L., Thompson, A., McClintock, P., Reid, Y.A.
and Hay, R.J. (1991) 'An effective method for
establishing human B lymphoblastic cell lines using
Epstein-Barr virus' J. Tiss. Cult. Meth. 13, 39-44.

Carney, D.N., Gazdar, A.F., Bepler, G., Guccion, J.G.,
Marangos, P.J. et al. (1985) 'Establishment and
identification of small cell lung cancer cell lines
having classic and variant features' Canc. Res. 45,
2913-2923.

Friedman, H.S., Burger, P.C., Bigner, S.H.,
Trojanowski, J.Q., Brodeur, G.M., et al. (1988)
'Phenotypic and genotypic analyses of a human
medulloblastoma cell line and transplantable
xenograft (D341 Med) demonstrating amplification of
c-myc' Amer. J. Pathol. 130, 472-484.

Gazdar, A.F., Carney, D.N., Nau, M.M. and Minna, J.D.
(1985) 'Chaaracterization of variant subclasses of
cell lines derived from small cell lung cancer
having distinctive biochemical, morphological and
growth properties' Canc. Res. 45, 2924-2930.

Gazdar, A.F., Linnoila, R.I., Kurita, Y., Oie, H.K.,
Mulshine, J.L. et al. (1990) 'Peripheral airway
cell differentiation in human lung cancer cell
lines' Canc. Res. 50, 5481-5487.

Gilbert, D.A., Reid, Y.A., White, C., Hay, R.J. and
O'Brien, S.J. (1990) 'Application of DNA
fingerprints for cell line individualization'.
Amer. J. Hum. Gen. 47, 499-514.

Hamburger, A.W, Reid, Y.A., Pella, B., Milo, G.E.,
Noyes, I., Krakauer, H. and Fuhrer, J.P. (1985a)
'Isolation and characterization of a monoclonal
antibody specific for epithelial cells' Canc. Res.
45, 783-790.

Hamburger, A.W., Reid, Y.A., Pelle, B.A., Breth, L.A., Beg, N. Ryan, U. and Cines, D.B. (1985b) 'Isolation and characterization of monoclonal antibodies reactive with endothelial cells' Tiss. and Cell 17, 451-459.

Hay, R.J., Williams, C.D., Macy, M.L. and Lavappa, K.S. (1982) 'Cultured cell lines for research on pulmonary, physiology available through the American Type Culture Collection. Amer. Rev. Resp. Dis. 125, 222-232.

Hay, R.J. (1984) 'Problems of specificity from the cell banking perspective: Colon cell lines at the American Type Culture Collection'.Prog. Cancer Res. Ther. 29, 3-21.

Hay, R.J. 'Preservation and characterization'. Chapter 4 ed. R.I. Freshney, in Animal Cell Culture: A Practical Approach, IRL Press, Washington, D.C. pp. 71-112.

Hay, R.J. (1988) 'The seed stock concept and quality control for cell lines'. Anal. Biochem. 171, 225-237.

Hay, R.J., Macy, M., Chen, T.R., McClintock, P. and Reid, R.A. (1988). American Type Culture Collection Catalogue of Cell Lines and Hybridomas, 6th edition, Rockville, MD.

Hay, R.J.; Macy, M., Chen, T.R. (1989), 'Mycoplasma infection of cultured cells'. Nature 339, 487-488.

Hay, R.J., Macy, M., Chen, T.R., McClintock, P. and Reid, R.A. (1990). American Type Culture Collection Update Cell Lines and Hybridoma, Rockville, Md.

Lawlor, D.A., Ward, F.E. Ennis, P.D., Jackson, A.P. and Parham, P., (1988). 'HLA-A and B polymorphisms predate the divergence of humans and chimpanzees' Nature 135, 268-271.

Logsdon, C.D., Moessner, J., Williams, J.A. and
    Goldfine, I.D. (1985) 'Glucocorticoids increase
    amylase mRNA levels, secretory organelles, and
    secretion in pancreatic acinar AR42J cells' J. Cell
    Biol. 100, 1200-1208.

Reid, Y.A., Gilbert, D.A. and O'Brien, S.J. (1990)
    'The use of DNA hypervariable probes for human cell
    line identification'. ATCC Newsletter 10 (4), 1-2.

Ruoff, N.M. and Hay, R.J. 'Metabolic and temporal
    studies of pancreatic exocrine colonial aggregates
    in vitro (1979). Cell and Tiss. Res. 204,
    243-252.

Seva, C., Scemama, J.L., Bastie, M.J., Pradayrol, L.,
    and Vayasse, N. (1990). 'Lorglumide and loxiglumide
    inhibit gastrin-stimulated DNA synthesis in a rat
    tumoral acinar pancreatic cell line (AR42J)' Canc.
    Res. 50, 5829-5833.

Thompson, A.A., Dilworth, S. and Hay, R.J. (1985)
    'Isolation and culture of colonic epithelial cells
    in serum-free medium' J. Tiss. Cult. Meth. 9, 117-
    122.

RECENT ADVANCES IN ANIMAL CELL BIOTECHNOLOGY.

R.E. Spier

Wolfson Cytotechnology Laboratory,
School of Biological Sciences,
University of Surrey,
Guildford, Surrey, GU2 5XH, UK.

1. PRODUCTS.

It is generally agreed that from the the products of the new biotechnology, (that involving genetic engineering and/or cell fusion), the value of products made from animal cells in culture accounts for about half of the total revenue which accrues to all the new biotechnology products. (Table 1).

Table 1.
The Approximate Value in 1990 of Animal Cell Derived Products in relation to all the Products derived from the New Biotechnologies

Product	Cell	Revenue ($/annum)
Erythropoietin	CHO	400,000,000
Erythropoietin	Mouse Cells	
h-Growth Hormone	CHO	
h-Growth Hormone	Mammalian	
t-Plasminogen Activator	CHO	200,000,000
t-Plasminogen Activator	Mouse Cells	
Factor VIIIC	CHO	
Interferon-alp	Namalwa	
Interferon-alpha 2	Namalwa	
Interferon-alpha	Human	
Interferon-alpha	Buffy Coat Cells	
Interferon-alpha N1	Lymphoblasts	
Interferon-beta	Human	90,000,000
Hepatitis B Vaccine	CHO	
Fibriscint-MAB	Hybridoma	
Myoscint-MAB	Hybridoma	
Othoclone OKT3-MAB	Hybridoma	10,000,000
Interleukin-2	CHO	
G-CSF	E.Coli	
EGF	E.Coli	
Interferon-gamma	E.Coli	
Interleukin-2	E.Coli	
Interferon-alpha	E.Coli	
TNF	E.Coli	
Interferon-beta	E.Coli	
h-Insulin	E.Coli	160,000,000

R. Sasaki and K. Ikura (eds.), Animal Cell Culture and Production of Biologicals, 41–46.
© 1991 *Kluwer Academic Publishers.*

42

Interferon-alpha 2a	E.Coli	90,000,000
Insulin-like GF	E.Coli	
h-GH	E.Coli	300,000,000
Superoxide Dismutase	Pichia Pastoris	
TNF	Yeast	
G-M CSF	Yeast	
Hepatitis B Vaccine	Yeast	100,000,000
TNF	Pichia pastoris	
h-Insulin	Sacc. Cerevisae	

Total Value Animal Cell Derived Products $ 700,000,000
Total Value all New Biotechnology Products $1,350,000,000

In addition to the particular products delineated above there are numerous other commercialisable applications of animal cells in preparation. Such other future developments include the use of animal cells in replacement organ therapies, (liver, pancreas, skin, adrenals) or for the testing of the toxicity and/or activity of materials with putative pharmaceutical applications. Cells themselves may become therapeutic products as in the LAK cells, (lymphocyte activated killer cells) or TIL cells, (tumour infiltrating lymphocytes). Finally, there is likely to be a resurgence of virus vaccines based, this time, on recombinant viruses such as Vaccinia or Polio.

2. AREAS FOR PROGRESS.

In reviewing the subject as a whole it is possible to identify a number of areas where attention is focused; these constitute a cursory description of the advanced work-face of the discipline. They are summarised in Table 2.

Table 2.

Topic Areas of Focal Concern in Animal Cell Biotechnology.

* Can we design a feasible 1000 Kg/annum Production Unit.

* The Control of Glycosylation and Glycoforms

* Genetic Engineering of Cells , Phase II. (Site specific insertion etc)

* Packed Bed versus Fluidised Bed Bioreactor

* Batch Processing versus Continuous Processing

* Bioprocess Optimisation operating at the levels of

       # the Cell
       # the Bioreactor
       # the Production Plant
       # the Factory
       # the Society

* Bioprocess Safety; The Regulatory Aspects

* Immortalisation of Secretory Cells

* Large Molecular Weight Products versus Small Molecular Weight

* The Influential Interactions between Cells and Surfaces

* The Control of Secondary Metabolism in Animal Cells in Culture

* The Impact of Transgenic Animals and Plants on Animal Cell Opportunities

* Oxygenation of Cell Cultures.

Many of these topics have been dealt with elsewhere, (1, [Transgenic Animals], 2, [Fluidised Bed versus Packed Bed], 3, [Glycosylation], 4, [Secondary Metabolism], 5, [Safety of Cell Substrates],6, [Small and Large Molecules]), while others need further consideration from viewpoints which have not been explored in the existing literature. This paper will therefore deal with those issues whose continuing development is necessary to maintain progress in the field.

3. THE 1000KG A YEAR PLANT.

There is a view that as Animals Cells in Culture have not been used in operations which yield tonnage quantitities of cell derived product that it is useless to do the necessary exploratory research which could result in the requirement to make a product which would be needed in quantities in excess of 1000kg per annum. Clearly this attitude can easily become self-fulfilling. However, it is necessary to face this issue and to begin to delineate the design parameters which would be engendered by the construction of such a plant. Were these to appear reasonable and effectable then it is hoped that research will be undertaken which could avail itself of such a facility.

In order to produce 1000KG per annum of a product which is biosynthesised by animal cells in culture the following parameters offer some guidelines;

> # 300 days a year of operations
> # 30pg/cell/day productivity
> # cell concentration 10^6-10^8 cells/ml
> # medium utilisation rate of 500L/10^{12} cells/day.
> # the plant to run as a continuous culture

This leads to the conclusions as in table 3.

Table 3.

The Scale of the Problems for the 1000KG/annum production unit.

Cell Concentration (cells/ml)	10^6	10^7	10^8
Bioreactor Volume (L)	100,000	10,000	1,000
Medium Consumption (L/day)	100,000	100,000	100,000

The implication of the figures in Table 3 are that at present so far as this author is aware animal cell bioreactors do not operate at the scales required. However there are bioreactors operating at the 20,000L scale and reactors at 10,000L are not uncommon for cells held at concentrations of 10^6/ml. When we come to consider bioreactors operating in the commercial domain at 10^7 cells/ml there are a few reports of experimental systems under development at the 100-750L scales, (the volume referred to in reference 4). The latter type of bioreactors have to operate with medium perfusion with cell retention and involve some form of cell separation device which has to operate at relatively high cell concentrations. Such devices have received much attention in recent times and range from sedimentation units through centrifuges to membrane separators either inside or outside the main cell containing chamber. Clearly the long term operation of such devices in the commercial work-place has to be established. Thus in both of the previous cases the current state of the art is about an order of magnitude away from that which would be necessary for the 1000KG/ annum plant. Obviously it would be possible to set up an operation with 10 unit bioreactors, but this is not normally regarded as the preferred way of proceeding largely due to logistic problems, (the number of breakdowns is proportional to the number of units available to breakdown), and the heavier manning requirements. With regard to operations at the 10^8 cells/ml concentration, there are a number of bioreactor types which can function at or slightly above this level. These are those based on bundles of hollow fibres or on three-dimensional microcarriers. While the former system can be scaled up in units of the order of 1L in volume the latter system has been run at the 24L scale. Such systems are about 2 orders of magnitude away from the objective.

The difficulties which are inherent in the scale-up of the high concentration systems, (over 10^8 cells/ml), puts a greater onus on the development of the lower concentration systems. These systems in their turn tend to be limited by the amount of oxygen which can be supplied to the cells when the latter are held in a unit process system at the scales necessary to achieve the 1000KG/annum plant.

4. OXYGENATION OF LARGE-SCALE CELL CULTURES.

Cells require some 1-10pg oxygen/cell/hr, (7). It is possible to obtain oxygen transfer rates (K_La) of 100 hr^{-1}, (in the authors laboratory using bubbles of air with diameters in the sub millimetre range; a 5 fold increase in K_La would be obtained were oxygen used). It has also been shown that bubbles can be used for the aeration of animal cell cultures when the medium contains such materials as Pluronic F68, (8). With such transfer rates it can be calculated that with cells respiring at a rate of 3pg oxygen/cell/ hour that a system with a K_La of 100 hr^{-1} would support some 10^8 cells/ml. (A system gassed by oxygen would support 5 times this number of cells). However, this calculation is based on a respiration rate which could be sub-optimal and with the maximum driving force for oxygen transfer. Were the latter halved and the former doubled then a conservative estimate would be that $2*10^7$ cells/ml could be supported by aerating with air bubbles. One way to increase the amount of oxygen transferred is to increase the aeration rate. The limitation to this is the consequential formation of a foam. (Diagram 1 shows the relationships between the various factors involved in the aeration of a cell culture and constitutes a model of the process).

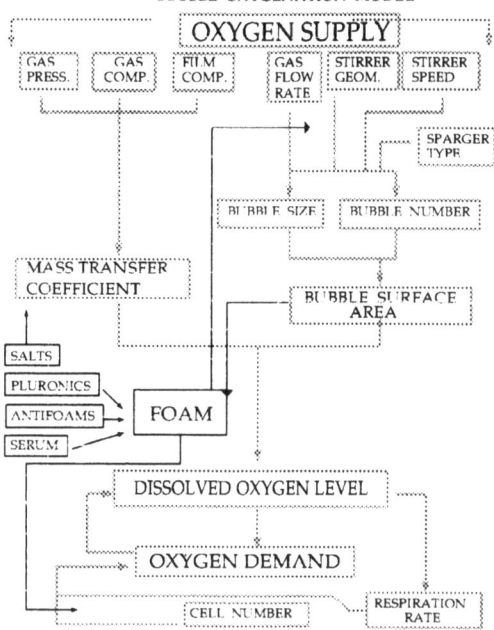

BUBBLE OXYGENATION MODEL

Both the quality and quantity of the foam is involved in the determination of the consequences of that foam. When the quality of the foam is stable and viscous, then there seems to be little danger of cell damage whereas when the foam is unstable and limpid there is significant danger of bubble damage at the time of bubble disengagement. Should the

quality of the foam be excessive then this could lead to the blockage of air filters with consequential danger of contamination. Also there is the possibility of cell loss as a result of cell entrainment in the foam. When attempts to control the foam lead to the overuse of antifoam there is also an effect on cellular metabolism.

It is of interest to note that materials such as antifoam and/or Pluronic F68 do not have a deleterious effect on the oxygen transfer coeficient (9), although serum may lower the oxygen transfer from small bubbles.

5. CONCLUSIONS

Animal Cell Biotechnology is entering a phase of activity which will yield both conventional products and, as a result of the incremental advances which have been made over the last 30 years , we can confidently expect to see products whose form and function has not been predicated on past operations. Additionally when we convince those with the responsibility for initiating research projects that it is possible to practically manufacture materials made from animal cells in culture at the levels of 1000KG/annum or more both reliably and reproducibly then we may expect an additional requirement for efficient processes based on animal cells in culture.

6. REFERENCES.

1. R.E. Spier, 1990
   Introduction to Volume 4; Contemporary Issues in Animal Cell Biotechnology.
   In Animal Cell Biotechnology Volume 4,
   Ed. R.E. Spier and J.B. Griffiths
   Academic Press, London, pp 1-13

2. R.E. Spier and N. Maroudas, (1991)
   Microcarriers for Animal Cell Biotechnology:An Unfulfilled Potential
   In Animal Cell Bioreactors,
   Ed C.Ho and D. Wang
   Butterworh-Heinemann, Guildford.

3. R.E. Spier, (1989)
   Animal Cell Culture: From Obscurity to Prominence to....
   In the Proceedings of the Eighth australian Biotechnology Conference
   University of New South Wales pp 41-51.

4. R.E. Spier, 1991
   Animal Cells Make Secondary Metabolites Too !
   In Production of Biologicals from Animal Cells in Culture
   Ed. R.E. Spier, J.B.Griffiths, B. Meignier
   Butterworths, London,

5. R.E.Spier, 1991
   Meeting Report in Vaccine
   To be published.

6. R.E. Spier, 1990
   Editorial in Enzyme and Microbial Technology,
   EMT v.12, 562-3

7. R.E. Spier and J.B. Griffiths, (1983)
   An Examination of the Data and Concepts germane to the Oxygenation of Cultured
   Animal Cells
   Dev. Biol. Standard. v. 55   pp 81-92

8. A. Handa-Corrigan, N. Emery and R.E. Spier (1989)
   Effect of gas -liquid interfaces on the growth of suspended mammalian cells:
   mechanisms of cell damage by bubbles.
   Enzyme and Microbial Tech. Vol 11., No. 4. pp 230-235.

9. R.E. Spier, (1990)
   The Oxygenation of Animal Cell Cultures by Bubbles
   In Animal Cell Biotechnology volume 4
   Ed. R.E. Spier and J.B. Griffiths
   Academic , London, pp 134-149.

CRITICAL ANALYSIS OF PROCESS DEVELOPMENT ON IN-VITRO GROWTH OF CHICK-EMBRYO

A. VENKATARAMAN, J. ANJANI KUMARI, P.S.R. BABU & T. PANDA*
Division of Biochemical Engineering
Department of Chemical Engineering
Indian Institute of Technology
Madras 600 036, INDIA.

ABSTRACT The propagation of animal cells in artificial media may be the basis for large scale biosynthesis of commercial animal proteins. This is equally applicable for the in-vitro development of chick-embryo. The major problems of in-vitro growth of chick-embryo are artificial environment for the growth of cells, suitable reactor configuration for cultivation and the mode of growth. In-vitro growth of chick-embryo can be better understod in stationary flask, agitated flask, continuous oscillatory glass reactor, and intermittent oscillation in glass reactor and continuous rotary reactor cf the reactor and methods developed for general cell growth. A survey of its advantages and disadvantages are summarized.

Introduction

The propagation of animal cells in artificial media can be the basis for large scale biosynthesis of commercial animal proteins. For the large scale cultivation of cells that will adhere to surfaces, one of the most promising techniques is to grow the cells on the surface of micro-carriers suspended in media [1]. Improvements in micro carrier technique have primarily been confined to media formulations and inoculation procedures [2,3]. Although these reactors have been used successfully for mammalian cell cultivation, no detailed analysis is available, which compares and determines the best reactor design and agitation procedures. Cultivation of animal cells e.g. chick-embryo is difficult because many of the cell types of interest exist as dense packing of similar or dissimilar cells. The challenges faced in the submerged cultivation of these cells is to provide an acceptable environment for the growth of such system independently. Attempt has been made to elucidate different aspect of chick-embryo cultivation in in-vitro [4]. This study summarizes the application of different configurations of reactors which can be studied in the process development of in-vitro chick-embryo.

47

R. Sasaki and K. Ikura (eds.), Animal Cell Culture and Production of Biologicals, 47–52.
© 1991 *Kluwer Academic Publishers.*

Existing reactor configuration for mammalian cell growth

Suspension cells are typically cultivated in flasks or bottles at the laboratory scale. They can be cultivated in a mechanically agitated vessels. This is equally applicable for the large scale application of both suspension and micro-carrier culture [5,6]. For the anchorage dependent cells, scale up depends on the increase in total surface area. Several attempts are exercised on new development of reactor configuration on increasing the growth surface area per unit vessel volume viz. packed-glass beads, stacked-plate, rotating multiple tubes and roller bottles with spiral film inside [5]. Modified stirred tank bioreactors, micro-carrier devices and hollow fiber reactors are also developed to meet these demand [5,7,8]. However, these devices can not meet the demand of chick-embryo development.

Reactors configuration for chick-embryo development

Various cultivation mode are practiced in our laboratory for the development of in-vitro growth of chick-embryo. A few of them are to be tried for this purpose.

STATIONARY FLASK MODE

This is consisted of a typical glass flask which is kept unstirred in a temperature and humidity controlled chamber (Fig.1).

Figure.1 Arrangement of stationary flask reactor

AGITATED FLASK SYSTEM

The glass flask, as it is used previously, is also used in this case. The slow agitation is, generally, caused by swirling motion and/or vibration in reaction phase. No mechanical agitation device in reaction fluid is advised for the development of chick-embryo. Temperature, humidity and gas conditions are necessary to be maintained during entire phase of development.

OSCILLATORY REACTOR

A typical arrangement of the reactor is shown in Fig.2. This is undulated continuously or intermittently for 180° per day.

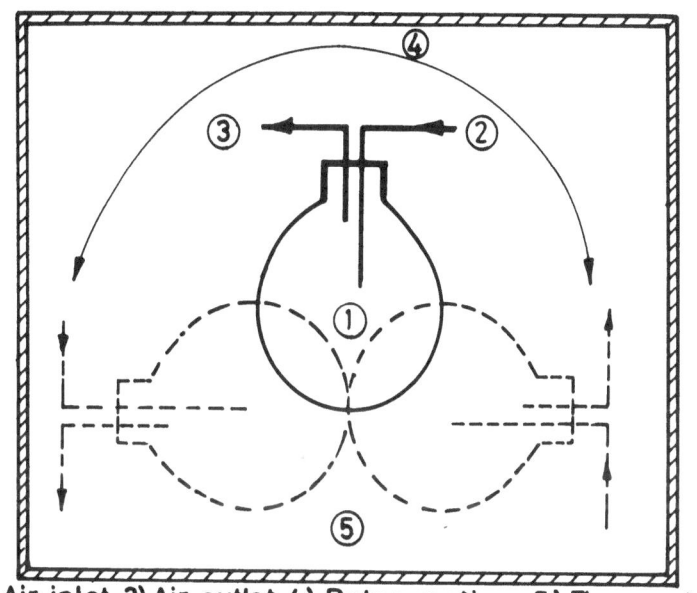

1) Reactor 2) Air inlet 3) Air outlet 4) Rotor motion 5) Thermostat

FIG. 2. OSCILLATORY REACTOR ASSEMBLY.

The arrangement of gas supply is as per Fig.3. This device does not contain any mechanical agitator in the reaction fluid.

CONTINUOUS ROTARY REACTOR

Another device of rotary reactor is also promising. This reactor is housed in a controlled environment chamber (Fig.4), having gas entry and exit arrangement as described in Fig.3. The reactor is allowed to rotate in both direction at 360°/day.

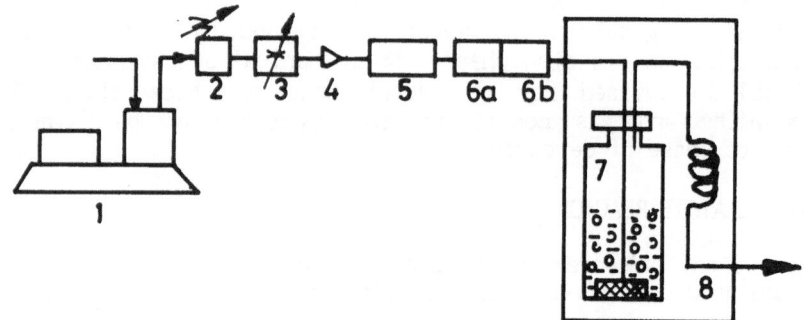

1)Compressor 2) Pressure reduction station 3) Finer pressure adjustment
4) Rotameter 5) CO_2 separator 6a) Activated carbon filter 6b) Microbial
filter 7) Gas washer cum humidifier 8) Thermostatic chamber.

FIG. 3.   AIR INLET ASSEMBLY

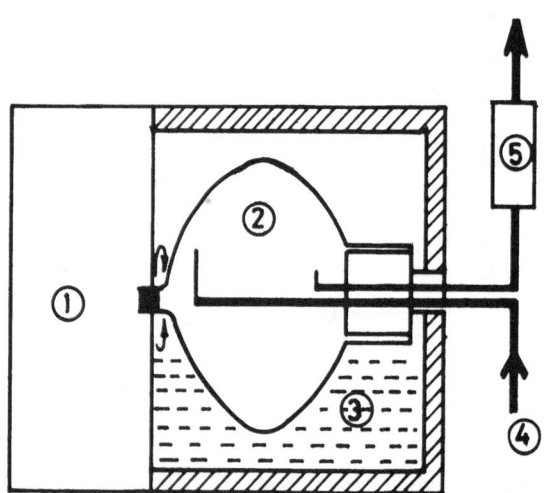

1) Rotor assembly and control units   2) Reactor 3) Thermostat
4) Air inlet, 5) Air outlet filter

FIG·4· ROTARY REACTOR ASSEMBLY·

Discussion

The studies on stationary reactor (i.e. conical and circular types) have been made in batch mode. The circular configuration has shown better performance compared to the conical geometry which requires more volume of medium. On the other hand, it has higher simplicity in operation whereas the survival rate is very low. This might require a minimum agitation in both circular and conical configuration. The bottom geometry of conical reactor has higher spreading compared to spherical type. The experiments have motivated in this direction. This has shown an important criteria in this aspect.

For further studies, spherical/oval reactors have been used with mild agitation in different devices viz: oscillatory-continuous and intermittent, continuous rotary. The continuous rotary reactor has advantage because it could accommodate series of reaction simultaenously.

The viability of chick-embryo has been observed to be less in stationary reactor. The agitated flask/spherical device has shown higher degradation due to vibration/agitation. For this purpose, the process has been performed in continuous/intermittent oscillatory reactor. It has been observed that large scale development of chick-embryos can not be made. However, oval reactor/continuous rotary type) has the prospect of large scale development of chick-embryo. The viability of chick-embryo has been found to be more or less same in oscillatory and continuous rotary reactors. Further studies are in progress to resolve some more problems.

References

1. Croughan, M.S., Hamel, J.-F., and Wang, D.I.C.(1987)'Hydrodynamic effects on animal cells grown in microcarrier cultures', Biotechnology and Bioengineering, 29, 130-141.

2. Smiley, A.L., Hu, W.-S, and Wang, D.I.C. (1989) 'Production of human immune interferon by recombinant mammalian cells cultivated on microcarriers', Biotechnology and Bioengineering, 33, 1182-1190

3. Lauffenburger, D. and Cozens, C.(1989), 'Regulation of mammalian cell growth by autocrine growth factors : analysis of consequences for inoculum cell density effects', Biotechnology and Bioengineering, 33, 1365-1378.

4. Venkataraman, A., Babu, P.S.R., and Panda, T. (1990) 'Studies on the in-vitro, development of chick-embryo' in Proc. APBioChEC'90, Kyungju, Korea, pp 149-151.

5. Hu, W.-S. and Dodge, T.C. (1985) 'Cultivation of mammalian cells in bioreactors' Biotechnology Progress, 1(4), 209-215.

6. Croughan, M.S., Hamel, J.-F.P. and Wang, D.I.C. (1988) 'Effects of microcarrier concentration in animal cell culture' Biotechnology

and Bioengineering, 32, 975-982.

7. Chresand, T.J., Gillies, R.J. and Dale, B.E. (1988) 'Optiomum fiber spacing in a hollow fiber bioreactor' Biotechnology and Bioengineering, 32, 983-992.

8. Cherry, R.S. and Poputsakis, E.T. (1988) 'Physical mechanisms of cell damage in microcarrier cell culture bioreactors' Biotechnology and Bioengineering, 32, 1001-1014.

IMPROVED METHOD FOR INOCULATION OF A CELL SUSPENSION INTO A HOLLOW FIBER BIOREACTOR

George Avgerinos*, Sean Downing**, Beth Gefre, Harvey
Freedman, and Christina Pelletier***
Amicon Division, W.R. Grace & Co.-Conn.
72 Cherry Hill Drive
Beverly, MA 01915
U.S.A.

ABSTRACT. Monoclonal antibody (MAb) production was evaluated with three different murine hybridoma cell lines in a "mini" hollow fiber bioreactor. Optimized production rates between 17-25 mg/day could be achieved with medium consumption of 2 liters per week. Antibody product was harvested at 3-4 mg/ml up to five times per week. Bioreactor productivity was enhanced twofold when cartridges were inoculated by direct ultrafiltration of cells in their conditioned medium as opposed to centrifugation, resuspension and inoculation in a smaller volume of fresh medium. Control reactors were started by centrifuging 2-2.5 x 10^8 cells and resuspending in 3 ml of fresh medium with 5-25% fetal bovine serum (FBS) prior to inoculation into the 7 ml extracapillary space (ECS) of the bioreactor. Alternatively, the same total number of cells were inoculated into the ECS directly from up to 100 ml of conditioned medium by ultrafiltration. The twenty to sixty minute ultrafiltration procedure concentrates various serum and cellular growth factors and potentially better distributes the cells around the fibers. It resulted in the decrease of the lag phase from twelve to sixteen days to three to six days and surprisingly, increased rates of monoclonal antibody production throughout the course of runs of up to sixty days duration.

Introduction

Hollow fiber bioreactors possess a number of inherent advantages for MAb production. These include recovery of relatively pure, highly concentrated MAb and minimization of serum requirements [1]. However, we have found that inoculation methodology has a major influence on realization of the bioreactor's full potential. In this report we characterize a direct ultrafiltration inoculation method which significantly enhances MAb production over other methods described in the literature

*Present affiliation: BASF Bioresearch Corp., Cambridge, MA
**Present affiliation: Rochester Institute of Technology, Rochester, NY
***Present affiliation: Serono Laboratories, Randolph, MA

R. Sasaki and K. Ikura (eds.), Animal Cell Culture and Production of Biologicals, 53–60.
© 1991 *Kluwer Academic Publishers.*

[2]. In addition to enhanced production, this method decreases start-up time from approximately fourteen days to about four days. A significant reduction of media consumption and an increase in monoclonal antibody titers over the course of the runs were also observed.

Materials and Methods

Mini Flo-paths (Amicon part #83130, Beverly, MA) with a total of 400 cm^2 of 30,000 molecular weight cut-off anisotropic polysulfone fibers were used to evaluate MAb production and inoculation procedures with three cell lines: BIPI, BD166, and CRL1606. A total of ten "minis" were run: three with BIPI, four with BD166, and three with CRL1606. CRL1606 is a P3X murine hybridoma which produces an IgG_1 antibody against human fibronectin. BD166 and BIPI are both P3X hybridomas producing proprietary IgG_1 antibodies. The cell lines were maintained in WRC 935 medium (Amicon, Beverly, MA). In T-flasks, BIPI and BD166 were grown with 10% added FBS and produced approximately 200 and 100 $\mu g/ml$ respectively, when grown to saturation. CRL1606 was grown with the addition of 2% FBS and produced 200 $\mu g/ml$.

Approximately 100 ml of T-flask culture medium at 2-2.5 x 10^6 cells/ml was used to inoculate three "minis" with an ultrafiltration inoculation procedure. Seven other "minis", inoculated using a standard inoculation procedure with 2-2.5 x 10^8 cells served as controls. Cell count and viability were determined just prior to inoculation into the ECS. Cell viability was typically greater than 90%. The "minis" were kept in a warm room at 37°C and medium was continuously recirculated at 100 ml/min from a 1L reservoir, through the fiber lumen, and returned to the reservoir.

STANDARD INOCULATION

In our standard inoculation method, 2-2.5 x 10^8 cells are inoculated into the ECS. Cells are harvested from T-flasks and gently spun and resuspended in a 3 ml volume of fresh medium containing 10% FBS. Prior to injection of cells into the cartridge, 10 ml of medium with 5-10% FBS is flushed over the fibers. A syringe containing the cells in fresh medium is placed on the inlet port and gently pushed into the bioreactor. Approximately 2-2.5 ml of medium with 5-10% of serum is placed into the reactor following inoculation to chase cells remaining in the inlet port into the ECS.

ULTRAFILTRATION INOCULATION

The ultrafiltration inoculation procedure also uses 2-2.5 x 10^8 cells, but in a 100 ml volume of conditioned medium without centrifugation. The dilute cell suspension is directly placed in two 60 cc syringes and attached to both the inlet and the outlet ECS ports (see Figure 1). The tubing just above and just below the cartridge is clamped off and the flushing line above the cartridge is routed to the pump. When the pump is turned on it creates a negative pressure inside the cartridge

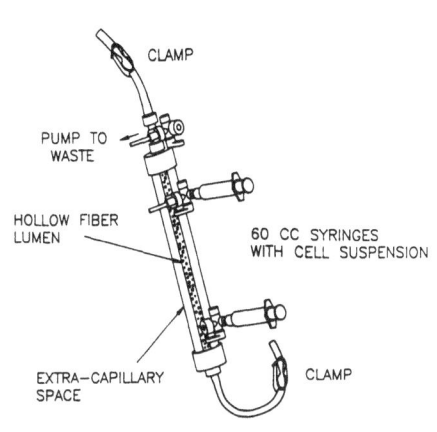

CLAMP

PUMP TO WASTE

HOLLOW FIBER LUMEN

60 CC SYRINGES WITH CELL SUSPENSION

EXTRA—CAPILLARY SPACE

CLAMP

Figure 1. Schematic illustration of bioreactor inoculation methodology.

which draws the conditioned medium and cells into the ECS and discharges a cell-free permeate at very low protein concentration through the fibers. The procedure concentrates the cells, as well as higher molecular weight growth factors, and serum proteins in the ECS.

OPERATION OF THE BIOREACTORS

"Mini" bioreactors were set up and rinsed with 1L of distilled water according to operating instructions. After inoculation as described above the Mini Flopath's 1L medium reservoir was filled with fresh WRC 935 medium. Recirculation media samples were taken daily and the pH was measured. The one liter medium bottle was replaced on a twice week basis. After two weeks of operation the WRC 935 medium was supplemented with 1 ml of 1N NaOH at each change. Typically, the recirculation medium reached a pH of 6.8 at the time of change.

All of the "mini" bioreactors were started with 5% FBS in the recirculating medium. In the case of cell line 1606, this FBS level was rapidly reduced to 0% by the third medium change. In the case of cell line BIPI, the serum level was reduced to 2½% FBS while cell line BD166 was maintained at 5% FBS in the recirculating medium.

Harvesting of MAb from the ECS was done on a daily basis with 10 ml being taken each weekday. The MAb concentration was determined by radial immunodiffusion (RID) on mouse IgG_1 plates (code #RN273.4) supplied by The Binding Site Ltd. (San Diego, CA).

Results and Discussion

The new inoculation procedure described uses the potential of Amicon's high flux anisotropic hollow fibers to ultrafilter 100 ml of conditioned medium and cells into the ECS. Not surprisingly, however, the time required for inoculation is related to the percentage of serum contained within the conditioned medium and the amount of cells present. Cell line BIPI was inoculated in 10% serum with a cell concentration of 2.5×10^6 cells/ml in 100 ml. The inoculation took sixty minutes to complete. BD166 which also had 10% FBS, but only a cell concentration of 2.0×10^6 cells/ml, took fifty minutes. However, the 1606 cell suspension also at 2.0×10^6 cells/ml, containing only 2% FBS, was inoculated into the bioreactor in just twenty minutes.

The results of the Mini Flo-path bioreactor runs are summarized in Table 1. These runs lasted from 44 to 89 days and produced from 280 to 1250 mg of MAb. Production of MAb over time for each run is shown in Figures 2-4. With the ultrafiltration inoculation method, one gram of MAb was produced in forty to sixty days with each cell line tested. Data comparing the performance of the standard and new inoculation methods with each cell line is shown in Table 2.

TABLE 1. Summary of Mini Flo-path experiments with various inoculation methods

RUN NUMBER	CELL LINE	INOCULATION METHOD	DAYS RUN	TOTAL MAb PRODUCED (g)	TOTAL MEDIUM USED (Liters)
1	BD166	Std.	60	0.5	16
2	BD166	Std.	60	0.6	17
3	BD166	Std.	80	0.89	21
4	BD166	UF	55	0.97	16
5	BIPI	Std.	65	0.28	15
6	BIPI	Std.	70	0.5	15
7	BIPI	UF	59	0.99	18
8	CRL1606	Std.	69	1.25	21
9	CRL1606	Std.	89	0.96	35
10	CRL1606	UF	44	1.02	13

TABLE 2. Comparison of Mini Flo-path performance with standard and ultrafiltration (UF) inoculation methods

	Std.	UF
A) Cell Line BD166		
Average MAb Productivity (mg/day	9.8	17.6
Average MAb Harvest Conc. (mg/ml)	1.8	2.9
MAb Yield (mg/L medium used)	36.3	60.6
Start-Up Time (days)	12.3	3.0
B) Cell Line BIPI		
Average MAb Productivity (mg/day)	5.8	16.8
Average MAb Harvest Conc. (mg/ml)	0.9	2.7
MAb Yield (mg/L)	21.6	55.0
Start-Up Time (days)	15.5	6.0
C) Cell Line CRL1606		
Average MAb Productivity (mg/day)	14.5	23.2
Average MAb Harvest Conc. (mg/ml)	2.2	3.7
MAb Yield (mg/L)	43.5	78.5
Start-Up Time (days)	12.5	5.0

The average productivity of the standard inoculation method in three runs for BD166 was 9.8 mg/day and an average harvest concentration of 1.8 mg/ml was observed. With the ultrafiltration inoculation procedure

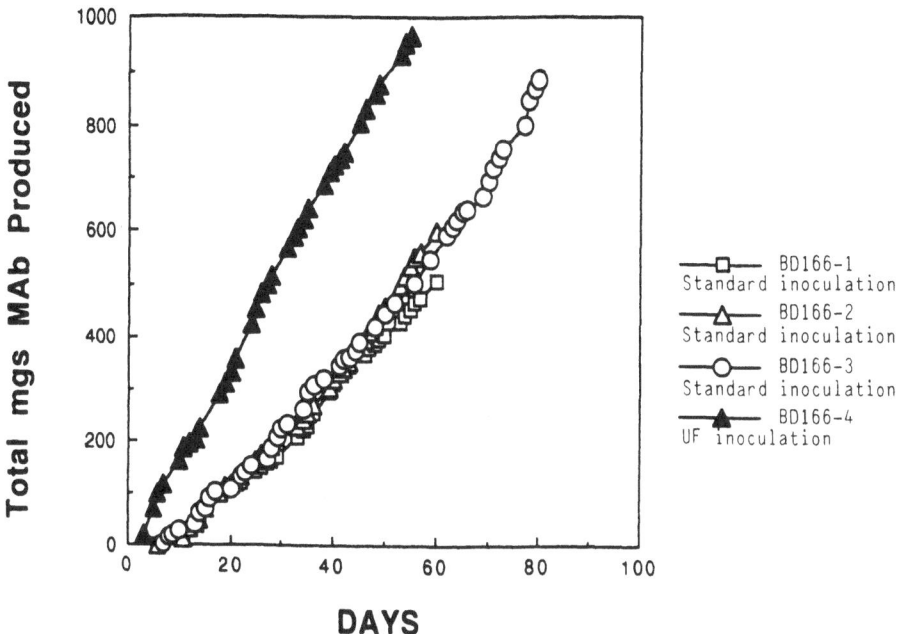

Figure 2. Effect of inoculation procedure on MAb production by cell line BD166

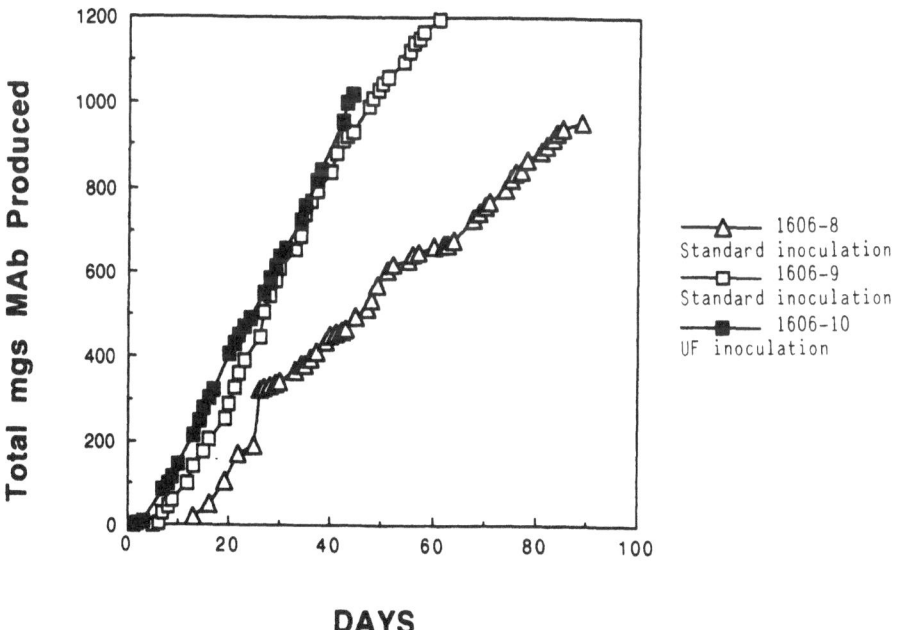

Figure 3. Effect of inoculation procedure on MAb production by cell line CRL1606

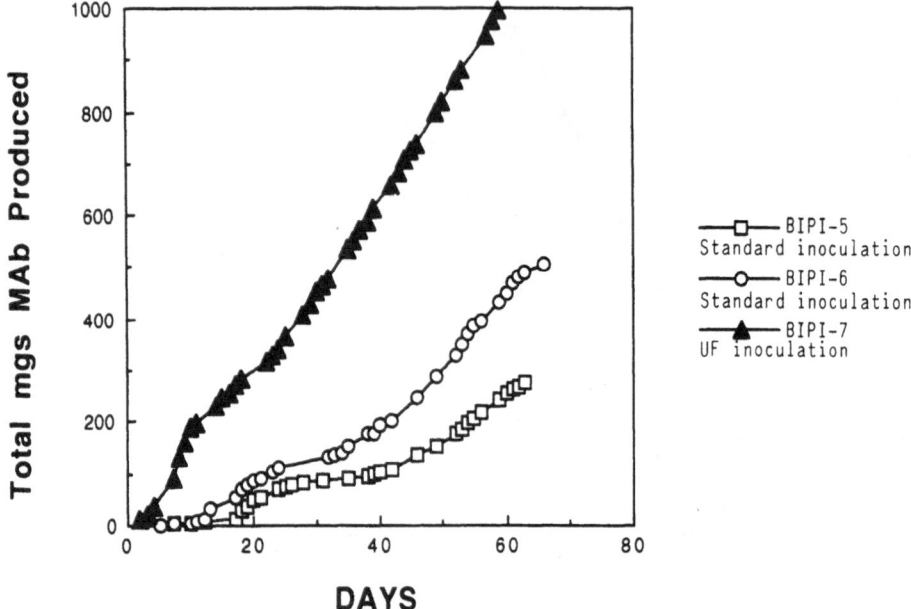

Figure 4.   Effect of inoculation procedure on MAb production by cell line BIPI

almost double the average productivity was observed at 17.6 mg/day with an average harvest concentration of 2.9 mg/ml.

Using the standard inoculation procedure, BIPI had an average productivity of 5.8 mg/day and an average harvest concentration of 0.9 mg/ml in two runs. The ultrafiltration inoculation method had an average productivity of 16.8 mg/day, which is over double that of the standard method. The ultrafiltration inoculation procedure also had an average harvest concentration of 2.7 mg/ml.

The "minis" containing CRL1606 inoculated with the standard procedure, in two runs had an average productivity of 14.5 mg/day and an average harvest concentration of 2.2 mg/ml. The ultrafiltration inoculation procedure had an average productivity of 23.2 mg/day and an average harvest concentration of 3.7 mg/ml.

The ultrafiltration inoculation procedure resulted in a 60-250% increase in the average daily production of MAb with each cell line tested. The average daily MAb produced was determined by dividing the total amount of antibody produced over the course of a run by the number of days of the run. Due to the increase in daily MAb production with the ultrafiltration inoculation procedure, the average MAb concentration in the harvest increased 61-200% as well. The start up times of the bioreactors were compared by determining the time required to achieve the average MAb productivity of the run. These start up times were reduced by seven to ten days with each cell line. Due to the improvements in start up time and MAb productivity, the medium required to achieve each mg of MAb was also significantly reduced. The ultrafiltration inoculation procedure increased MAb yield (mg/L medium used) by 67-155% with the cell lines tested.

These data demonstrate the effectiveness of an ultrafiltration inoculation procedure, towards enhancing monoclonal antibody production in Amicon's Mini Flo-path. This could have resulted from the concentration of serum and growth factors in the extra capillary space and/or the better distribution of cells about the fibers throughout the entire bioreactor. Investigators have dealt with the problem of cell distribution during inoculation by rotating or shaking the cartridge after inoculation [3] or by transferring cells in one ECS volume of conditioned medium by pressurization of a transfer vessel which is connected to the ECS with a tube [4]. Patankar and Oolman used a method in which a single hollow fiber was sealed at one end, connected to a vacuum source at the other and immersed in a suspension of cells. After the cells were drawn to the fiber it was placed in fresh medium [5]. This method draws the cells into the spongy layer of the fiber but does not concentrate high molecular weight factors from the conditioned medium.

The ultrafiltration method we have demonstrated concentrates over ten ECS volumes of cells with their conditioned medium factors into the ECS of the bioreactor. The high water flux and porosity structure of the anisotropic fibers enables this concentration which may not be feasibly done with many isotropic fiber types. The result of this method with two of the cell lines, BD166 and BIPI was an increased average MAb productivity over two fold. On the other hand, a 50% increase was shown for CRL1606. This smaller improvement may be due to the fact that 1606 has the lowest serum dependence of the lines tested.

Summary

In conclusion, an ultrafiltration inoculation method effectively enhances rapid start up and growth of cells in the bioreator. The sustained impact of this ultrafiltration inoculation method on monoclonal antibody productivity over the course of the "mini" runs was a surprising finding. Use of this procedure would significantly reduce the operating cost associated with MAb production in the "Mini" Bioreactor. Data, not reported, has shown that the productivity and start-up enhancement of this inoculation method is translated to hollow fiber systems two to twenty fold larger in scale as well.

References

1. Hopkinson, John (1985) 'Hollow fiber cell culture systems for economical cell-product manufacturing', Biotech. 3, 225-230.
2. Ku, K., Kuo, M.J., Delente, J., Wildi, B.S., and Feder, J. (1981) 'Development of a hollow-fiber system for large scale culture of mammalian cells', Biotech. Bioengin., 23, 79-95.
3. Piret, J.M. and Cooney, C.L. (1990) 'Mammalian cell and protein distributions in ultrafiltration hollow fiber bioreactors', Biotech. Bioengin., 36, 902-910.

4. Knazek, R.A., Wu, Y.W., Aebersold, P.M., and Rosenbery, S.A. (1990) 'Culture of human tumor infiltrating lymphocytes in hollow fiber bioreactors', j. Immunol. Meth., 127, 29-37.

5. Patankar, Dhananjay and Oolman, Timothy (1990) 'Wall-growth hollow-fiber reactor for tissue culture: 1. preliminary experiments', Biotech. Bioengin. 36, 97-103.

STRATEGIES TO INCREASE THE EFFICIENCY OF MEMBRANE AERATED AND PERFUSED ANIMAL CELL BIOREACTORS BY AN IMPROVED MEDIUM PERFUSION

H.D. BLASEY and V. JÄGER
Gesellschaft für Biotechnologische Forschung mbH
Mascheroder Weg 1
D-3300 Braunschweig
Germany

ABSTRACT. A bioreactor system based on internal membrane aeration and perfusion was used to cultivate a variety of mammalian and insect cell lines for the production of recombinant proteins, monoclonal antibodies or biomass. The efficiency of this reactor system is evaluated with respect to maximum cell concentration, productivity and membrane lifetime in order to give some fundamental data for comparison with other stirred tank perfusion bioreactors. Due to the membrane capacities for aeration and perfusion, eight to thirty fold increases in cell concentration compared to batch cultivation in culture flasks could be obtained routinely with unchanged cell viabilities.

For further increase of the efficiency of the perfusion system several strategies have been developed. Different membrane materials and pore sizes have been investigated and the membrane lifetime has been prolonged by backflushing and membrane precoating.

1. Introduction

High density perfusion culture in stirred tank bioreactors has been shown to be an attractive alternative to conventional batch or fed-batch processes for the large-scale cultivation of animal cells. Medium exchange in these systems can be performed either internally by spin filters (Himmelfarb et al. 1969, Tolbert et al. 1981, Varecka et al. 1987), by inserted membrane modules (Takazawa et al. 1988) or by settling zones (Tokashiki et al. 1988) or externally by hollow fiber or flat-plate cross flow filtration systems (Velez et al. 1988) or by dynamic filters (Rebsamen et al. 1987). In our research group, we have developed an internal perfusion system based on microfiltration via moving hollow fiber membranes (Lehmann et al. 1987, Lehmann et al. 1988).

Generally, maximum cell concentration is influenced by the cell specific consumption rates for oxygen as well as for other nutrients and it is limited by aeration and perfusion, respectively. In addition, attention has to be focussed on the accumulation of toxic metabolic by-products which also can be controlled by the perfusion rate. Therefore, for comparative purposes, data have to be interpreted with respect to oxygen transfer rates (OTR) and medium exchange rates. OTR of this system have already been investigated by Lehmann et al. (1987), and the usual medium exchange rates are summarized here.

To optimize the perfusion process several strategies have been developed to reduce membrane fouling and increase membrane lifetime.

61

R. Sasaki and K. Ikura (eds.), Animal Cell Culture and Production of Biologicals, 61–73.
© 1991 *Kluwer Academic Publishers.*

2. Material and Methods

2.1. CELL LINES AND MEDIA

For the evaluation of the bioreactor system various cell lines were used including hybridomas (mouse-mouse, rat-mouse, rat-rat, human-human-mouse), recombinant cells (BHK-21, L-929) and insect cell lines. These are summarized in TABLE II. A variety of different serum supplemented, serum-free (DIF) (Jäger et al. 1988) or protein-free media (Lucki-Lange et al. 1991) were used for the cultivation of these cells, according to their individual growth requirements.

2.2. BIOREACTOR SYSTEMS

Stirred tank bioreactors in a scale from 1 to 5 liters were used. They were equipped with moving microporous polypropylene hollow fiber membranes (S6/2) (Enka, Wuppertal, Germany) for aeration and additional membranes of the same type for medium perfusion (Lehmann et al. 1988). Usually 3 meters of membranes per liter reactor volume were used for aeration and for medium perfusion, respectively.

For optimization of the perfusion system, several other hydrophilic or hydro-philized microporous membranes were used as shown in TABLE I. The length of these membranes was kept within a range from 20 to 40 cm to shorten the time until membrane fouling starts to influence flux and transmembrane pressure. The lengths of the different hollow fiber membranes were adjusted to get comparable membrane surface areas. Transmembrane pressure was measured with autoclavable piezoresistive pressure gauges (Kistler, Switzerland). A schematic diagram of the reactor system is shown in Fig. 1.

TABLE I. Hollow fiber membranes used for optimization of the perfusion system.

Type:	Q3/2	R5/2	S6/2	S6/4	C3/2	F L O W
Manufacturer:	Enka	Enka	Enka	Enka	Enka	Flow Lab.
Material:	Poly-propylene	Poly-propylene	Poly-propylene	Poly-propylene	Nylon PA6	Polyether-imid
Pore size [μm], nominal/maximal:	0.2/0.65	0.2/0.7	0.2/0.6	0.4/0.85	0.2/?	0.8-1.0
Bubble point (Water) [hPa]:	3100	3100	3500	2500	-	-
Inner Diameter [μm]:	600	1200	1800	1800	600	2000
Wall Thickness [μm]:	200	300	400	400	200	800
Max. Pressure during Operation at 40°C [hPa]:	?	1400	1400	800	?	?
Porosity [%]:	75	75	75	75	?	?

2.3.   ANALYTICAL METHODS

Cell number was estimated by trypan blue exclusion with a hemocytometer.
Protein content in fouled hollow fiber membrane was measured by the Kjeldahl method.

The lipids bound to the membrane were extracted and separated by high performance thin layer chromatography (HPTLC). Phospholipids were quantified as described by Eibl et al. [5].

Nucleic acids were extracted from the membranes and quantified photometrically at 260 nm. The size of the nucleic acids was estimated by gel electrophoresis. For destinction between RNA and DNA a DNA-microassay was used [6].

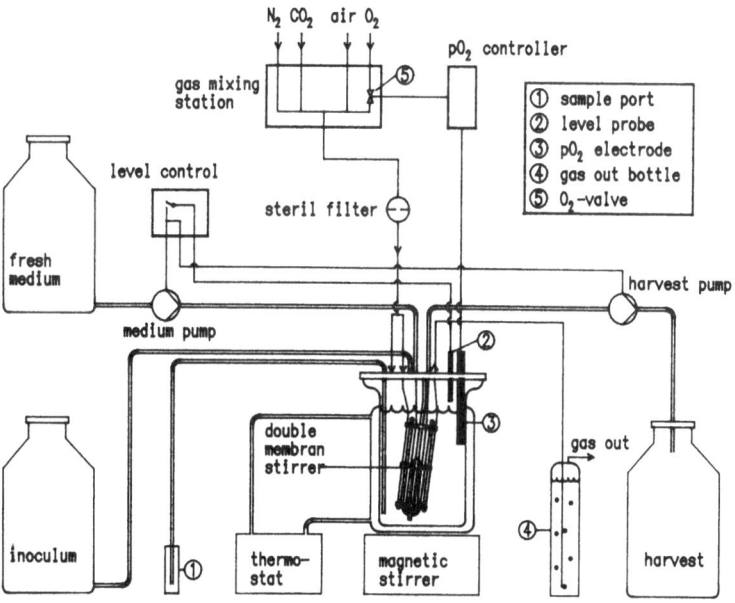

Fig.1: Schematic diagram of a 1.4 liter perfusion bioreactor system.

TABLE II. Comparison of maximum viable cell concentrations of animal cell lines in culture flasks, in batch and in continuous perfusion culture.

Cell line	Type	Product	Maximum viable cell concentrations / ml			Medium exchange rate (V_R/d)	Cultivation method
			in culture flasks	in batch culture	in continuous perfusion culture		
CB-hahE	human hybridoma	human IgG_2	$3.0 \cdot 10^5$	$5.4 \cdot 10^5$	$4.8 \cdot 10^6$	0.6	Suspension
CB-maHIV 4-1	murine hybridoma	mouse IgG_{2a}	$1.5 \cdot 10^6$	$3.2 \cdot 10^6$	$3.0 \cdot 10^7$	3.6	Suspension
CB-maHIV 13-5	murine hybridoma	mouse IgG_1	$1.2 \cdot 10^6$	$1.4 \cdot 10^6$	n.d.		Suspension
LA/C4	murine hybridoma	mouse IgG_1	$8.0 \cdot 10^5$	$1.0 \cdot 10^6$	$7.9 \cdot 10^6$	5.0	Suspension
561	murine hybridoma	mouse IgG_1	$1.1 \cdot 10^6$	$1.4 \cdot 10^6$	n.d.		Suspension
2125	murine hybridoma	mouse IgG_{2a}	$1.2 \cdot 10^6$	$1.2 \cdot 10^6$	$7.5 \cdot 10^6$	2.5	Suspension
H28	rat-mouse hybridoma	rat IgG	$7.7 \cdot 10^5$	$1.6 \cdot 10^6$	$1.4 \cdot 10^7$	5.5	Suspension
412	rat-mouse hybridoma	rat IgG	$7.0 \cdot 10^5$	$1.3 \cdot 10^6$	$1.1 \cdot 10^7$	5.5	Suspension
187.1 (ATCC HB58)	rat-mouse hybridoma	rat IgG_{2c}	$1.2 \cdot 10^6$	$1.7 \cdot 10^6$	$1.9 \cdot 10^7$	5.0	Suspension
YL 1/2	rat hybridoma	rat IgG	$1.2 \cdot 10^6$	$1.6 \cdot 10^6$	$1.4 \cdot 10^7$	1.6	Suspension
BHK-21 pSVIL2	syrian hamster kidney	r hu IL-2	$2.0 \cdot 10^6$	$2.0 \cdot 10^6$	$2.8 \cdot 10^7$	4.0	Suspension
BHK-21 pSVIL2	syrian hamster kidney	r hu IL-2	$2.0 \cdot 10^6$	$3.0 \cdot 10^6$	$8.0 \cdot 10^6$	3.0	Microcarrier
BHK-21 pODa cl39	syrian hamster kidney	r hu PDGF-AA	$1.7 \cdot 10^6$	$1.9 \cdot 10^6$	$7.0 \cdot 10^7$	5.0	Susp. (Spheroids)
BHK-21 pODa cl39	syrian hamster kidney	r hu PDGF-AA	$1.7 \cdot 10^6$	$1.7 \cdot 10^6$	$1.6 \cdot 10^7$	2.0	Microcarrier
BHK-21 pSVIFN-ß	syrian hamster kidney	r hu IFN-ß	$1.7 \cdot 10^6$	$1.7 \cdot 10^6$	$2.0 \cdot 10^7$	2.0	Microcarrier
L-M(TK)pSVIFN-ß	mouse connective tissue	r hu IFN-ß	$1.0 \cdot 10^6$	$5.5 \cdot 10^6$	$1.2 \cdot 10^7$	1.0	Microcarrier
L-M(TK)pSVIL2	mouse connective tissue	r hu IL-2	$1.0 \cdot 10^6$	$2.0 \cdot 10^6$	$8.0 \cdot 10^6$	1.0	Microcarrier
IPLB-Sf-21	fall armyworm ovary	-	$1.2 \cdot 10^6$	$6.6 \cdot 10^6$	$2.1 \cdot 10^7$	2.5	Suspension
Kc	drosophila embryo	-	$4.0 \cdot 10^6$	$6.2 \cdot 10^6$	$1.6 \cdot 10^7$	0.6	Suspension
Schneider-2	drosophila embryo	-	$5.0 \cdot 10^6$	$1.2 \cdot 10^7$	$3.1 \cdot 10^7$	0.5	Suspension

3.   Results and Discussion

3.1.   EFFICIENCY OF THE BIOREACTOR SYSTEM

To compare the efficiency of our perfusion bioreactor with other systems, data from various cultivation processes have been summarized in TABLE II. Numerous hybridomas of different origin and insect cell lines have been grown as suspension cultures, and several recombinant cell lines have been grown either in suspension, on microcarriers or as spheroids. The maximum viable cell density which could be reached during batch culture was up to 100 % higher in the bioreactor when compared with conventional culture flasks. The population doubling time in the bioreactor was also reduced, indicating that the controlled and more homogeneous conditions combined with a better oxygen supply and a gentle mixing with extremely low shear forces provided a superior environment for animal cell lines. In continuous perfusion mode this cell density could be increased again by a factor of 8 to 35. The medium exchange rates necessary to reach these cell densities were quite different and depended upon the individual consumption rates or minimum dilution rates to reduce toxic metabolic by-products. Normally, perfusion rates were not increased to the theoretical maximum of more than 5 reactor volumes per day. Membrane lifetime could be significantly increased to more than 6 weeks by not exceeding a rate of 2 reactor volumes per day.

Due to the higher cell densities obtained in the reactors, the volumetric productivity also was generally higher. Moreover, some cell lines significantly increased their specific productivity up to a factor of 5, as shown in TABLE III.

TABLE III.   Comparison of cell specific productivity in culture flasks and in batch fermentation

Cell line	Cell specific productivity (mg $\cdot$ 10^{-9} $\cdot$ d^{-1})	
	in culture flasks	in batch fermentation
H28	7.0	10.4
412	4.6	4.4
BHK-21 pSVIL2	4.0	21.0

3.2.   STRATEGIES TO INCREASE THE MEMBRANE LIFETIME

For further improvement of the perfusion system several parameters could be optimized: the hollow fiber membrane, the filtration mode, the culture medium and treatment of membranes to reduce fouling.

3.2.1. *Restoration of the membrane with proteases.* Due to the fact that the perfusion system was integrated into the bioreactor, it was impossible to exchange membranes in place after they are clogged. Alternatives to increase the duration of a process were a restoration of the membrane in place or an increase of the membrane lifetime by reducing membrane fouling.

A complete restoration of the membrane in situ could be carried out by a short term treatment with proteases such as trypsin, but this process was associated with degradation of product, especially when cells were grown in serum-free or protein-free media which contain no trypsin inhibitors. Normally, such procedures could not be accepted in view of the product quality.

3.2.2. *Characterization of fouling materials.* In developing strategies for an increase of membrane lifetime it is necessary to identify and characterize the substances responsible for membrane fouling. Substances found in clogged Accurel[R]-polypropylene hollow fiber membranes are summarized in TABLE IV. As expected, proteins with a total of 6.8 g per m^2 were the major class of fouling substances. No DNA but 6 mg per m^2 of RNA, with a very characteristic size of 800 bp, was found.

A very important factor seemed to be the different lipid fractions (phospholipids, glycolipids, etc.). It can be assumed that these compounds are able to bridge between the hydrophobic surface of the membrane and hydrophilic substances and thus enhance the fouling potential of proteins.

TABLE IV. Fouling materials in clogged Accurel[R] hollow fiber membranes

Material:	Concentration [mg/m^2]:
Proteins	6800
DNA	Not detected
RNA (800bp)	6
Phospholipids	400
Glycolipids	+
Other lipophilic substances	+
Particles ($<1\mu m$)	+
Whole cells	Not detected

3.2.3. *Screening for an optimum perfusion membrane.* In Fig. 2 the hydrophilized S6/2 polypropylene membrane which was normally used for perfusion in our bioreactor system was compared with two hydrophilic membranes. Flux through the membranes has been standardized for better comparison:

$$(Flux_n = Flux / Flux_{max})$$

The polypropylene membrane showed only a slight reduction in flux whereas the fluxes of the hydrophilic membranes decreased rapidly. Nylon membranes have a high affinity for proteins, which were partially denatured after adsorption. A comparison of polypropylene membranes with different diameters but an identical pore size of 0.2 μm gave results which were very similar, but again the S6/2 membrane proved slightly superior (Fig. 3). The S6/4 membrane has the same diameter as the S6/2 but a larger pore size of 0.4 μm, which results in a better flux as shown in Fig. 4. However, the maximum transmembrane pressure of the S6/4 is significantly lower thgan that of the S6/2 (800 hPa compared to 1400 hPa) and in a critical range during operation (Fig. 5), thus limiting the use of this membrane, especially in combination with backflushing (see below).

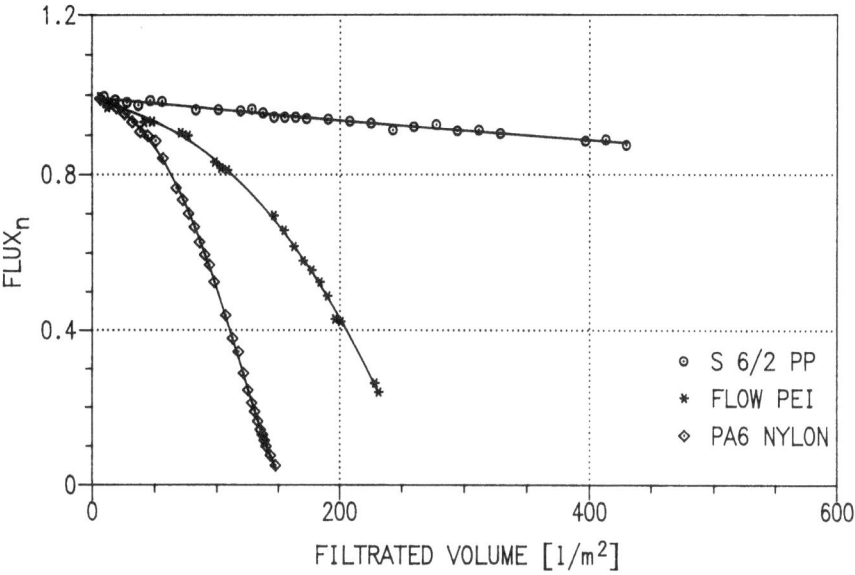

Fig. 2.   Comparison of the hydrophilized S6/2 polypropylene membrane with hydrophilic nylon and polyetherimid membranes.

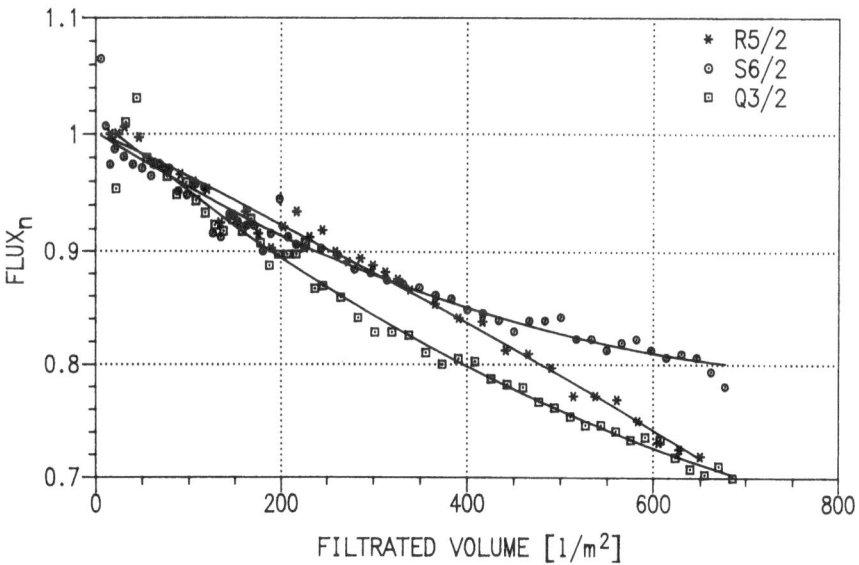

Fig. 3.   Comparison of three 0.2 μm hydrophilized polypropylene membranes with different diameters.

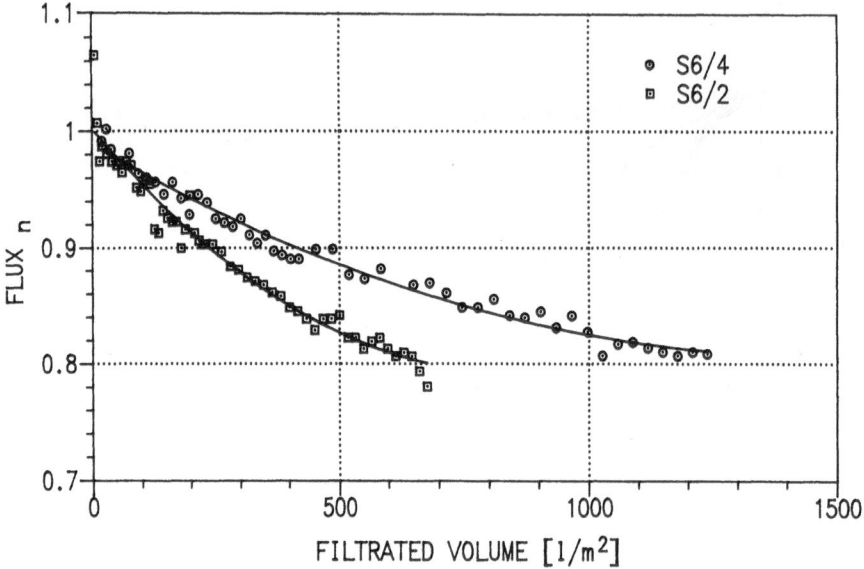

Fig. 4.   Comparison of two hydrophilized polypropylene membranes with different pore sizes (0.2 μm and 0.4 μm).

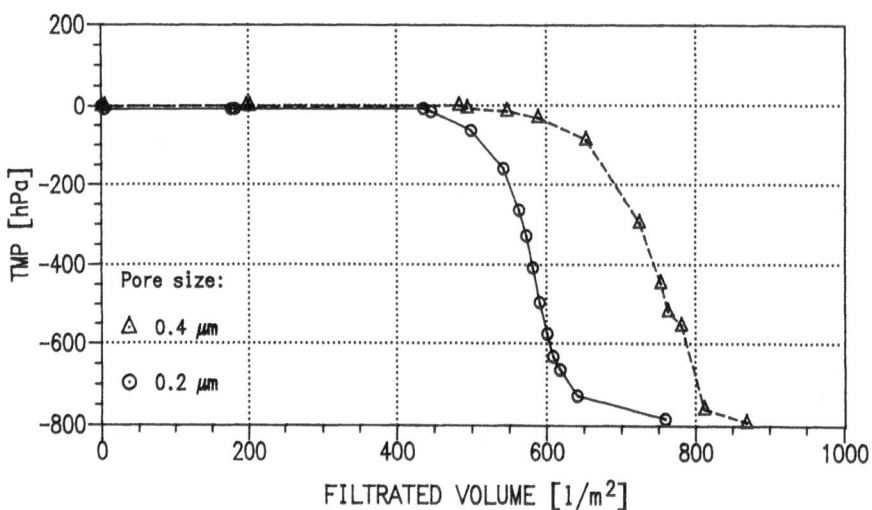

Fig. 5.   Transmembrane pressure during operation of hydrophilized polypropylene membranes with a pore size of 0.2 or 0.4 μm.

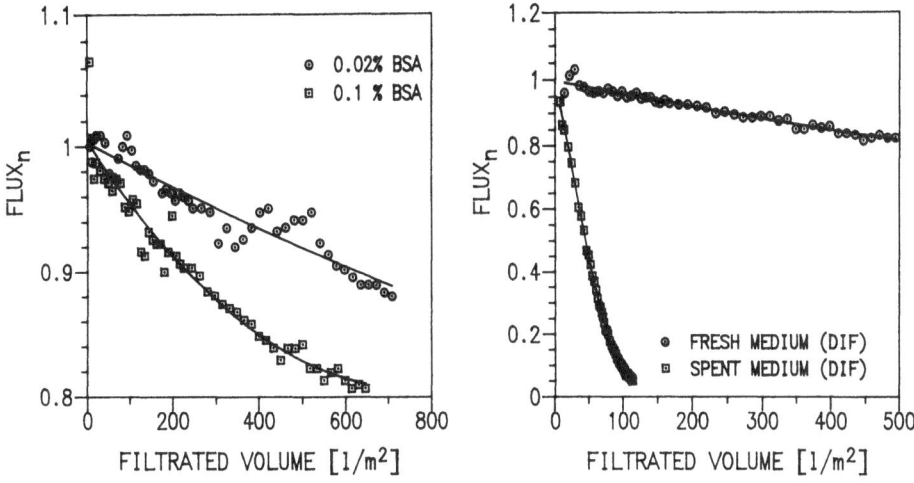

Fig. 6.   Influence of the protein concentration of the culture medium on the flux of S6/2 membranes.

Fig. 7.   Comparison of fresh medium and supernatant for their potential for membrane fouling.

3.2.4. *Influence of cell culture media and supernatants on membrane fouling.* The reduction of the protein content of the medium increased the membrane lifetime significantly (Fig. 6). However, compromises have to be made since most cell lines require several proteins as growth factors, nutrient carriers, attachment factors, etc. and can be cultivated much better at higher protein concentrations. In addition, cell derived proteins, phospolipids, membrane vesicles and cellular debris, which proved to have much influence on membrane fouling (Fig. 7) could not be eliminated by low protein culture media.

3.2.5. *Comparison of different perfusion modes.* Medium exchange via the membrane perfusion system could be performed in three different modes as shown in Fig. 8. In mode a.) supernatant is filtered out of the bioreactor without backflushing and fresh medium is added directly. In mode b.) pumps were run alternatively as described previously (Lehmann et al. 1988). Fresh medium was used for backflushing of the membrane. However, approx. 30% of the fresh medium was lost in this mode when the feed pump stopped and the harvest pump started to remove medium out of the membrane. In mode c.) the harvest pump was running continuously and the fresh medium was added directly into the bioreactor. A third pump was run in cycles backflushing the membrane with filtered supernatant.

a. Dead end filtration   b. Backflushing with fresh medium

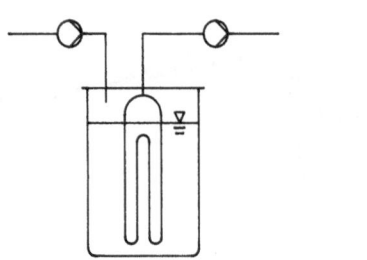 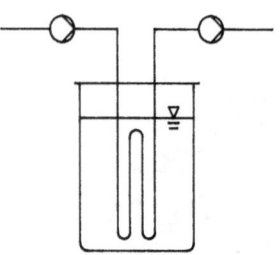

c. Backflushing with filtrated supernatant

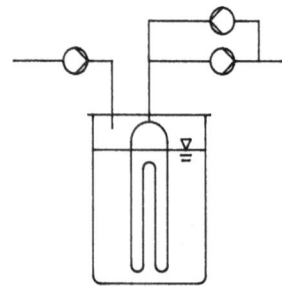

Fig. 8.   Different operation modes for medium perfusion via hydrophilized microfiltration membranes.

3.2.6. *Optimized backflushing.* In order to achieve a positive effect, backflushing volume and velocity have to be optimized. To keep the volume which is filtered through the membrane as small as possible, the backflushing volume should be very small. However, approx. 20 % of the supernatant was required for back-flushing because of expansion of the flexible tubes with commensurate increase in volume, especially when the transmembrane pressure was already increasing as a result of membrane fouling. The backflushing velocity should be much higher than the velocity of filtration. Backflushing velocities in the range from 20 to 25 liters per $m^2 \cdot min$ have proved to be ideal for the S6/2 membrane (Blasey et al. in prep.).

3.2.7. *Membrane pretreatment.* Membrane precoating is a procedure which was shown to reduce fouling of ultrafiltration membranes as well as microfiltration membranes. Non-ionic surfactants (Fane et al. 1985) have been used as well as non-ionic polymers (Kim et al. 1988). For use in animal cell technology these compounds have to be nontoxic to cells. We have used Pluronic[R] F68 as a non-ionic surfactant and polyethyleneglycol (PEG) as non-ionic polymer. Both of these substances have been used as media supplements (Mizrahi 1975, Shintani et al. 1988, Blasey and Winzer 1989). Pretreatment of the membranes with Pluronic[R] F68 resulted in a small increase in the membrane lifetime. Better results were obtained after pretreatment with PEG (Fig. 9).

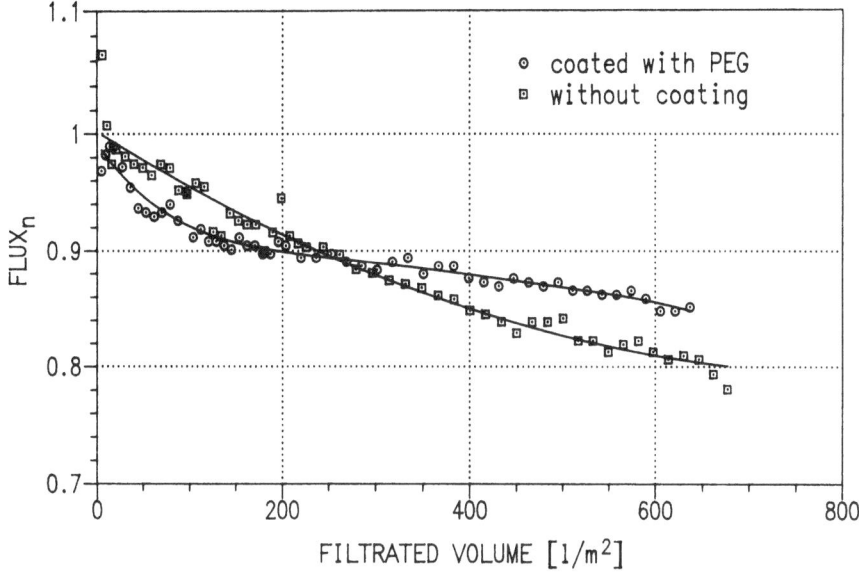

Fig. 9.   Effect of precoating with PEG on the flux of S6/2 membranes.

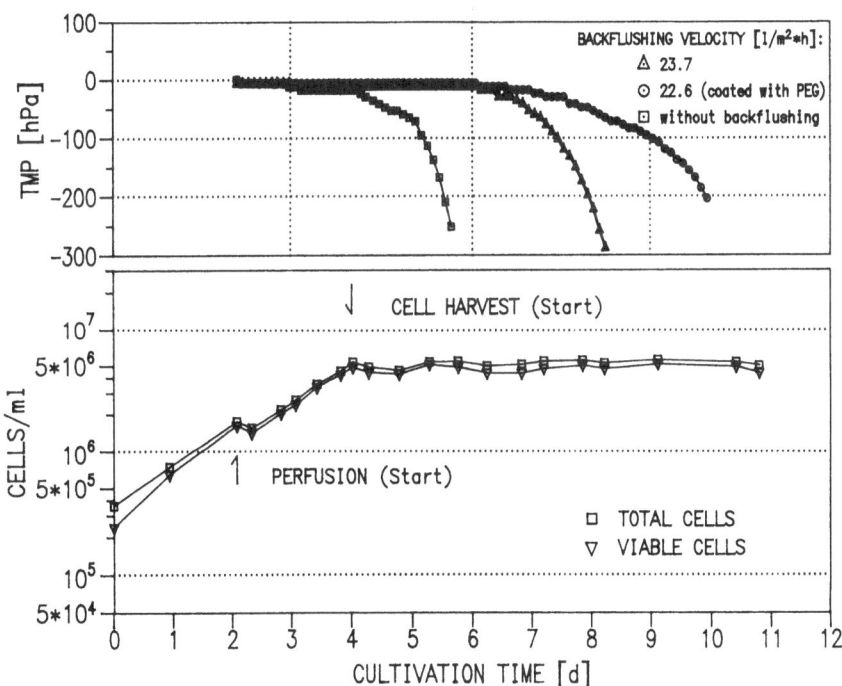

Fig. 10.   Continuous perfusion culture of 187.1 hybridoma cells (ATCC HB58) running three membrane perfusion modes in parallel: dead end filtration without backflushing, dead end filtration with high velocity backflushing and PEG-precoating together with backflushing.

3.2.8. *Use of different perfusion modes for the cultivation of hybridoma cells.*
To prove the efficiency of backflushing and precoating in perfusion culture of
mammalian cells, a 1.4 liter bioreactor was equippped with three perfusion membranes of 40 cm length. All membranes were run separately. One membrane was
used without backflushing. The other two membranes, one of which was pretreated
with 1.5% PEG, were used with backflushing. The results of this comparative
expe-riment are shown in Fig. 10. Perfusion was started after two days of
cultivation at a rate of 11.3 $l/m^2 \cdot h$. After 4 days cell bleeding was started to
keep the cell concentration at $5 \cdot 10^6$ cells per ml. The membrane which was run
in a simple dead end filtration mode was clogged after 3.5 days. With backflushing, the uncoated membrane clogged after 6 days and with additional
precoating after 8 days of perfusion. The filtrated net volume could be increased
by 40% with backflushing and again by 40% with a precoated membrane (Fig.
11). With a membrane length of 3 meters per liter of reactor volume, which was
normally used for media perfusion in our bioreactors, it should be possible to
run the system for up to two months before membrane fouling leads to termination of the process.

4. **Conclusion**

The internal perfusion system with microporous hollow fiber membranes appears
to be one of the most efficient systems for medium exchange and cell retention
in continuous culture of animal cells. Due to the membrane capacities for aeration

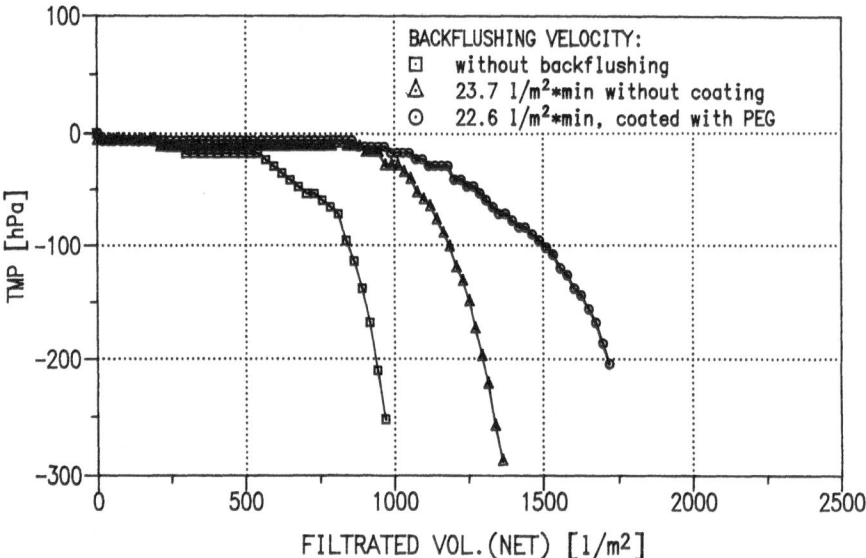

Fig. 11.   Comparison of the net filtered supernatant during continuous perfusion
culture of 187.1 hybridoma cells running three membrane perfusion
modes in parallel: dead end filtration without backflushing, dead end
filtration with high velocity backflushing and PEG-precoating in
combination with backflushing.

and perfusion, 8 to 35 fold increases in cell concentration compared to batch cultivation in culture flasks could be obtained routinely with unchanged cell viabilities. The latter is a direct consequence of the cells not being exposed to mechanical stress by pumping at high velocity, which is a necessary procedure for external perfusion with cross flow membrane systems.

Hydrophilized polypropylene membranes have proved to be superior to hydrophilic microfiltration membranes for medium perfusion. The lifetime of these membranes could be increased significantly by backflushing with spent medium at high velocities and by precoating with PEG, resulting in a highly efficient perfusion system.

5. Acknowledgement

We are grateful to Dr. D. Monner for linguistic advice.

6. References

Blasey H.D. and Winzer U. (1989) 'Low protein serum-free medium for antibody production in stirred bioreactors', Biotechnol. Lett. 11, 455 - 460

Eibl, H. and Lands, W.E.M. (1969) 'A new sensitive determination of phosphate' Anal. Biochem. 30, 51 - 57

Fane , A.G.; Fell, C.J.D. and Kim, K.J. (1985) 'The effect of surfactant pretreatment on the ultrafiltration of proteins', Desalination 53, 37 - 55

Hill, B.T. and Whatley, S. (1975) 'A simple, rapid assay for DNA' FEBS Lett. 56, 20 - 23

Himmelfarb, P.; Thayer, P.S. and Roberts, D.W. (1969) 'Spin filter culture: The propagation of mammalian cells in suspension', Science 164, 555 - 557

Kim, K.J.; Fane, A.G. and Fell C.J.D. (1988) The performance of ultrafiltration membranes pretreated by polymers', Desalination 70, 229 - 249

Lehmann, J.; Piehl, G.W. and Schulz, R. (1987) 'Bubble free cell culture aeration with porous moving membranes', Develop. biol. Standard., Vol. 66, S. Karger, Basel, pp. 227 - 240

Lehmann, J.; Vorlop, J. and Büntemeyer, H. (1988) 'Bubble-free reactors and their development for continuous culture with cell recycle', in R.E. Spier and J.B. Griffiths (eds.) 'Animal Cell Biotechnology Vol. 3', Acad. Press, London, pp. 221 - 237

Lucki-Lange, M. and Wagner, R. (1991) 'Conditions for the production of recombinant IL-2 in stirred suspension culture using a protein-free medium', in R.E. Spier, J.B. Griffiths, B. Meignier (eds.), 'Production of Biologicals from Animal Cells in Culture. Research, Developments and Achievements', Butterworths Pub., Kent, in press

Mizrahi, A. (1975) 'Pluronic polyols in human lymphocyte cell lines cultures', J. Clin. Microbiol. 2, 11 - 13

Rebsamen, E.; Goldinger, W.; Scheirer, W.; Merten, O.-W. and Palfi, G.E. (1987) 'Use of a dynamic filtration method for separation of animal cells', Develop. biol. Standard. 66, 273 - 277

Shintani, Y.; Iwamoto, K. and Kitano, K. (1988) ' Polyethylene glycols for promoting the growth of mammalian cells', Appl. Microbiol. Biotechnol. 27, 533 - 537

Takazawa, Y.; Tokashiki, M.; Murakami, H.; Yamada, K. and Omura, H. (1988) 'High-density culture of mouse-human hybridoma in serum-free defined medium', Biotechnol. Bioeng. 31, 168 - 172

Tokashiki, M.; Hamamoto, K. and Takazawa, Y. (1988) 'Perfusion culture of hybridoma cells recycling higher-molecular-weight components', Hakkokogaku 66, 31 - 35

Tolbert, W.R.; Feder, J. and Kimes, R.C. (1981) 'Large-scale rotating filter perfusion system for high-density growth of mammalian suspension cultures' In Vitro 17, 885 - 890

Varecka, R. and Scheirer, W. (1987) 'Use of a rotating wire cage for retention of animal cells in a perfusion fermentor', Develop. biol. Standard. 66, 269 - 272

Velez, D.; Miller, L. and Macmillan, J.D. (1988) 'Use of tangential flow filtration in perfusion propagation of hybridoma cells for production of monoclonal antibodies' Biotechnol. Bioeng. 33, 938 - 940

Cultivation of Hepatocytes in a New Entrapment Reactor: a Potential Bioartificial Liver

Wei-Shou Hu,[1] Scott L. Nyberg,[1,2] Russell A. Shatford,[2] William D. Payne,[2] and Frank B. Cerra[2]
[1]Department of Chemical Engineering and Materials Science,
[2]Department of Surgery, University of Minnesota, Minneapolis, MN 55455 U.S.A.

INTRODUCTION

Liver failure is a major cause of mortality.[1] Most patients admitted to an intensive care unit in liver failure do not survive. Currently, the only available treatment for refractory liver failure is hepatic transplantation. Renal dialysis, which revolutionized the treatment of renal failure, does not have a hepatic equivalent. There is a critical demand to develop an artificial liver for short-term support of hepatic failure. Given the functional complexity of the liver, the most promising approach to the development of an artificial liver involves using cultivated hepatocytes to supply function.

The number of hepatocytes necessary for normal liver activity is unknown. We expect that a large number of hepatocytes will be needed to sustain a fraction of the normal liver activity. These cells shall be confined in a relatively small volume to ensure that the device is of a practical size. In addition, there must be mechanisms for supplying the cells with nutrients to sustain viability and function as well as of allowing exchange of molecules between the patient and the artificial liver.

We have previously developed a hollow fiber cell entrapment bioreactor for mammalian cell cultivation and protein production.[2] In this reactor, cells entrapped in collagen gel are located within the hollow fiber lumen; an open space between the collagen gel and the fiber wall allows medium to be continuously supplied to the cells, while metabolites and small molecular weight components of basal medium freely diffuse through the hollow fiber membrane (Figure 1). Such a bioreactor allows the continuous supply of nutrients and removal of metabolites to and from a very high concentration of cells packed in a relatively small volume. Extracellular matrix components, which have been reported to be important for

75

R. Sasaki and K. Ikura (eds.), Animal Cell Culture and Production of Biologicals, 75–80.
© 1991 Kluwer Academic Publishers.

differentiated hepatocyte function, can be easily incorporated into the collagen gel. Furthermore, the hollow fiber membrane separates the patient's blood from the xenogeneic hepatocytes, while the relatively large surface area allows for the rapid exchange of small and medium sized molecules across the bioreactor. Here we report our preliminary results of cultivating rat hepatocytes in this bioreactor.

MATERIALS AND METHODS

Description of the bioreactor

Our bioreactor employs hollow-fiber semipermeable membranes to separate the patient's blood from the cultured hepatocytes (Figure 1). The hepatocytes are suspended in a collagen solution at 4°C and inoculated into the lumens of hollow fibers. Subsequently, fiber formation of collagen molecules is induced by warming the bioreactor cartridge to 37°C. As a result, cells are entrapped in an insoluble, fibrous, and highly porous collagen matrix. The matrix then contracts, creating two phases within the lumen of the hollow fiber: a gel phase consisting of a dense network of polymer fibers and cells, and a separate liquid phase. Juandiced blood, plasma, or medium for detoxification flows through the extracapillary space, while the growth factor enriched medium for sustaining hepatocyte function flows in the intraluminal channel.

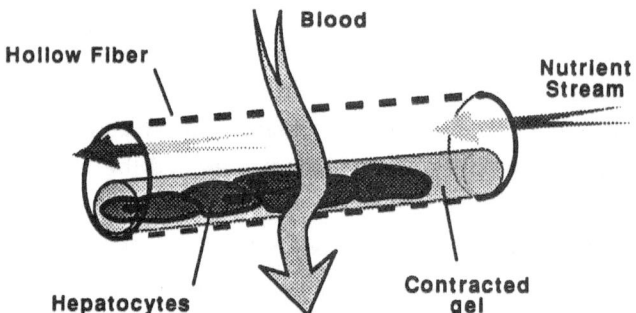

Figure (1) *Schematic of a contracted gel with entrapped hepatocytes inside a single hollow fiber. Molecular exchange occurs through pores in the hollow fiber wall.*

Cell harvest

Hepatocytes were harvested from 4-6 week old male Sprague Dawley rats, weighing 250 - 300 grams, by a modification of Seglen's perfusion technique.[3,4] Briefly, following intraperitoneal nembutal anesthesia and portal vein cannulation, *in vivo* perfusion was performed with a calcium-chelating balanced salt solution. Portal vein perfusion was continued with 0.05% collagenase (clostridiopeptidase A)

(Sigma #C-0130 Type 1, 200 units/mg) in a balanced salt solution. The liver was then gently combed to isolate hepatocytes. Following three washings in balanced salt solution, the hepatocyte pellet was resuspended in an isotonic type I collagen solution for hollow fiber inoculation. Rat harvests yield 2.0-3.0 x 10^8 hepatocytes with 80-90% viability by trypan blue staining.

Measurement of Bilirubin Conjugation

A concentrated bilirubin stock solution was prepared by dissolving unconjugated bilirubin powder in 0.1M NaOH. Concentrated stock solution was added to serum containing culture medium [William's E medium supplimented with Hepes buffer, L-glutamine (2 mM), insulin (10 mg/L), penicillin (100 U/mL), streptomycin (100 mg/mL), and 5% bovine serum] so that the concentration of unconjugated bilirubin is approximately 50 mg/L. Samples were collected throughout the bioreactor run and stored frozen at -80°C. The presence of bilirubin mono- and diglucuronide conjugates was later assessed by a high performance liquid chromatography (HPLC). Briefly, thawed samples were diluted 1:1 with dimethyl sulfoxide. Analysis of 150 µL aliquots was performed on a Varian 5000LC with a Merck (#50942) LiChrospher 100 RP-8 column. A 20 to 95% linear gradient of isopropranol in 0.1M sodium acetate, pH 4.0, containing 3 mM octane sulfonic acid was used. Data was acquired over the first 35 minutes of each 45 minute run. Absorbance detection was at 440 nm.

RESULTS AND DISCUSSION

Contraction of collagen gel

An important feature of our bioreactor is the contraction of collagen gel after cell entrapment. Hepatocytes harvested from a rat were mixed with collagen solution and allowed to form a gel disk of approximately 2.0 mm thickness in 1.6 cm diameter tissue culture plates. Contraction began within a few hours and reached a stable minimum within a few days. The final diameter was approximately 40 to 60% of the original diameter. A contracted gel disk is shown in Figure 2. In the absence of viable cells, the collagen solution gelled, but failed to contract. The initial cell concentration weakly influenced the extent of gel contraction (data not shown).

78

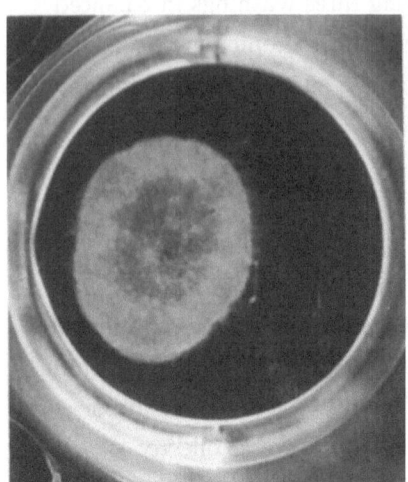

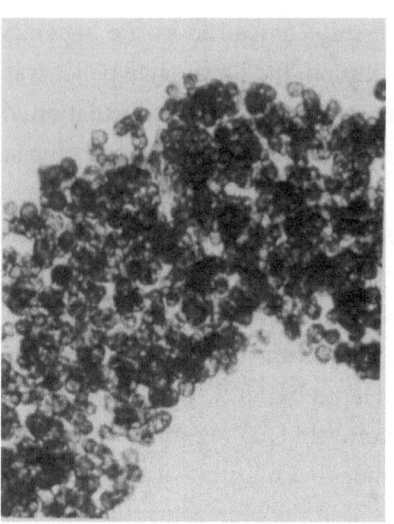

Figure (2) *A contracted collagen gel from a 7 day old hepatocyte culture. Well diameter, and initial gel diameter, each measured 1.6 cm.*

Figure (3) *Hepatocytes within a contracted collagen gel fiber which was extracted from the hollow fiber bioreactor after 24 hours.*

Hepatocytes suspended in collagen solution (10^7/mL) were loaded into a hollow fiber bioreactor and medium was recirculated in the extracapillary space. The hollow fiber cartridge measured 20 cm and contained about fifty fibers. The inner diameter of the each fiber is 1.1 mm. After 24 hours, the bioreactor was dissembled and the collagen gel was removed from the bioreactor for observation. As shown in Figure 3, the gel maintained structural integrity; measurement of the gel diameter indicated that the diameter has reduced to about 60% of its original value.

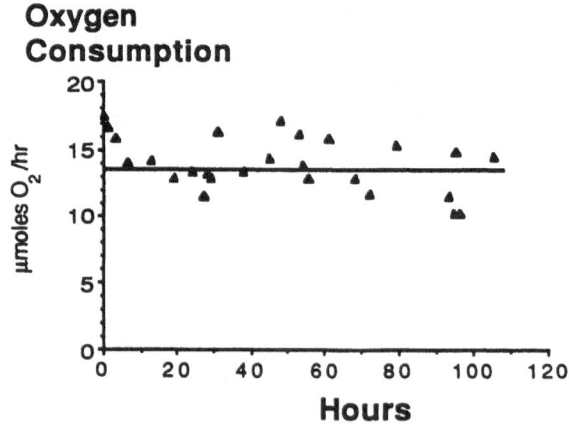

Figure (4) *Stable Oxygen consumption by hepatocytes within the hollow fiber bioreactor. Oxygen consumption was calculated from oxygen tension measurements from the inlet and outlet streams of the extracapillary space*

Viability and metabolic activity

The hepatocytes cultivated in this bioreactor remained viable and metabolically active after one week as illustrated by their ability to consume oxygen and glucose. The oxygen consumption rate, as determined by on-line measurement of inlet and outlet dissolved oxygen concentration through the extracapillary space, remained relatively constant over the cultivation period (Figure 4).

We examined the viability of the hepatocytes entrapped in the collagen gel using a fluorescein diacetate/ethidium bromide stain (FDA/EB).[5] Following FDA/EB staining, viable cells appear green, while the nuclei of dead cells appear orange under epifluorescent microscopy. The viability of the cells entrapped in the collagen gel exceeded 50% after a seven day cultivation.

Bilirubin Conjugation

Bilirubin is a toxic and poorly water soluble end product of red blood cell heme degradation. Bilirubin conjugation, catalyzed by the hepatic enzyme UDP-glucuronyl transferase, changes bilirubin into a water soluble molecule capable of elimination in aqueous bile. We thus chose bilirubin conjugation as an essential liver-specific measure of hepatic function in our bioreactor.

Hepatocytes were cultivated in the bioreactor for two days with the medium being continuously recirculated through the extracapillary space. On day 2, the medium was replaced with fresh medium containing 50 mg/L of unconjugated bilirubin. The reactor was wrapped with aluminum foil and the incubation room was kept dark to prevent photodegradation of bilirubin. The conjugated bilirubin was demonstrated by HPLC analysis. As shown in Figure 5, the conjugated form of bilirubin increased over time.

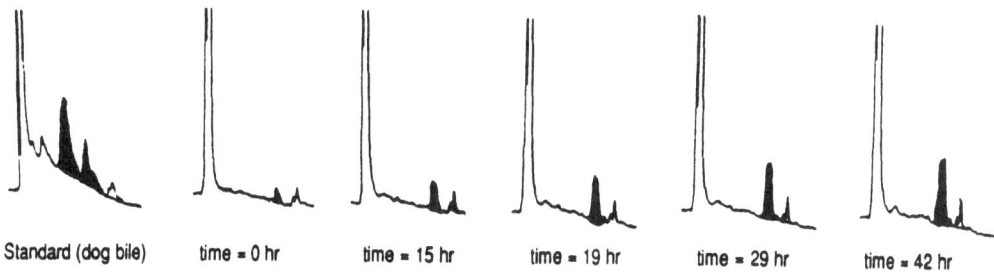

Standard (dog bile)   time = 0 hr   time = 15 hr   time = 19 hr   time = 29 hr   time = 42 hr

Figure (5) *HPLC chromatogram of a dog bile standard and hepatocyte culture fluids The bile standard demonstrates both mono- and diconjugate forms of bilirubin from dog bile. Increasing (mono-) conjugate (shaded peak) is demonstrated during 42 hours of continuous medium recirculation.*

CONCLUSION

A major need exists for an artificial liver. A hybrid design utilizing metabolically active hepatocytes to provide essential hepatic function has been explored in our laboratory. Hepatocytes cultured at high density within a continuously perfused hollow fiber bioreactor retain viability and UDP-glucuronyl transferase activity throughout the bioreactor run. Bioreactor optimization and scale-up is essential before this design may be tested in an animal model, or possibly in clinical trials.

ACKNOWLEDGEMENTS

This work was supported in part by a grant from the National Science Foundation (BCS-8915307) and the National Institutes of Health (DK34931). SLN was supported by a National Institutes of Health Grant (CA09024-15). We also wish to thank Larry D. Bowers and Doug A. Scheeler from the Department of Laboratory Medicine and Pathology, University of Minnesota, for HPLC analysis of bilirubin.

REFERENCES

1.   Blake JE, Compton KV, Schmidt W, Orrego H. (1988) Accuracy of death certificates in the diagnosis of alcoholic liver cirrhosis. *Alcoholism* (NY); 12(1): 168-72.

2.   Scholz MT, Hu W-S. (1990) A two-compartment cell entrapment bioreactor with three different holding times for cells, high and low molecular weight compounds. *Cytotechnology*; 4: 127-37.

3.   Seglen PO. (1974) Preparation of isolated rat liver cells. In Prescott DM, ed. *Methods in Cell Biology*. New York: Academic Press; 29-83.

4.   Keller GA, West MA, Cerra FB, Simmons RL. (1985) Modulation of hepatocyte protein synthesis by endotoxin activated Kupffer cells. *Annals of Surgery* 201:87-94.

5.   Nikolai TJ, Peshwa MV, Göetghebeur S, Hu W-S. (in press) Improved microscopic observation of mammalian cells on microcarriers by fluorescent staining. *Cytotechnology* .

CONTINUOUS CULTURE WITH CELL PRECIPITATION FOR RECOMBINANT PROTEIN PRODUCTION

H. R. KIM, B. H. CHUNG, C. H. KIM and I. S. CHUNG
Department of Genetic Engineering
Kyung Hee University
Suwon, KOREA

ABSTRACT. A continuous bioreactor using a spinner flask equipped with cell precipitator was applied to cultivate insect cells. Maximum viable density of insect cells in this bioreactor system reached 1.38×10^6 cells/ml at the dilution rate of 0.005 hr^{-1}. Recombinant protein production by cells infected with a genetically-modified baculovirus was also demonstrated. The maximum β-galactosidase synthesis of 7500 units per reactor volume was achieved at the dilution rate of 0.011 hr^{-1}.

INTRODUCTION

Protein expression systems based on the Autographa californica nuclear polyhedrosis virus (AcNPV) have wide applicability as an alternative to prokaryotic or other eukaryotic expression system (1-5). However, the application of the baculovirus expression systems has been limited by difficulties in insect cell culture scale-up (6-7). Therefore, it is quite natural that scale-up of insect cell culture would be a challenging scientific and technological undertaking to ensure efficient production of various significant products for human health and medicine.

The major obstacle to scale-up involves obtaining high cell density in culture vessels. The most effective way to achieve high cell densities is to recycle insect cells in the bioreactor by suitable methods.

Cell recycle can be problematic, especially the cells are separated outside the cultivation vessel. Filtration and centrifugation steps may pose problems of contamination and cell lysis. The separation of the aqueous medium from the cells within the cultivation vessel would be advantageous as this would eliminate the need for pumping the fragile cells and should limit the problems with aseptic operation. Internal cell recycle has been achieved by several investigators (8-9) through cell precipitation or filtration. Cell precipitation is particularly attractive for insect cells because of the large cell diameters and the tendency of the cells to clump, enhancing settling.

In this study, Spodoptera frugiperda cells were cultured in a continuous bioreactor equipped with cell precipitator in order to obtain information for future development of large scale suspension culture. This work also

R. Sasaki and K. Ikura (eds.), Animal Cell Culture and Production of Biologicals, 81–85.
© *1991 Kluwer Academic Publishers.*

investigated the feasibility of recombinant protein production by cells infected with a genetically-modified baculovirus.

MATERIALS and METHODS

Cell line and culture conditions

The cell line used in this study was <u>Spodoptera frugiperda</u> insect cells. The cells were maintained in 25 cm^2 and 75 cm^2 tissue culture flasks to provide cells for continuous reactor. The medium used was Grace's insect medium, which was supplemented with 50 μg/ml gentamycin sulfate, 2.5 μg/ml fungizone, 0.35 g/l sodium bicarbonate and 5 % (v/v) fetal bovine serum.

Continuous culture

Continuous culture was conducted in a siliconized spinner flask under the conditions of the initial pH 6.2, 28 °C, 80 rpm agitation and surface aeration.

Continuous culture with cell precipitation

The set-up for continuous culture with cell precipitation was basically the same as that continuous culture with an additional provision of cell precipitator. Cell retention was done by natural settling of insect cells with a tube side port. The system was designed in such a way that cell settling occurs while the culture effluent passes through the side port. The aggregation property of <u>Spodoptera frugiperda</u> cell was helpful in achieving nearly complete cell retention over an experimental range of dilution rates. The insect cell in the effluent leaving the culture system was less than 1 % of that in the culture vessel. Other conditions were the same as those described in continuous culture.

Analytical methods

The cell number was counted with a hemacytometer under the microscope. The cell viability was determined by the dye exclusion method with 0.4 % trypan blue solution. The recombinant protein production by cells infected with a genetically-modified baculovirus was estimated by the β-galactosidase assays given by Miller (10). One unit of activity is defined as 1.0 mM of ONPG cleaved per minute at 37 °C and pH 7.3. One mg of pure β-galactosidase contains approximately 300 units of activity (Sigma, G-6008).

RESULTS and DISCUSSION

Continuous culture

The 500 ml spinner reactor was inoculated with 150 ml of complete medium containing 3 x 10^5 cells/ml. When the cell concentration reached 1.3 x 10^6 cells/ml after 5 days of batch growth, the feed containing 5 % fetal bovine serum was started at the dilution rate of 0.006 hr^{-1} and the reactor volume was maintained at 150 ml. The dilution rate was varied from 0.006 hr^{-1} to 0.024 hr^{-1}.

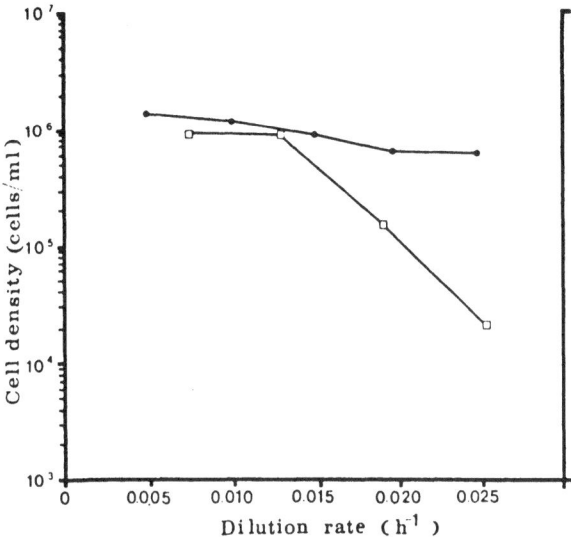

Figure 1. Effect of cell precipitation on cell growth.
□: continuous culture, •: continuous culture with cell precipitation

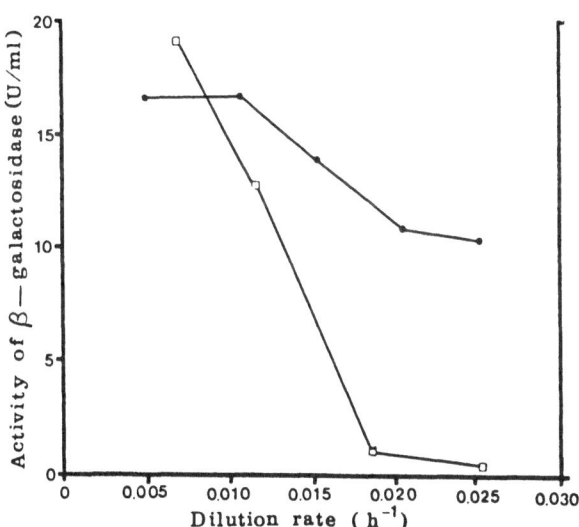

Figure 2. Effect of cell precipitation on β-galactosidase production.
□: continuous culture, •: continuous culture with cell precipitation

84

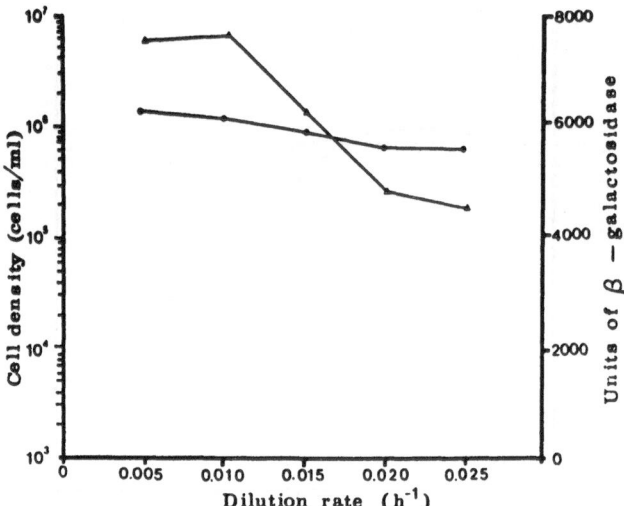

Figure 3. Cell growth and total β-galactosidase production in continuous
culture with cell precipitation.
● : Cell density, ▲ : β-galactosidase

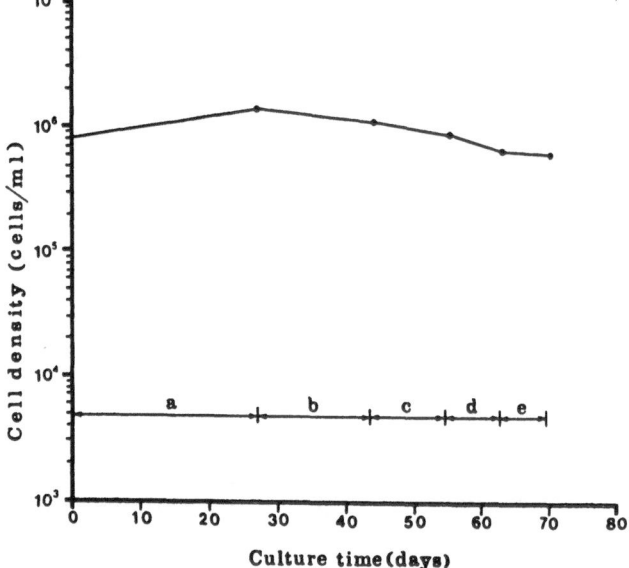

Figure 4. Time course changes of viable cell concentration in continuous
culture with cell precipitation. Perfusion rate : a, 65ml/day ;
b, 117ml/day ; c, 175ml/day ; d, 233ml/day ; e, 292ml/day

At each dilution rate representative samples were taken from the spinner reactor for viable cell counts and quantification of β-galactosidase activity. The steady state cell concentration and recombinant protein (β-galactosidase) production are shown in Figures 1 and 2 as a function of dilution rate. The viable cell content and β-galactosidase production decrease monotonically as the dilution rate increases.

Continuous culture with cell precipitation

Continuous culture with cell precipitation was started 5 days after inoculation and the perfusion rate (cited as the dilution rate in this paper) was increased stepwise (Figures 1, 2 and 3). As shown in Figures 1 and 2, even when the dilution rate reached the highest point of 0.024 hr^{-1}, viable cell concentration did not decrease dramatically compared with that of continuous culture. Also the amount of β-galactosidase synthesis was maintained high. But as the dilution rate increased, the reduction of β-galactosidase synthesis was more noticeable than that of viable cell concentration. The amount of β-galactosidase synthesis per reactor volume was highest at the dilution rate of 0.011 hr^{-1} (Figure 3). In a continuous culture with cell precipitation, viable cell concentration was maintained above 6.3 x 10^5 cells/ml over a period of 70 days (Figure 4).

REFERENCES

1. Luckow, V.H. and Summers, M.D. (1988) 'Trends in the development of baculovirus expression vectors', Bio/Technology, 6, 47-55.
2. Maiorella, B., Inlow, D., Shauger, A. and Harano, D. (1988) 'Large-scale insect cell-culture for recombinant protein production', Bio/Technology, 6, 1406-1410.
3. Miller, L.K. (1988) 'Baculoviruses as gene expression vectors', Ann. Rev. Microbiol., 42, 177-199.
4. Smith, G.E., Summers, M.D. and Fraster, M.J. (1983) 'Production of human beta-interferon in insect cells infected with a baculovirus expression vector', Mol. Cell. Biol., 3, 2156-2165.
5. Smith, G.E., Ju, G., Ericson, B.L., Moschera, J., Lahm, H.W., Chizzonite, R. and Summers, M.D. (1985) 'Modification and secretion of human interleukin-2 in insect cells by baculovirus expression vector', Proc. Natl. Acad. Sci., 82, 8404-8408.
6. Murhammer, D.W. and Gooche, C.F. (1988) 'Scale up of insect cell cultures: protective effects of pluronic F-68', Bio/Technology, 6, 1411-1418.
7. Tramper, J., Williams, J.B. and Joustra, D. (1986) 'Shear sensitivity of insect cells in suspension', Enz. Microb. Technol., 8, 33-36.
8. Kitano, K., Shintani, Y., Ichimori, Y., Tsukamoto, K., Sasai, S. and Kida, M. (1986) 'Production of human monoclonal antibodies by heterohybridomas', Appl. Microbiol. Biotechnol., 24, 282-286.
9. Takazawa, Y. and Tokashiki, M. (1989) 'High cell density perfusion culture of mouse-human hybridomas', Appl. Microbiol. Biotechnol., 32, 280-284.
10. Miller, J.H. (1972) 'Assay of β-galactosidase', in Experiments in Molecular Genetics, Cold Spring Harbor Lab., New York, pp. 352-355.

ELIMINATION OF MICROORGANISMS FROM CELL CULTURE MEDIUM USING
REGENERATED CELLULOSE HOLLOW FIBER (BMM)

S. Manabe[*1], M. Umeda[*2], A. Kono[*3], I. Togo[*1], S. Fukada[*1],
K. Yamaguchi[*4]
*1 Asahi Chemical Ind. Co. Ltd., The Imperial Tower 18F, Uchi-
   saiwaicho 1-chome, Chiyoda-ku, Tokyo 100, Japan
*2 Yokohama City University, Mutsugawa 3-122-21, Minami-ku,
   Yokohama 232, Japan
*3 National Kyushu Cancer Center, Notame 3-1-1, Minami-ku,
   Fukuoka 815, Japan
*4 Institute of Laboratory Animals, Yamaguchi University
   School of Medicine, Ube, Yamaguchi 755, Japan

ABSTRACT. We intended to show microbe removal by cuprammonium regenerated cellulose hollow fiber (BMM) from cell cultrure media with a high
recovery rate of its components. The MEM or D-MEM solution varying in
FBS content from 0% to 100% was used. BMM modules with mean pore sizes
of 35, 40, 50, and 75 nm were tried. Dead-end type filtration was done
with a constant filtration rate. The cytotoxicity of the filtrate was
evaluated through the cell growth rate and viable cell ratio for five
cell lines. The microbes larger than 40 nm in size could be removed to
less than $1/10^5$ of the original concentration by BMM with a mean pore
size of less than 40 nm. The filtration rate and filtration capacity
strongly depended on the content of FBS in medium such as more than 1000
$1/m^2$ for non-FBS medium, 100 $1/m^2$ for 10% FBS medium, and 8 $1/m^2$ for 100
% FBS in the case of BMM having 40 nm mean pore size. The filtrate did
not show any cytotoxicity. We concluded that BMM can be used to prepare
microbe-free cell culture medium.

1. INTRODUCTION

   The accelerated progress of biotechnology encourages the industrial
production of bioactive substances. Because of the possibility of infection by viruses that may exist in the source material or in the environment during production, the manufacture and testing of the substances
such as monoclonal products for human use should be controlled by the
various international health authorities. The strategy to get high se-
curity of prohibition from virus infections or oncogenes that may occur
by transfusion of drugs derived through biotechnology or from the contaminated circumstances of manufacturing can be classified into three cla-
sses:
(1) prevention of virus contamination of the feed materials to be used
    in cell culture systems,
(2) prevention of scattering virus from bioreactor into its operational

87

environment,
(3) virus removal from products.

As for the former two terms, not only virus removal but also virus inactivation can be used.[1] The third term must be virus removal since the bioactivity of the product should be kept during the treatment. From the regulatory viewpoint of biotechnological products and product liability, there must be some precautions against viruses and dangerous genes.

In this paper, we intend to demonstrate the use of BMM to prevent contamination in cell cultrure media. Here, BMM has been developed so as to remove viruses from blood plasma and plasma products. For this purpose, at first we will investigate (1) removability of microbes that may contaminate cells or surroundings, (2) permeability of protein, (3) filtration characteristics, and (4) cytotoxicity, and then design a novel commodity optimizing cost performance in obtaining microbe-free culture medium.

2. EXPERIMENTAL

2.1 Materials: (1) BMM module; effective filtration area of a module of 6 cm^2 (referred to as S-module), 300 cm^2 (referred to as M-module) and 600 cm^2 (Mycocut module) were supplied by Asahi Chemical Ind. Co. Ltd. in Japan. The module was constructed with BMM having mean pore sezes of 35, 40, 50, and 75 nm. We named the module being constructed with BMM of 35 nm in mean pore size BMM35 and so on. (2) filtrand; solutions with a wide range of concentration of fetal bovine serum (FBS in short, manufactured by Gibco Co. USA) were prepared by diluting with D-MEM (manufactured by Gibco Co. USA). (3) mycoplasma; three kinds of mycoplasma of A. laidlawii, M. orale, and M. hyorhinis were kindly supplied by the Hakko Institute (Japan). After cultivating them in the modified Heyblick liquid medium, the solution was diluted ten-fold using the medium containing 7 vol.% (25°C) of FBS. (4) virus; Human immunodeficiency virus (HIV-1), cytomegalovirus (CMV), and Japanese encephalitis virus (JEV) were supplied by Prof. N. Yamamoto (Tokyo Medical and Dental University School of Medicine, Japan). (5) a mixture of HIV-1 and mycoplasma; The supernatant of the cell line of HIV-1 was filtered using commercially available plane type membrane having a mean pore size of 0.45 um. After the filtration, the filtrate was cooled to 4 °C and was left in a safety cabinet without a cover to be contaminated by mycoplasma existing in the environmental air.
2.2 Measurement: (1) ultrafiltration; Dead-end type filtration was done under a constant transmembrane pressure of 200 mmHg at 20 °C. The filtration volume was 17 1/m^2 in usual case but some modules were used with the volume of 50 1/m^2 for the sake of the observation of microbes in BMM by electronmicroscope. (2) Assay; Ultraviolet dying method using Vero (ATCC CCL 81) as an indicator cell and a colony-forming method using modified Heylick agar-agar medium were used with the mycoplasma. The MT-4 cells were infected with the causative agent of adult-T-cell leukemia and transformed in advance. These cells showed cytopathetic effect with HIV-1. The HIV-1 concentration was assayed by a plaque-forming method using these cells. With CMV and JEV, the TCID$_{50}$ method was used. (3)

electron microscopy; The BMM used in filtration was fixed by 2% glutar-aldehyde and paraformaldehyde and post-fixed by osmic acid. After de-hydration and embedding using epoxy resin, the samples to be observed were sliced into ultrathin sections of 80 nm thickness and then were stained using lead citrate and uranium acetate.

3. RESULTS AND DISCUSSION

Table 1 summarizes the values of the microbe (mycoplasma or virus) logatithmic rejection coefficient Φ for various BMM. Here, Φ is de-fined as

$$\Phi = \log (N_o/N_f) \tag{1}$$

where N_o and N_f are the total number of microbes in 1 ml of feed solu-tion and filtrate, respectively. When the size of the microbe increases and/or mean pore size of BMM decreases, the value of Φ increases. When the particle size of the microbe coincides with mean pore size, the value of Φ is more than 4.

Table 1 Virus or mycoplasma logarithmic rejection coefficient Φ for BMM35, BMM40, BMM50, and BMM75

Microbe/medium	BMM35	BMM40	BMM50	BMM75
HIV-1/human plasma	>5	>5	>5	>5
HIV-1/FBS10% D-MEM	>6	>6	>6	>6
CMV/FBS10% D-MEM	>5	>5	>5	>5
T-4/FBS10% D-MEM	>8	>8	7.8	6.5
VSV/FBS10% D-MEM	>12	>12	>12	NT*
Sindbis/FBS10% D-MEM	>8	>8	NT*	NT*
DHBV/duck plasma	6	5	NT*	NT*
JEV/FBS10% D-MEM	5.5	4.5	2.3	1.8
HBV/human plasma	6	5	2.4	1.9
ϕx174/FBS10% D-MEM	9	8	4	3
CJD/saline	>6	4	3.5	2.2
A. laidlawii/FBS7% D-MEM	>5	>5	>5	>5
M. orale/FBS7% D-MEM	>5	>5	>5	>5
M. hyorhinis/FBS7% D-MEM	>5	>5	>5	>5

*NT; not tested

Figure 1 shows an electron microphotograph of a cross section of BMM75 after filtration. Both HIV-1 and mycoplasma (probably M. orale) can be observed near the same position in BMM indicating that mycoplasma can pass more easily than expected from its size of 200 nm in smallest diameter. Anyhow BMM75 can remove both HIV-1 and mycoplasma from the culture medium.

In Table 2 is shown the recovery rate of protein when the filtrand

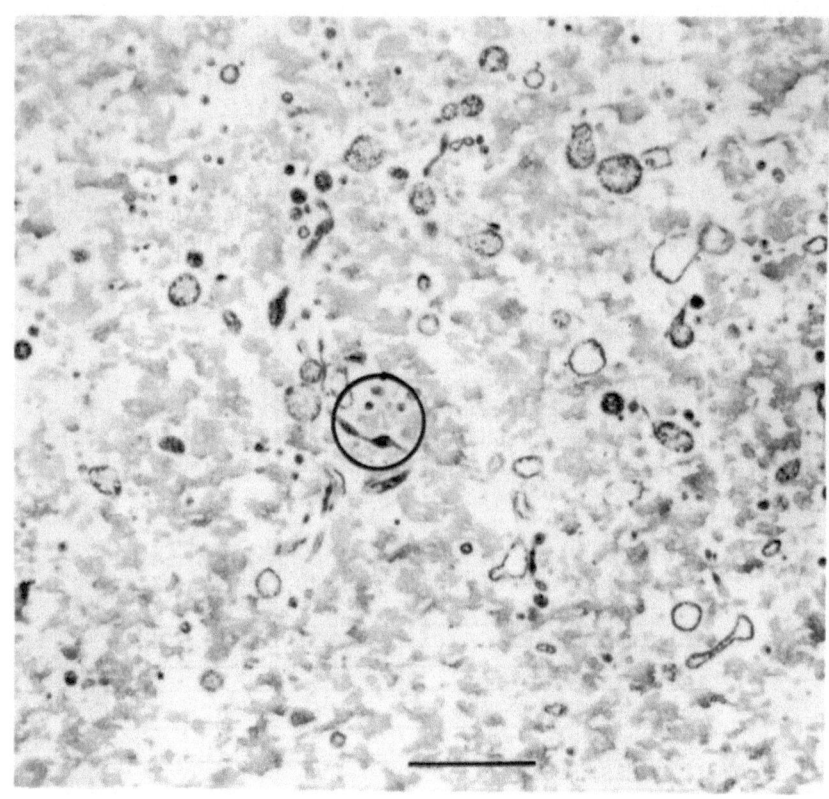

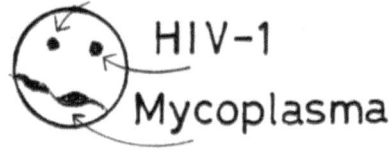

HIV-1

Mycoplasma

Figure 1. Transmission electron micrograph of cross section of BMM75 after filtration of culture medium of HIV-1 contaminated with mycoplasma. Scale bar is 1 μm in length. Gray area is source material (regenerated cellulose) and white area is pore. Mycoplasma and HIV-1 particles can be observed in the circle (represented by abstract patterns in the lower circle) near the center.

is the protein 1 wt% aq. solution. Most of the protein component can pass through BMM easily. The reason for the easy permeation of protein may be the good hygroscopic property of the source material, cuprammonium regenerated cellulose.

Filtration rate and filtration capacity, which is defined as the total filtration volume available before the transmembrane pressure becomes 1 atm. under a constant filtration rate, are nearly proportional to the effective filtration area. Then, we can represent these two characteristic values with the reduced values of unit filtration area such as $ml/min \cdot m^2$ and ml/m^2, respectively.

Figure 2 shows the composition dependence of the filtration capacity; when the content of FBS decreased the filtration capacity increased. For example when the content was 100% the capacity was 8 l/m^2 and 580 l/m^2 for the solution of 3 wt% of FBS.

Figure 3 shows the filtration volume dependence of the transmembrane pressure at the constant filtration rate of 7.2 $l/min \cdot m^2$, 11.7 $l/min \cdot m^2$, and 13.3 $l/min \cdot m^2$ for BMM75. The initial transmembrane pressure at these filtration rates for D-MEM solution ware 0.3, 0.5, and 0.7 atm. in order of the above filtration rate. The pressure did not increase even when the filtration volume was more than 1700 l/m^2.

The large filtration rate and capacity of BMM for the solution containing protein are promising for the use of BMM modules for microbe removal, not only in the preparation of culture medium to biotechnical products.

Table 2. Recovery rate of plasma protein in aq. solution (25°C)

Mean pore size (nm)	35	40	75
Albumin	100 %	100 %	100 %
Globulin	100	100	100
Factor V	100	100	100
Factor VIII	92	95	100
Factor IX	100	100	100

Transmembrane pressure; less than 760 mmHg
Filtration capacity; more than 30 $l/m2$
Filtration rate; more than 50 $l/m2$

If there remains the possibility that some kind of component being secreted into the filtrate shows a decrease in growth rate of cells and causes death of cells, then, the cytotoxicity must be checked in advance. Eight kinds of cells including delicate cells such as Molt-4 and P3U1 were investigated. The degree of cytotoxicity was evaluated from the relative growth rate (RG) of cells and the viable cell ratio (VR) in whole cells comarid with the control medium that was not filtered. Table 3 summerizes both values of RG and VR for various cell cultures. All filtrate showed more than 90% in RG and VR values, indicating that the filtrate obtained by BMM did not show cytotoxicity.

In the course of designing a new commodity for preparing microbe-free culture medium, we must take into account the following conditions:

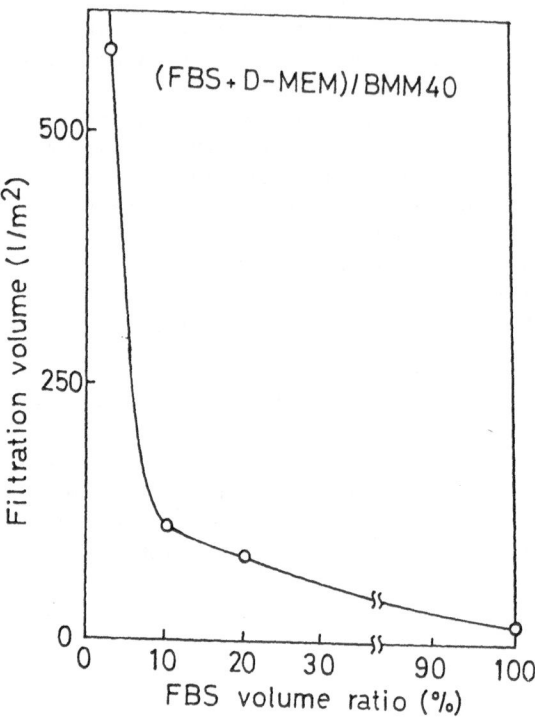

Figure 2. Composition dependence of filtration capacity for Mycocut module composed with BMM40.

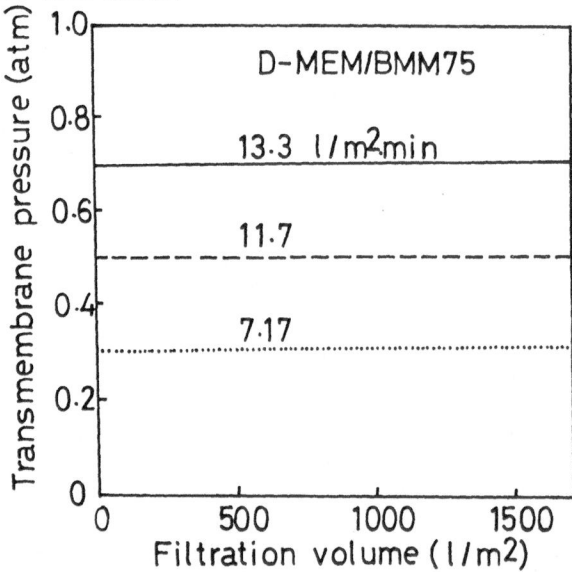

Figure 3. Filtration volume dependence of transmembrane pressure at constant filtration rate for Mycocut module composed BMM75: Filtrand; D-MEM

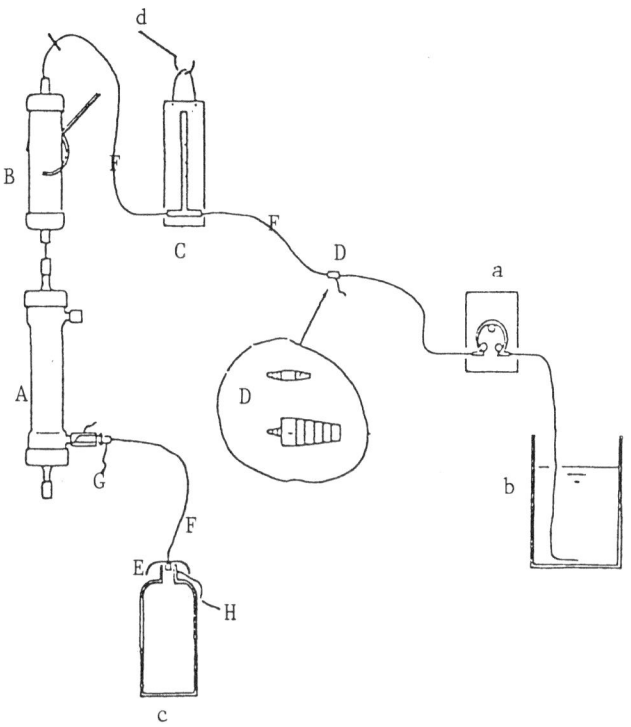

Figure 4. Schematic representation of outlook of Mycocut A :
A; BMM40 module, B; air trap chamber filled with pure water, C; pressure gauge, D; connector, E; cover with hemispheric shape, F; tubes, G;
plastic needle, H; adaptor, a; tubing pump, b; bottle with medium before
filtration, c; bottle for filtrate, d; clamp etc. for setting a pressure
gauge. Mycocut A does not contain a,b,c and d.

(1) The solution to be filtered contains serum or does not.
(2) The microbes to be removed include viruses or non-viruses.
(3) The volume of solution to be used is less than 50 l (laboratory usage) or more than 100 l (industry usage).
(4) The filtration should befinished within 3 hr.

Asahi Chemical Ind. Co. has developed two types of filtration set named Mycocut A and Mycocut B. The former is used for virus removal from cell culture medium containing serum and the later for microbes more than 80 nm in size. Both modules have effective filtration areas of 600 cm^2. Their mean pore sizes for Mycocut A and B are 40 nm and 75 nm, respectively. Their filtration performances are nearly same to those described above. Figure 4 shows the schematic representation of outlook of Mycocut A.

Table 3. Values of RG and VR of filtrate obtained by BMM40 for various cell culture

Cell culture		RG		VR	
		1 day	3 days	1 day	3 days
Human lymphocyte strain	MF-4	99 %	NT	94 %	98%
	MOLT-4	102	100 %	100	99
	P3HR-1	98	100	75	93
	U937	NT	101	NT	106
Islet of Langerhans cancer cell		NT	99	NT	101
Chinese hamster lung cell		NT	99	NT	100
Mouse myeloma cell P3U1		NT	100	NT	100
Mouse fibroblast cell		NT	100	NT	103

NT; not tested

REFERENCE

1) Harbour, C and Woodhouse, G. (1990) 'Viral contamination of monoclonal antibody preparation: Potential problems and possible solution', Cytotechnology, 4, 3-12.

EFFECTS OF SHEAR STRESS ON THE GROWTH OF HYBRIDOMA CELLS CULTIVATED IN SERUM-FREE MEDIUM COUPLED WITH AMMONIA REMOVING SYSTEM.

F. R. Nayve Jr., M. Matsumura and H. Kataoka
Institute of Applied Biochemistry
University of Tsukuba, 305, JAPAN

ABSTRACT

Serum-free perfusion cultures of mouse-mouse hybridoma TO-405 producing HBs monoclonal antibody were done in spinner flasks coupled with an ammonia-removing system. Ammonia in the culture broth was effectively maintained below inhibitory concentrations for cell growth. Maximum viable cell density levels similar to that in serum supplemented cultures as well a as higher of percentage viability were obtained using this system. Variation of the cross-flow rate through the ceramic filter between 0.29 m/s to 0.59 m/s did not cause immediate cell damage, but the shear stress generated in the system repressed the growth of the cells at the higher flow rate.

INTRODUCTION

Mouse-mouse hybridoma TO-405 cells have been reported to be especially susceptible to ammonia inhibition in serum-free media. The maximum density of cells cultivated in serum-free perfusion cultures was only about 70% of that in serum supplemented cultures. [1] Production of HBs monoclonal antibody (HBs-MAb) by mouse-mouse hybridoma TO-405 cells, however, have been also reported to be higher in serum-free medium than in serum-supplemented culture. It was thought that if ammonia could be selectively removed from the culture broth, then the maximum viable cell density and consequently the productivity in serum-free cultures could be improved. In this study, the ammonia inhibition of hybridoma cell growth was investigated. A system to selectively remove ammonia from the culture broth was developed and the effects of shear stress associated with the system on cell growth were examined.

R. Sasaki and K. Ikura (eds.), Animal Cell Culture and Production of Biologicals, 95–101.
© 1991 *Kluwer Academic Publishers.*

MATERIALS AND METHODS

Cell line and culture medium.
Hybridoma TO-405 cells, which secrete monoclonal antibody against hepatitis B surface antigen (HBs-MAb), were used in this study. The cells were maintained routinely at 37°C in 5% CO_2 atmosphere in E-RDF medium (Kyokuto, Japan) supplemented with growth factors: transferrin, Na-selenite, ethanolamine, insulin, and sodium bicarbonate. For the perfusion cultures, the above medium was further supplemented with 1 g/l bovine serum albumin (BSA).

Perfusion culture with ammonia removal system.
The experiments were done in a 500-ml stirred reactor (250 ml working volume) equipped with pH, DO, temperature, and liquid level controls (Fig. 1). Agitation was provided at 35 rpm. Surface and bubble-free aeration with 5% CO_2 in air and/or pure oxygen maintained the DO around 2 mg/l. The pH was controlled at around 7.2 with 1 N NaOH. The cells and HBs-MAb accumulate inside the reactor. A peristaltic pump controlled by the level sensor withdraws cell-free and product-free culture broth through a UF-membrane (Diaflo YM10, Amicon) fixed at the bottom of the reactor.

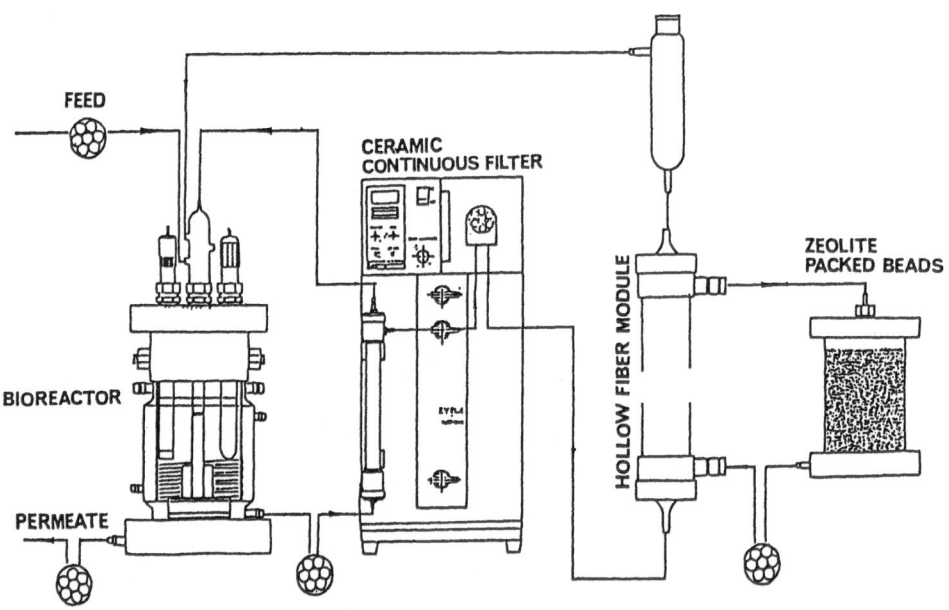

Fig. 1. Experimental set-up for perfusion culture coupled with ammonia removal system.

The culture broth was circulated through a ceramic cross-flow module (Al_2O_3, pore size: 0.2 μm) with backwashing. The cell-free filtrate from ceramic filtration was dialyzed through a hollow fiber module (AM-Neo, Asahi) at a flow rate of 3.4 l/day. Ammonia was then selectively removed from the culture broth by circulating the dialyzed permeate through 200 ml of packed zeolite A-3 beads (bead size: 0.997 mm, max. pore size: 0.42 nm, Tosoh, Japan) at a flow rate of 1.5 l/day.

Perfusion of serum-free medium at a dilution rate of 1 day^{-1} and ammonia separation were started when the viable cell concentration was about 1×10^6 cells/ml. The culture broth was circulated by a peristaltic pump through the ceramic cross-flow module at two linear velocities, v_s; 0.59 and 0.29 m/s.

RESULTS AND DISCUSSION

Zeolite selectivity for ammonia.

The selectivity of 3 synthetic and 2 natural zeolites were evaluated.[2] Pretreated zeolites were incubated in e-RDF medium supplemented with 10% FCS and 1.46 mmol/l ammonium chloride for 14 h at 37°C. The equilibrium concentrations of ammonia and amino acids in the medium are listed in Table 1

Table 1. Removal of ammonia and amino acids from E-RDF medium by different types of zeolites (percentage of original conc.).

zeo-lite	zeolite A-3	zeolite F-9	silica-aluminium ZCP-50	clinoptilolite	mordenite			
type	A-type zeolite	X-type zeolite	Y-type zeolite	ditto	ditto	compo-nent	concentration (m-mol/m³)	
origin	synthetic	synthetic	synthetic	natural	natural			
NH₃	−51.76	−68.69	−55.48	−49.19	−58.58	NH₃	1462.	
ASP	− 0.41	− 0.66	+ 3.13	− 4.33	− 5.46	ASP	73.48	
GLU	+ 1.56	+ 4.69	+ 0.19	+ 2.54	+ 4.41	GLU	108.0	
CYS	− 0.43	−10.61	− 2.37	−10.48	−15.60	CYS	93.55	
PHE	− 0.92	+ 0.04	− 0.80	+ 6.62	− 5.88	PHE	109.7	
GLN	− 6.34	−10.43	−10.78	−16.31	−19.26	GLN	1522.	
TYR	+ 2.19	+ 1.96	− 2.16	− 1.26	− 3.03	TYR	113.8	
MET	+ 1.68	− 8.60	− 4.52	−11.87	−12.48	MET	70.89	
ILE	+ 0.99	+ 2.19	− 0.81	− 1.94	− 2.72	ILE	281.8	
LEU	+ 1.84	+ 1.93	− 0.01	− 3.56	− 3.95	LEU	296.1	
ALA	+ 0.91	+ 0.76	+ 2.34	+ 5.09	−12.43	ALA	65.04	
HIS	−12.69	−24.32	− 1.64	−22.40	−20.29	HIS	85.62	
LYS	− 8.87	−68.67	−73.23	−20.70	−35.69	LYS	272.0	
ARG	− 1.58	−20.56	−61.30	−13.70	−32.77	ARG	646.4	

as percentages of their original compositions. The ion-exchange capacity for ammonia were almost the same but the uptake of some amino acids particularly lysine and arginine, varied significantly between zeolites. Since the molecular weights of amino acids are about 10 times that of ammonia, the high selectivity of zeolite A-3 seems to be due to the molecular sieving effect. For this study, therefore, zeolite A-3 was selected.

Characteristics of MAb production and sensitivity to ammonia.
    HBs monoclonal antibody production has been previously demonstrated to be growth-associated and proportional to the viable cell density.[1] Although the maximum viable cell densities are lower in serum-free cultures, HBs-MAb production is higher than that in serum cultures.
    The sensitivity of TO-405 cells to ammonia inhibition was investigated by doing perfusion cultures in e-RDF medium supplemented with 10% FCS at different DO levels. The cell viability appeared to be inhibited when ammonia accumulated to about 5 mmol/l regardless of the DO levels [Fig. 2]. Previous results also showed that although the final ammonia concentrations were similar in serum-supplemented and serum-free cultures, the rate of accumulation was faster with the latter.

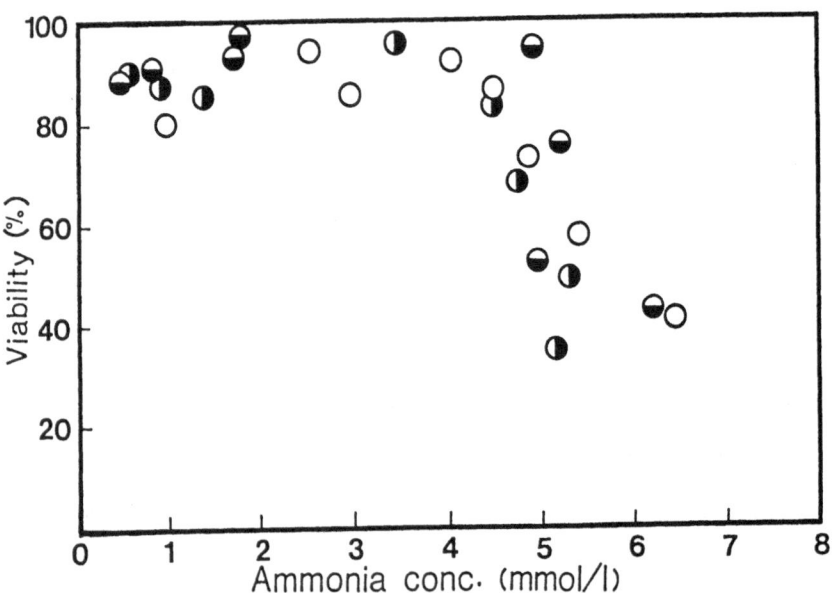

Fig. 2.   Effect of ammonia conc. on the viability of the TO-405 cells during perfusion cultures with E-RDF + 10% FCS medium.

Perfusion culture with ammonia removal system.

Perfusion culture without ammonia removal was done as a control experiment. The results are shown in Fig. 3. Although the cells continued to grow, the viable cell density did not improve an ymore from about 6.6 x 10^6 cells/ml. This was considered to be caused by ammonia toxicity.

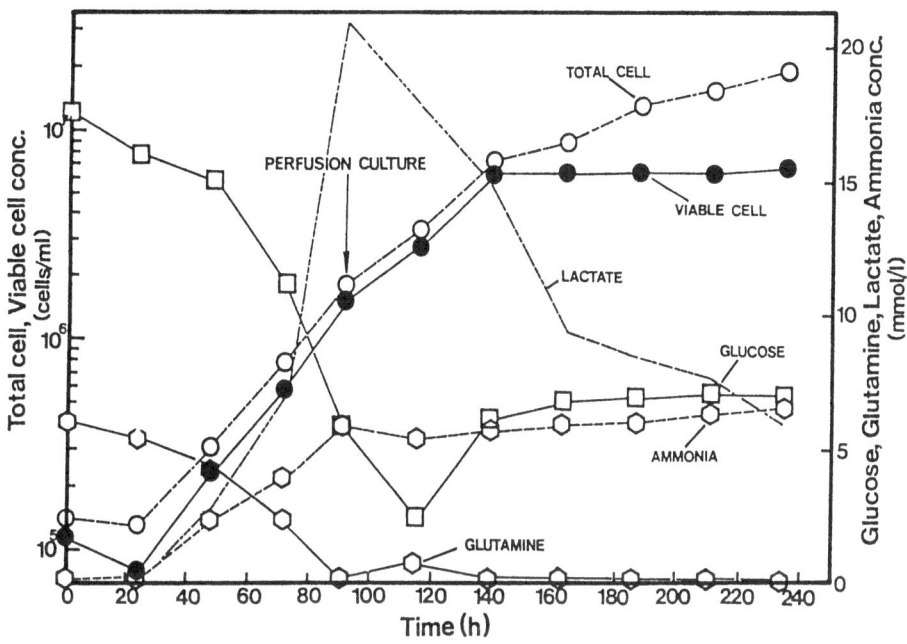

Fig. 3.   Perfusion culture of TO-405 hybridoma cells.

The effects of selective removal of ammonia on the growth and viability of hybridoma TO-405 cells producing HBs monoclonal antibody were investigated. The effect of hydrodynamic shear stress on the growth of the hybridoma cells was investigated by varying the cross flow rates of the broth through the ceramic filter module.

The cell growth was completely repressed in cultures with a circulation linear velocity of 0.59 m/s (Fig. 4). The cell viability did not change for more than 60 h after the start of the circulation. The inhibition of the cell growth was therefore considered to be due to shear stress generated by the high flow rate of the culture broth through the ceramic module. Ammonia concentration in the culture broth, however, was maintained below 1 mmol/l, which is much lower than the reported inhibition level of 5 mmol/l for this cell line.

100

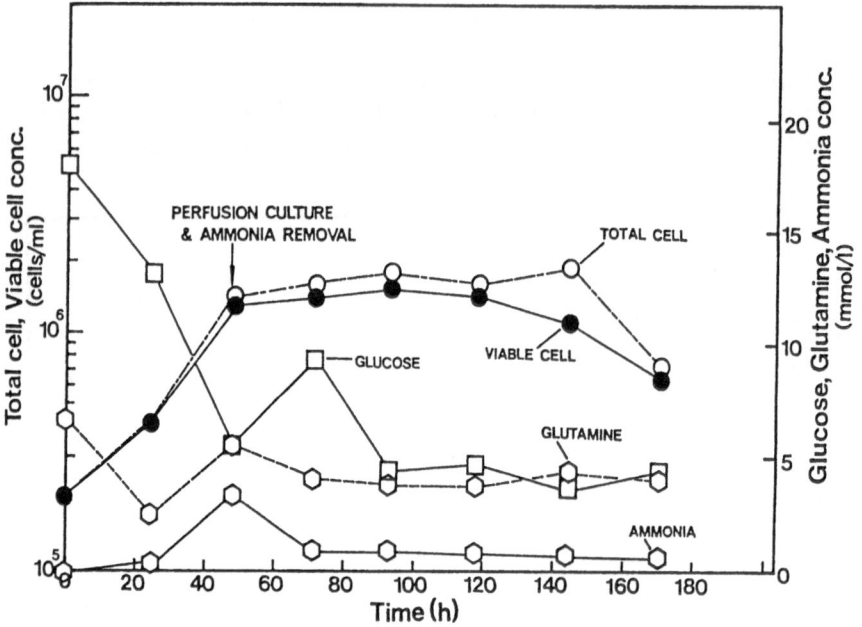

Fig. 4. Perfusion culture of TO-405 hybridoma cells with ammonia
removal (ceramic circultion velocity = 0.59 m/s).

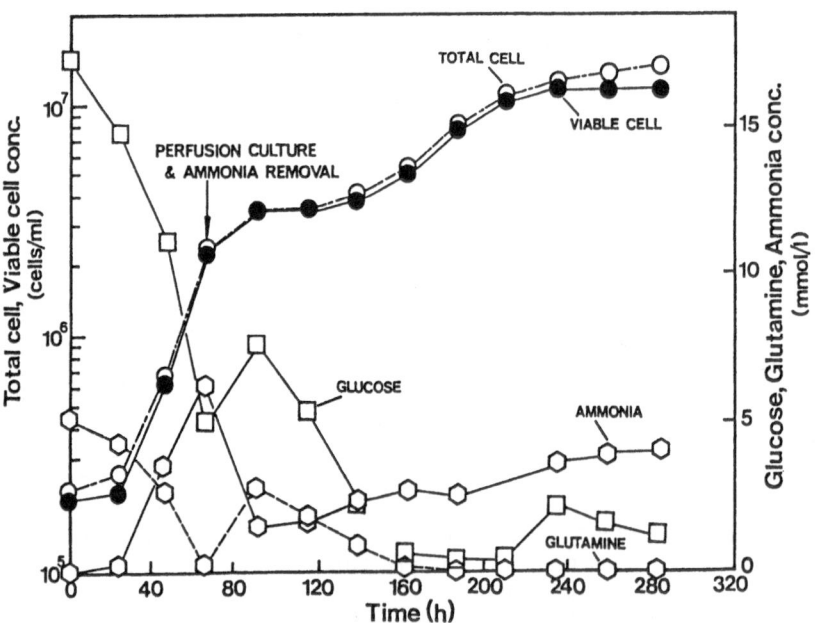

Fig. 5. Perfusion culture of TO-405 hybridoma cells with ammonia
removal (ceramic circultion velocity = 0.29 m/s).

Under lower circulation rate of 0.29 m/s, a viable cell density of 1×10^7 cells/ml, similar to that in serum culture was attained (Fig. 5). The percentage of viable cells was also maintained high above 90%. This result was attributed to the removal of ammonia inhibition from the culture system. The cell growth, however, seemed to be affected by shear stress generated during the ammonia removal even at this linear velocity. The low specific growth rate, however, may not be only due to shear stress but also to cell accumulation at the inlet of the module. Improved module construction is presently being tested.

Nevertheless, ammonia concentration in both cultures was successfully maintained far below the inhibitory level. The viability was much better compared to that in perfusion culture without ammonia removal (Fig. 3). Since HBs-MAb production is growth-associated and proportional to the viable density [1], improved viability means improved MAb productivity.

REFERENCES

1. Matsumura, M., Misato, T., Kataoka, H., Tanaka, T. and Mayumi, M. (1989) 'Production of HBs monoclonal antibody by perfusion culture in membrane reactor', in H. Murakami (editor), Trends in Animal Cell Culture Technology, Kodansha Ltd., Tokyo, pp. 145-148.

2. Motoki, M., Matsumura, M. and Kataoka, H. (1988) 'Removal of ammonia from animal cell culture broth using zeolite', Abstract of the Annual Meeting of Chemical Engineers, Sendai, pp. 4.

GMP PRODUCTION OF BIOPHARMACEUTICALS USING HIGH DENSITY, FLUIDIZED-BED CELL CULTURE TECHNOLOGY

Dr. Peter W. Runstadler
Mr. Michael W. Young

Verax Corporation
Lebanon, New Hampshire U. S. A.

ABSTRACT

Biopharmaceutical products from mammalian cell culture are currently being produced in cGMP facilities using high density, immobilized, perfusion culture. Animal cells, including hybridomas, natural, and recombinant cells, are being immobilized in collagen microsponges. Process enhancement of the physical, chemical, and physiological properties of the culture optimize cell growth, productivity and product quality. The use of collagen microsponges (Verax Microspheres) enables high density cell culture ($> 1.0 \times 10^8$ viable cells/mL matrix) at high perfusion rates and high cell specific productivities. Serum-free medium in combination with high density, perfused culture reduces production costs.

R. Sasaki and K. Ikura (eds.), Animal Cell Culture and Production of Biologicals, 103–119.

Introduction - Fluidized Bed Bioreactors: The Verax System

Verax Corporation has developed proprietary mammalian cell fluidized bed culture systems for the manufacture of therapeutic proteins. The Verax System uniquely optimizes the local microenvironment surrounding natural, genetically-engineered, or hybridoma cell lines, leading to high cell density, cell viability and most important of all, enhanced cell-specific productivity compared to conventional approaches. This process leads to low cost of product and high product quality. The Verax process also facilitates the use of serum-free media which offers a significant cost advantage: savings are realized by avoiding the high cost of fetal bovine serum, minimizing product degradation, and reducing downstream purification costs. This process is being used in cGMP facilities for the production of biopharmaceuticals.

The Verax System is the result of synergistic effects deriving from the interaction of several unique components, including: sponge-like, weighted, collagen microspheres for the immobilization of cells to achieve high cell densities (> 1.0×10^8 viable cells per mL matrix) and high productivities; and a fluidized bed perfusion bioreactor designed to deliver the oxygen and other nutrients required, and to efficiently remove the secreted therapeutic protein products and waste products generated.

The Extra Cellular Matrix (ECM) and The Cell Microenvironment

The cell culture microenvironment affecting each cell is the local biochemical, physiological, and physiochemcial milieu directly affecting the growth of the cells and their metabolism. The behavior of cells, for example attachment, spreading, morphology, and biosynthetic capacity, are influenced by their immediate surroundings. It is well documented that in culture mammalian cells retain or lose their morphological and biochemical differentiation, and thus, their protein synthetic and secretory capabilities depending on the chemical composition and structure of the culture substrate.

For example it is known that the extra cellular matrix (ECM) exerts physical and chemical influences on the cell through transmembrane receptors and that these influences alter regulation of the cell by changing the association of the cytoskeleton with mRNA and the interaction of chromatin with the nuclear matrix (Bissell, Hall and Barry (1982)). The cell membrane attaches to ECM glycoproteins and other cells through integrin receptors. The intercellular domains of the integrins interact with the cytoskeleton (Hynes (1987)). Current workers in the field hypothesize that the ECM communicates through the cytoskeleton with connections to the nuclear matrix medicated by cell shape changes. These communications assist in the regulation of gene expression at all levels: transcription, mRNA processing, translation, post - translational modifications, secretion and extracellular organization (Bissell and Barcellis - Hoff (1987)).

Typically, the sequence of events in the establishment of a productive cell/substrate interaction involves (1) cell attachment, mediated initially by the interaction of a family of structurally related proteins including fibronectin, with their cell receptors, the integrins, (2) expression and transport of cellular fibronectin, vitronectin, laminin, proteoglycans and other proteins, leading to the formation of an extracellular matrix (ECM) in the presence of appropriate substrate molecules such as native collagen, (3) development of a basal surface in contact with ECM and an apical layer, coincident with the appearance of a highly organized cytoskeleton in the cell, and (4) establishment of a pattern of gene expression that results in both morphological and biochemical phenotypes that are correlated with high levels of protein synthesis and secretory capacity as well as with low rates of DNA synthesis and cell replicative activity. As discussed above, independent researchers have proposed that there is a direct mechanochemical transduction of information from the extracellular matrix environment through the cytoskeleton to the cell's nuclear matrix that is responsible for the organization of these events.

In considering immobilized mammalian cell culture in the three-dimensional collagen microsphere matrix, the Verax microenvironment can be described as consisting of three compartments: the contact environment, the diffuse environment, and cell junctions. Each of these three aspects regulate cell behavior by specific molecular mechanisms, acting synergistically to provide a given overall condition for cell behavior. The combined effect of these three components can be optimized by appropriate and judicial choice of cell culture substrates and bioreactor design and operation to achieve the optimum conditions for mammalian cell growth and productivity.

The Contact Environment. The contact environment of cell tissues is provided primarily by the extracellular matrix and an insoluble meshwork of protein, carbohydrate and other matrix-bound molecules synthesized and assembled by the cells themselves. Although the ECM varies in composition among tissues, it generally consists of basement membrane and connective tissue proteins such as collagen and elastin, of adhesion surface glycoproteins such as fibronectin, laminin, and vitronectin, and of proteoglycans and gylcosaminoglycans.

Collagen has been employed extensively for animal cell culture in order to attempt to closely mimic natural tissue conditions. Collagen is immunologically benign, highly resistant to proteolysis, and is a natural substrate for cell adhesion. To reconstruct the natural microenvironment of cells as closely as possible, Verax's three dimensional microspheres are prepared from Type 1 collagen. These microspheres are manufactured using patented technology from bovine hide collagen, which is composed of 80% collagen and 20% other insoluble matrix components. This material provides a natural scaffolding for the attachment of cultured cells and cell-specific factors either produced by the cells themselves or provided by serum in media that can be used to precondition the microspheres prior to a culture run.

Once attached to the collagen substrate, cells enhance their contact environment by producing and organizing their extracellular matrix on the surface of the microsphere, including the production of cell-specific fibronectin. Eventually, the microspheres are coated with a complex spectrum of macromolecules and important other components such as adherent growth factors which have a very high affinity for matrix proteoglycans. The secretion of these factors also obviates the need for expensive bovine serum. Thus, the collagen microspheres provide an ideal, natural material for the development of the optimum contact environment for cultured cells.

The Diffuse Environment. The diffuse environment includes all molecules that are freely diffusible in the extracellular and intracellular tissue culture space, including hormones and growth factors (e.g., autocrine factors synthesized by the cells themselves and other cytokines), dissolved gases (e.g., oxygen and carbon dioxide), nutrients (e.g., glucose and amino acids), salts and ions (e.g., Na^+, K^+, Ca^{++}) and cellular products including secreted proteins as well as waste products such as lactate and ammonia In general, the diffusive substances can access the cell from either the basal or the apical surface. Many of the molecules in the diffuse compartment are internalized by cells either by active and/or passive transport systems, or by receptor mediated endocytosis.

The fluidized bed bioreactor optimizes the diffuse environment by two mechanisms: it causes fluid flow effects forcing culture medium into and through microspheres, and it facilitates control of nutrient, oxygen, and waste product concentrations by allowing a variable and high recycling of the culture medium and a rapid removal of the conditioned medium from the bioreactor. The tumbling motions of the fluidized bed superimposed on the medium flow around the microspheres, provides a thorough access of the recycling medium to all cells.

Cell Junctions. The formation of cell junctions is critical in the development of tissue-like aggregations of cells. Figure 1A shows a scanning electron micrograph illustrating the morphology of Chinese Hamster Ovary (CHO) cells in a microsphere loaded at low cell density. These cells have invaded the open pores of the microsphere and have formed contacts on the matrix surface. Cell/cell contacts have also been formed, which in time will result in the population of the microsphere at near

FIGURE 1
CELL MORPHOLOGY

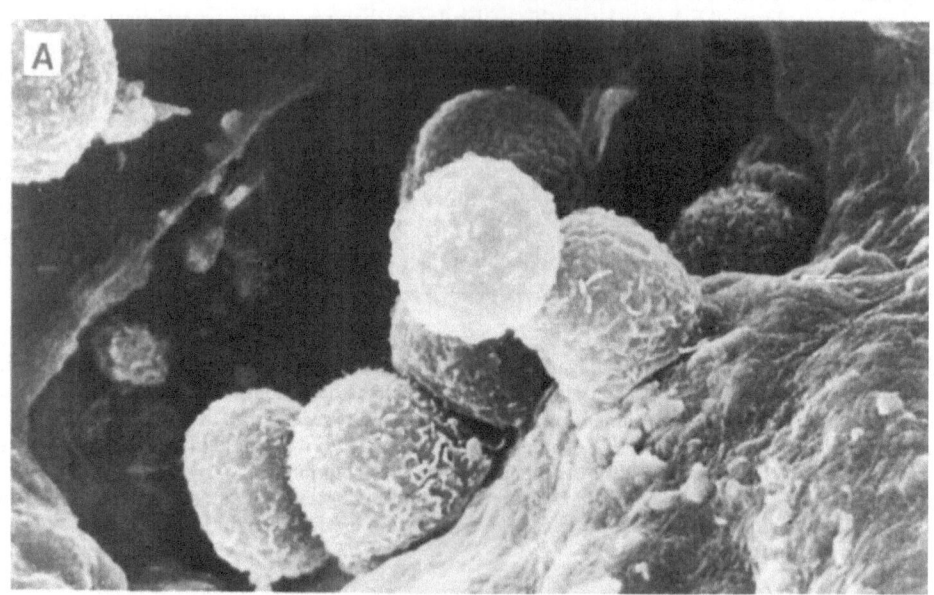

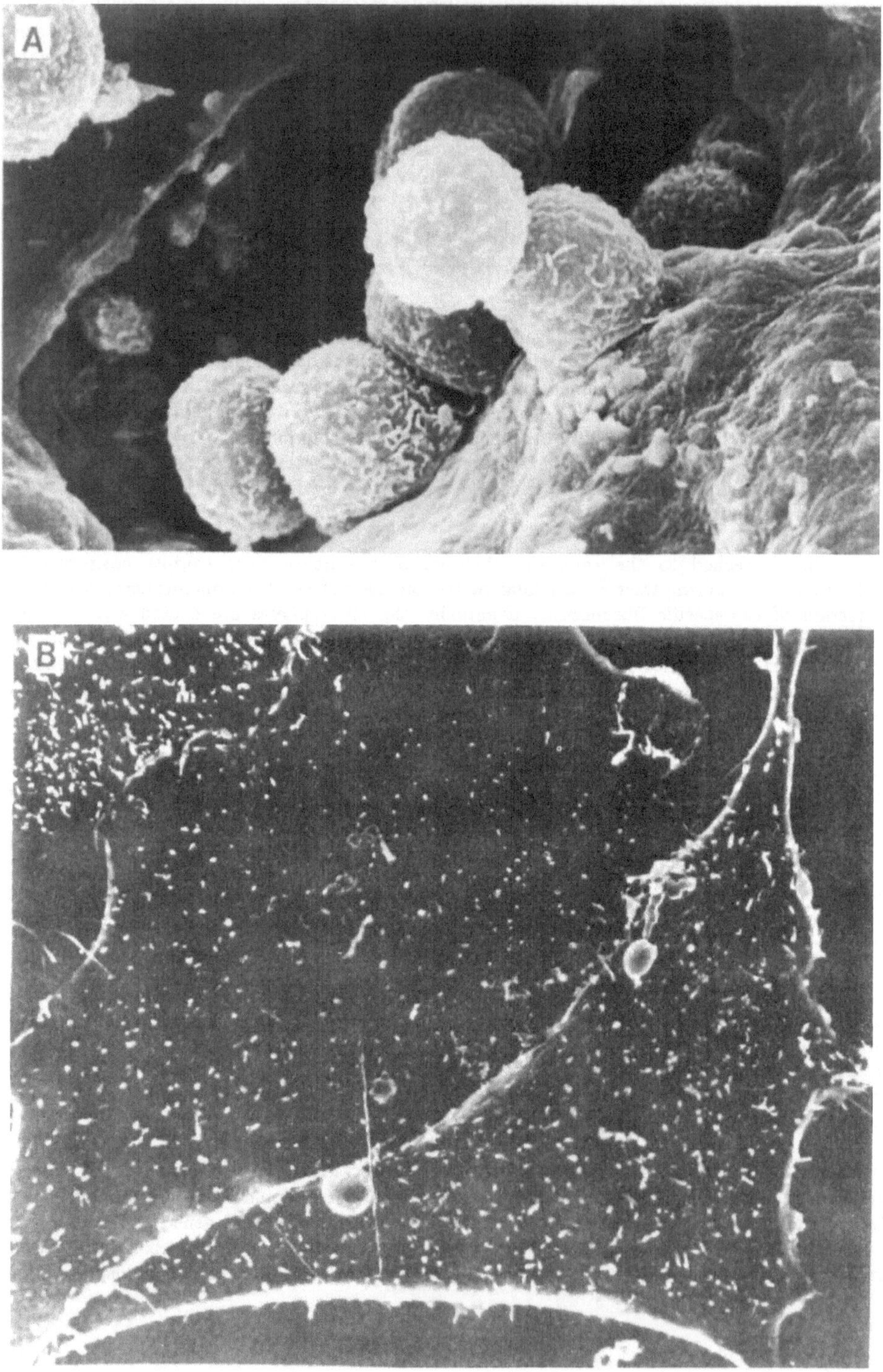

tissue-like cell densities. These cells have a much more normal and healthy appearance compared to cells attached to a two-dimensional substrate, typical of most of the other reactor systems, as seen in Figure 1B. As expected, cells grown in collagen microspheres are highly efficient and productive in generating the target therapeutic proteins.

Verax Collagen Microspheres

The heart of the Verax System is the sponge-like, bovine collagen microsphere, containing small metal particles that increase its density and, thus, allow it to be suspended in a fluidized bed bioreactor. The microspheres are manufactured having an average diameter of 500-600 microns, although this can vary. Microspheres contain interconnected pores and channels which allow cells to enter, attach and grow in the internal volume, which is approximately 85% open. The pore size, with a typical diameter of approximately 50 microns, can also be controlled during the manufacturing process. Pores are surrounded by leafy surfaces of collagen, available for the attachment and/or proliferation of cells. Figure 2 is a scanning electron micrograph indicating the morphological structure of a typical microsphere.

Collagen microspheres contain bovine fibronectin. Fibronectin, a natural cell adhesion molecule, provides a natural ligand available to promote the initial contact of attachment dependent cells having fibronectin surface receptors. It is known that many animal cells have such receptors and therefore recognize and readily adhere to the microspheres. Growth factors and other important molecules may also adhere to the microspheres and are, therefore, available in high local concentrations to support initial cell growth and viability.

Two important and unique properties of Verax microspheres, allowing them to function optimally in fluidized bed systems, are their porosity and flexibility. The open porous structure of microspheres enhances medium flow and nutrient product exchange. In addition, the flexibility property allows cells to populate microspheres in a manner that optimizes cell/cell and cell/matrix contacts. As a result, cells differentiate to a natural state optimal for biosynthesis, processing, and secretion of proteins.

More than 120 cell/product combinations, including natural secreting cells, attachment dependent and suspension cells engineered to produce recombinant proteins, and hybridomas producing monoclonal antibodies have been cultured in these microspheres in Verax bioreactors and have achieved high cell densities and viabilities. A partial list of the cell types cultured in the Verax Process is indicated in Table 1.

TABLE 1

CELL TYPES CULTURED IN COLLAGEN MICROSPHERES

Chinese Hamster Ovary	Human Hepatoma (HEP-G2)
African Green Monkey Kidney	Human Lymphocytes
Human Embroynic Kidney	Transformed HeLa Cells
Baby Hamster Kidney	Murine Myeloma
Rat Kidney: Normal or Transformed	Human Myeloma
Mouse Mammary Tumor (C-127)	Murine Hybridomas
Human Diploid Fibroblasts (MRC-5)	Murine Hybridomas

FIGURE 2
COLLAGEN MICROSPHERE

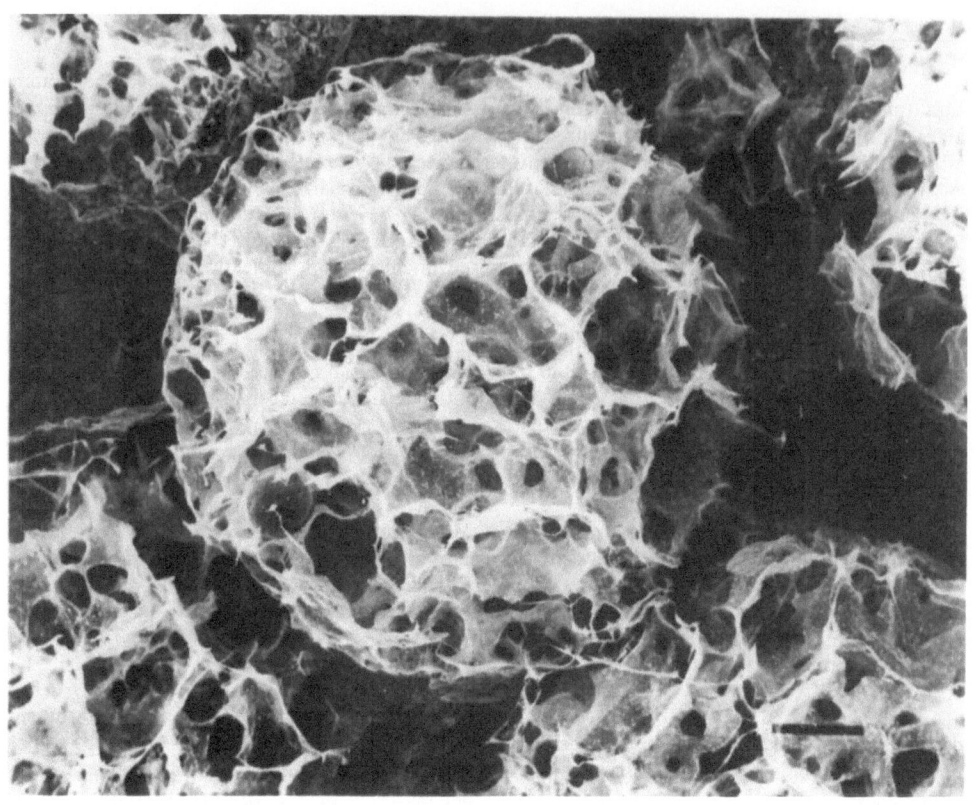

Table 2 summarizes some of the growth and viability data and illustrates the general applicability of the collagen microspheres for the culture of cell lines of commercial interest. High cell viabilities are usually observed over long periods of time in the fluidized bed bioreactor system. The majority of cells, usually more than 70%, remain viable and highly productive for periods of months in heavily cell populated microspheres.

TABLE 2

CELL DENSITY AND VIABILITY IN COLLAGEN MICROSPHERES

Cell Type	Product	Cell Density ($\times 10^7$ cells/ml)	% Viability
A. Hybridoma Cell Lines			
Human/Mouse	IgM	5.0	70
Human/Mouse	IgG	8.0	68
Rat/Mouse	IgG	15.0	72
B. Attachment-Dependent Cell Lines			
CHO	thrombolytic	18.0	79
CHO	thrombolytic	20.0	80
CHO	interleukin	26.0	92
CHO	thrombolytic	22.0	75
CHO	cardiovascular	20.0	80
CHO	hormone	25.0	72
CHO	anti cancer		
C127	thrombolytic	9.0	85
Myeloma	chimeric antibody	6.7	87
Human Kidney	thrombolytic	16.8	60

Verax currently manufactures microspheres from bovine collagen using a proprietary patented process. The structure and fluidization behavior of microspheres are controlled during manufacture. Microspheres are made in advance and are stored in a ready-to-use, sterile state. They are treated as a consumable with a useful lifetime of one cell culture run.

The Fluidized Bed Bioreactor

The central feature of the Verax fluidized bed bioreactor is a column containing the microspheres through which the culture medium flows upward at sufficiently high velocity so that the weighted microspheres do not rest on one another, but are suspended and move about within the vessel as a slurry. The collagen microspheres are subjected to medium flow dynamics that result in a pumping of medium throughout their interior as cells populate them to very high densities. The fluid dynamics of the rapidly recycling culture in Verax bioreactors generates pressure differences over the surface of the microsphere resulting in the forced flow of nutrients throughout the porous structure. This effect provides convective flow through the microspheres that leads to a greater uniformity in the distribution of medium components, such as oxygen and other cell culture nutrients, than is possible in other methods for mammalian cell culture. This property is uniquely attributable to the porous, flexible nature of the microspheres.

The flow dynamics described above that lead to the flow through the collagen microspheres are optimized by the vertical flow characteristics of the fluidized bed. The total impact of the bioreactor design is to help establish the formation of an optimal microenvironment around and in each of the densely populated microspheres.

The culture medium is continually removed and recycled through the bioreactor vessel and provides nutrients to the cells living inside the microspheres. Proteins secreted by the cells diffuse out of the microsphere and into the recycling culture medium. The weighted microsphere fluidized bed establishes a natural separation horizon near the top of the bioreactor vessel, providing the recycling culture medium with a zone of clear detachment from the slurry of microspheres. The recycle flow removes culture medium from the top, conditions it in the recycle loop, and returns it to the bottom of the bioreactor vessel. The microspheres, which contain the majority of the cells, do not enter the recycle loop but remain suspended.

The recycle loop (see Figure 3) consists of a pump, a membrane gas exchanger to add oxygen and to extract carbon dioxide from the recycling media, a heater, and sensors for monitoring pH, dissolved oxygen, temperature, and flow rate. A metering pump continuously supplies fresh nutrient medium to the culture from a chilled medium reservoir. A microcomputer and an analog-digital interface module allow electronic monitoring of the bioreactor and provide automatic and manual measurement, recording and control of the system. Control is maintained within operator-specified set points to provide near steady-state operation for long-term cultures of up to many months duration. The product proteins are harvested continuously at the same rate as the medium is added.

The key features that derive from the interactions of the elements of this system are:

- high mass transfer rates caused by the fluid dynamic properties of medium flow around and through individual microspheres resulting in efficient intramatrix transport of oxygen and other nutrients, and removal of cell products,

- highly viable immobilized cells, due to the affinity of cells for the natural collagen substrate of the microsphere, leading to high productivity,

- an optimal microenvironment, due to the cell/microsphere and cell/cell contacts and to the formation of a cell derived extracellular matrix, resulting in local concentrations of growth factors and other important constituents that combine to generate enhanced specific productivity, high product titer, and high product quality.

When in a fluidized bed system, each microsphere becomes an independent colony of cells, behaving effectively as an individual microreactor. The nutrient requirements and productivity of the total bioreactor are, therefore, the sum of those of the individual unit microreactors. This concept is illustrated in Figure 4 and is the basis for the scalability of the process. Verax's commercial systems have an increasing number of microspheres, going from the bench-top System One research bioreactor to the commercial pilot-scale System 200 to the large-scale production System 2000, as indicated in this figure.

Several unique features result from this cell culture system design. The bioreactor has inherent culture stability and scalability properties. Culture stability, in the absence of exogenous selective pressures, derives from the difficulty in propagating a mutation throughout the culture. In the Verax System, mutant cells originally appear on individual microspheres from which they must detach and then cross over a rapidly flowing medium stream in order to colonize a neighboring microsphere whose cell binding sites are usually fully occupied. In addition, the short mean residence time (on the order of 2-8 hours or less) of culture medium in this system facilitates rapid washout of mutant cells and limited exposure of protein product to the culture harvest, which may contain degradative enzymes.

FIGURE 3

FLUIDIZED BED BIOREACTOR RECYCLE LOOP

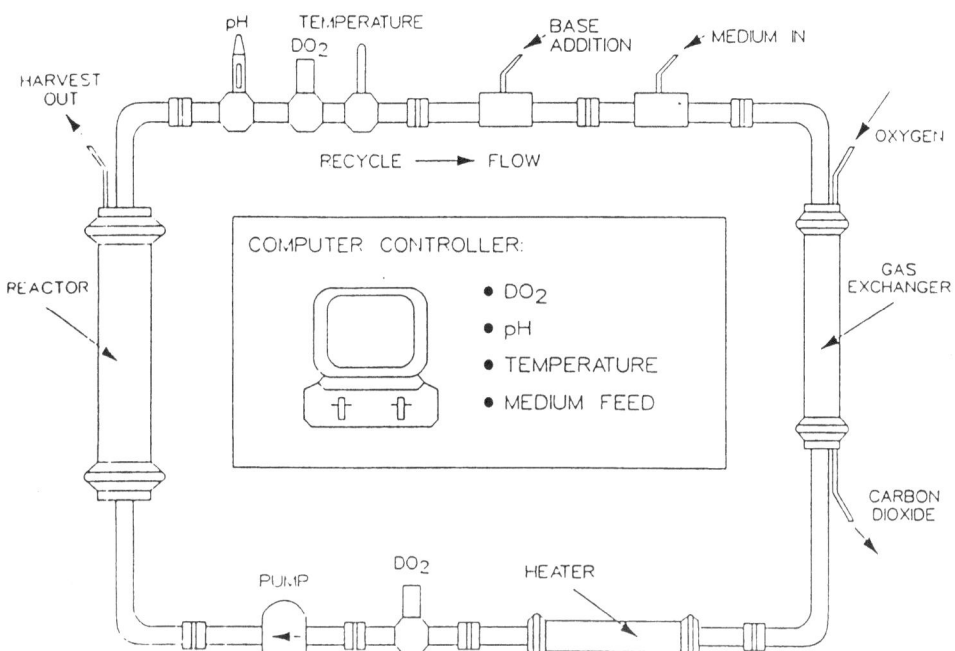

Figure 4: Scalability of a Fluidized Bed System

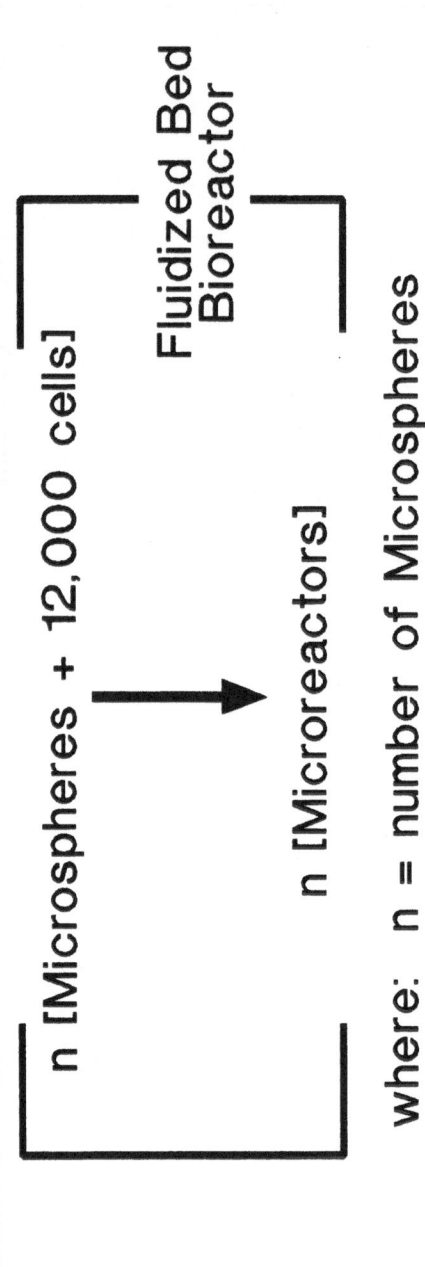

n [Microspheres + 12,000 cells]

→ n [Microreactors]

Fluidized Bed Bioreactor

where: n = number of Microspheres

	n	Viable Cells per Bioreactor
System One	80,000	$1.0 \times 10E9$
System 200	8,000,000	$1.0 \times 10E10$
System 2000	130,000,000	$1.5 \times 10E12$

The ultimate test of the effectiveness of the Verax tissue culture system for the production of biotherapeutic proteins is its scalability, its productivity, and the quality of the product it generates. Proteins produced by this process have excellent characteristics, including structural integrity, appropriate protein folding and glycosylation and specific activity. This is generally due to the total elimination of bovine serum from the medium, to the short residence time of the product in the culture medium after secretion, and to the impact of the system in creating a microenvironment to enable cells to achieve an optimally productive state, usually demonstrated by an increase in cell specific productivity i. e. in the amount of product produced per cell per unit of time.

Data showing an increase in cell specific productivity upon removal of fetal bovine serum from the growth medium, have routinely been obtained for a wide variety of recombinant cell lines cultured in Verax bioreactors. It has been observed, for example with CHO cell lines, that an "adaptation" of the cells occurs on the microspheres during the culturing processs. This is illustrated in Figure 5 for several CHO cell lines cultured in fluidized bed bioreactors. In each case, a significant increase in the product concentration is observed following removal of serum from the medium. In some cases increasing by 5X or more. Since in most cases the cell densities did not change significantly in these reactors following serum removal, it is concluded that protein specific productivity increased sharply. This protein specific productivity reflects an adaptive response of cells to the optimal microenvironment created in the microspheres during the growth (serum-containing) phase of the culture, and it is an important effect of the Verax process. Other cell types, including mouse mammary C-127 cell lines, exhibit similar, although somewhat less dramatic, behavior. It appears that most cells undergo an "adaptation" to Verax microspheres, which is enhanced by the fluidized bed bioreactor, leading to increased productivity and improved product quality.

Of equal importance, it is seen that the protein quality is improved after the serum was removed, with greatly reduced product degradatation as shown by Western Blot analysis (Figure 6). Removal of serum sharply reduces the presence of deleterious proteases in the culture medium, thus improving the likelihood that the product will remain intact during the culture process. In another instance, the specific activity of a hybridoma produced anti-IL-4 monoclonal antibody was approximately 10-fold more than the same hybridoma produced antibody made using conventional culture techniques (Finkelman et. al. (1988)). Results similar to these have been observed with most recombinant protein products, suggesting that considerable improvement in product quality is realized in the serum-free Verax Process.

Adjustment of bioreactor operating parameters can further improve productivity. Increases in the perfusion rate (medium feed rate) has resulted in dramatically-improved daily reactor output with hybridoma, natural kidney and recombinant CHO cell lines. The product concentration remains constant as more medium is perfused, thus leading to the observed improvements in protein-specific productivity and daily output. Perfusion rates of approximately 1000 liters per day have been used for the GMP production of recombinant protein for clinical studies using a Verax System-2000 bioreactor. In addition, it has been found that the ratio of oxygen transfer rate and glucose consumption rate affects the product concentration in recombinant CHO cell line cultures.

Scale-Up Performance and Predictability

A unique property of the microspheres and the fluidized bed bioreactor is the inherent scalability of the system. Scale-up performance of the Verax Process has been demonstrated by culturing the same cell line in both the bench-top System One and System 10 and either System 200 or System 2000, and in some cases in all three systems, and comparing results. It can be seen that a number of important parameters are either enhanced or remain constant by scaling up. For example, the data in Table 3 show results with a blood factor producing CHO cell line that achieved a factor of 135X in daily productivity for a 150X increase in fluidized bed volume in going from a System 10 to a System 2000, indicating an essentially constant normalized productivity per unit volume fluidized bed.

114

FIGURE 5

CELL PRODUCTIVITY AFTER REMOVAL OF BOVINE SERUM

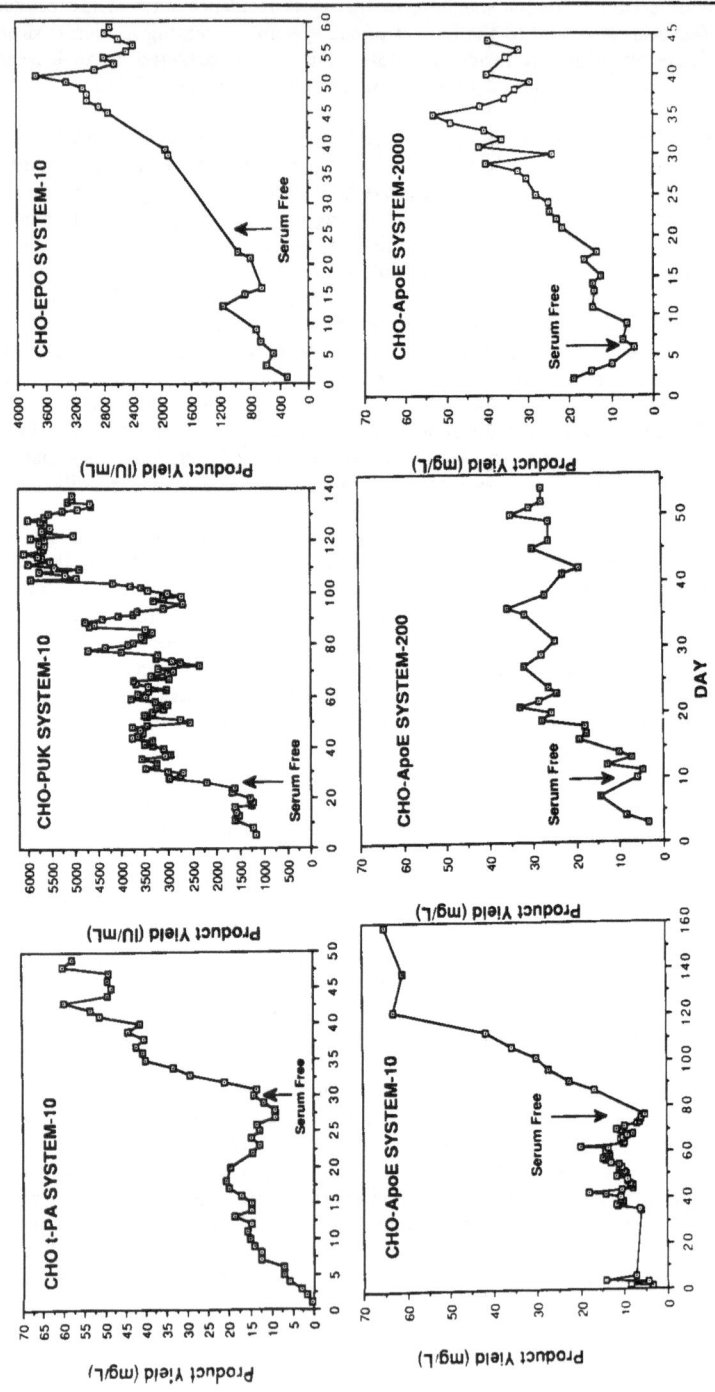

FIGURE 6

PRODUCT QUALITY

CARDIOVASCULAR THERAPEUTIC

A. SYSTEM-10 BIOREACTOR PERFORMANCE

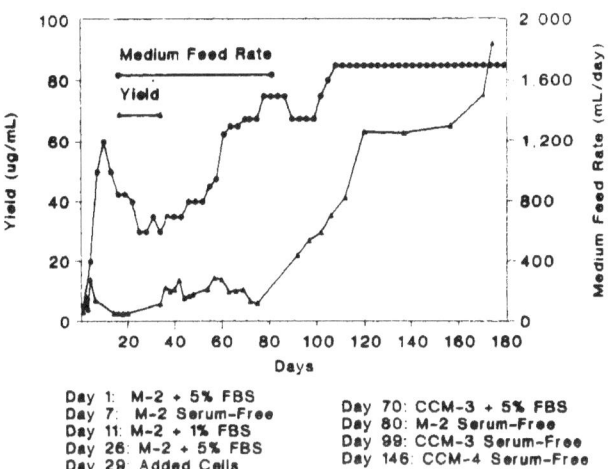

Day 1: M-2 + 5% FBS
Day 7: M-2 Serum-Free
Day 11: M-2 + 1% FBS
Day 26: M-2 + 5% FBS
Day 29: Added Cells

Day 70: CCM-3 + 5% FBS
Day 80: M-2 Serum-Free
Day 99: CCM-3 Serum-Free
Day 146: CCM-4 Serum-Free

B. PRODUCT QUALITY: WESTERN BLOT ANALYSIS

+ FBS   - FBS

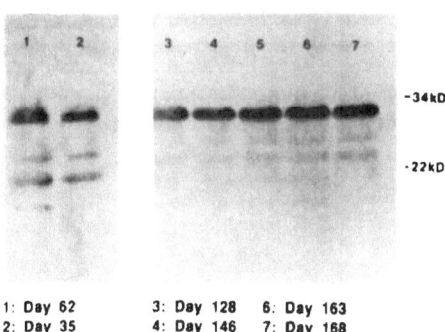

1: Day 62   3: Day 128   6: Day 163
2: Day 35   4: Day 146   7: Day 168
               5: Day 155

Also the amount of product produced as a function of glucose consumed remained nearly constant, as did the steady-state product concentration.

TABLE 3

DEMONSTRATION OF SCALE-UP CULTURE PERFORMANCE OF RECOMBINANT BLOOD FACTOR PRODUCING CHO CELL IN FLUIDIZED BED CULTURE PROCESS (a)

Culture Parameter	System 10 Bioreactor	System 200 Bioreactor	System 2000 Bioreactor	S-2000 S-10
Fluidized Bed Volume (L)	0.14	1.5	21.0	150.0
Culture Duration (days)	181	55	55	-
Medium Feed Rate (L/day)	1.6	20.0	191.0	119.0
Product Conc. (mg/L)	35.6	28.1	40.3	-
D-glucose Consumption (g/day)	4.3	48.4	544.2	126.0
Product Yield (mg/day)	57.0	562.0	7.7	135.0
Ratio: mg Product Produced/g of D-glucose Consumed	13.2	11.6	14.1	-

(a) Analysis of steady-state results

In the case of a tPA producing CHO cell (Table 4), the ratio of fluidized-bed volume was 300X and the observed daily productivity was 522X, indicating a significant increase in normalized productivity in going from the System 10 to the System 2000. Also, the product yield as a function of glucose consumption and product concentration increased in the System 2000.

Similar results have been obtained with hybridomas, and other cell types. Scale-up results are favorable in the Verax System, with positive increases observed in productivity.

TABLE 4

DEMONSTRATION OF SCALE-UP CULTURE PERFORMANCE OF
RECOMBINANT tPA PRODUCING CHO
CELL IN FLUIDIZED BED CULTURE PROCESS (a)

Culture Parameter	System 10 Bioreactor	System 2000 Bioreactor	S-2000 S-10
Fluidized Bed Volume (L)	0.08	24.0	300.0
Culture Duration (days)	81	27	-
Medium Feed Rate (L/day)	0.85	330.0	388.0
Product Conc. (mg/L)	45.30	60.8	-
Percent Single Chain Product (b)	> 95%	> 95%	-
D-glucose Consumption (g/day)	2.21	957.0	433.0
Product Yield (mg/day)	38.50	20,100.0	522.0
Ratio: mg Product Produced/g of D-glucose Consumed	17.40	21.0	-

(a) Analysis of steady-state results
(b) Based on densitomeric scan analysis of Western Blots.

Conclusions

Present production alternatives to the Verax Systems include roller bottles, various types of retained cell suspension systems, such as stirred tank systems with spin filter cell retention, adaptations of stirred tank fermenters, hollow fiber reactors, packed beds, and a number of microcarrier and encapsulation technologies. The Verax System possesses attributes which make it the most attractive alternative for use as a production system for the manufacture of biotherapeutics. Some of the unique features and benefits of the system are:

Continuous Culture with Cell Retention: Process Optimization. In contrast to batch fermentation, Verax's continuous culture system maintains steady-state operating conditions that are the best culture environment for process optimization. Under these conditions, where process variables (e.g., pH, temperature, cell density, glucose concentration, etc.) have reached constantly maintained values, one of these parameters at a time can be varied to arrive at the best process definition for high productivity and product quality. Cell immobilization permits the additional degree of freedom of variable nutrient feed rates, that cannot be achieved in a simple stirred tank, continuous suspension culture that operates without cell retention.

Immobilized Cells at High Density in a Collagen Matrix: High Cell Productivity. Cell immobilization in the unique Verax microspheres encourages the cells to grow to high densities. At high densities, the immobilized cells alter their microenvironment and metabolism and redirect their metabolic energy from growth to product formation. The microspheres have important characteristics of composition and morphology that are essential for optimal cell/matrix and cell/cell interactions. These interactions lead to high cell density and productivity through the dynamic modification of the microenvironment by cell-derived extracellular constituents. This results in local concentrations of growth factors and other important constituents that combine to generate enhanced cell productivity and high product titers. The microsphere/fluidized bed system provides the means to control and optimize the diffuse and contact components of the cells' microenvironment. The enriched environment has nutrients, hormones, growth factors, enzymes, and other cell factors, that cause the cells to productively respond in ways not observed at lower cell densities, resulting in higher cell-specific productivity than can be derived in alternate production systems.

Use of Serum-Free Media: Low Cost Media. It is most significant that cells can be cultured in this system using defined media without supplemental serum. A central feature of the microenvironment produced in the Verax System is the production by the cells themselves of cell-specific factors that provide the appropriate signals for growth and productivity obviating the use of serum to provide these factors. This also reduces downstream purification costs. The result is low process costs through the use of low cost media.

High Medium Feed Rates: High Reactor Productivity. Using the immobilized cell, continuous process, the cell culture's microenvironment is made optimal in part by controlling several factors including the rate of addition of nutrient substrate and other medium components fed to the culture, the concentration of beneficial constituents made by the cells (such as hormones and growth factors), and the rate of removal of inhibitory culture metabolites (waste or toxic products produced by the cells). Very high medium feed rates approaching 1000 liters per day have been used with the Verax Process without any reduction in product titer, leading to very high reactor productivity not seen in batch or non-cell retention alternative systems.

Culture Stability: Long, Continuous Runs. A unique feature of the bioreactor is its inherent culture stability properties. Culture stability, in the absence of exogenous pressure, derives from the difficulty in propagating a mutation throughout the culture. Mutations present in a living cell culture pose a potential problem of non-producer or low producer cells growing faster than high producer cells so that over time, a decline takes place in the productivity of the culture. In the Verax System, mutant cells originally appear in individual microspheres from which they must detach and cross over a rapidly flowing medium stream in order to colonize neighboring microspheres whose cell binding sites may be fully occupied. Also the short mean residence time (on the order of 2-8 hours or less) of culture medium in this system facilitates rapid washout of mutant cells and limited exposure of product protein to culture liquid, which may contain degradative enzymes.

Process Scalability: Process Continuity. Process scalability derives from the additive nature of the basic component of the system, the cell populated microsphere in the fluidized bed. The bioreactor design based on the fluidized bed of microspheres allows reliable scale-up and process continuity up to large production scale systems, while maintaining adequate oxygen and other nutrient supplies to the culture. Therefore, process results from small bench-top scale systems can be used confidently to derive process definitions for large, commercial production reactors.

Anchorage Dependent and Suspension Cells: System Works Well with All Cell Types. The Verax culturing process works with all cell types, whether they grow in suspension or are anchorage dependent. Anchorage dependent cells attach to the collagen morphology of the microsphere aided by constituents such as attachment factors incorporated in the microspheres, while suspension cells become entrapped in the collagen sponge matrix where they form cell colonies that proliferate to high density. Many of the new biopharmaceuticals have been recombinantly engineered into both attachment and suspension cells and this system works well with all types.

REFERENCES

1. Bissell, M. and Barcellos-Hoff, M.; The Influence of Extracellular Matrix on Gene Expression: Is Structure the Message?; J. Cell Sci. Suppl. 8, pp. 327 - 343, 1987.

2. Bissell, M., Hall, G. and Parry, G.; How Does the Extracellular Matrix Direct Gene Expression?; J. Theor. Bio. 99, pp. 31088, 1982.

3. Finkelman, F. D., et. al., IL-4 Is Required to Generate and sustain In Vivo IgE Response; Vol. 141, No. 7, pp. 2335 - 2341, October 1, 1988.

4. Hynes, R. O.; Integrins: A Family of Cell Surface Receptors; Cell 48, pp. 549 - 554, 1987.

5. Tung, A. S., Ray, N. G., Corace, R. A., Vournakis, J. N., Runstadler, P. W.; Cell Culture Technology: Making It Work; Sterile Pharmaceutical Manufacturing Applications For the 1990's, Volume 2, 43 - 92, Ed. Groves, Olson and Anisfeld, Interpharm Press, 1991.

ANCHORAGE-DEPENDENT ANIMAL CELL GROWTH IN POROUS MICROCARRIER CULTURE

N. Shiragami, Y. Ohira, and H. Unno
Department of Bioengineering, Tokyo Institute of Technology,
Nagatsuta, Midori-ku, Yokohama 227, Japan

Abstract

The effects of the agitation rate on cell growth in porous microcarrier cultures were discussed. The specific growth rate in porous microcarrier cultures was independent of agitation rates. By estimating the total surface area occupied by cells from the maximum cell density, it was found that not all the surface area of the porous microcarrier was usable for cell growth. The supply of oxygen or nutrient into the central region of the porous microcarrier seems to be improved by the enlargement of the pore size.

Introduction

For the large-scale culture of anchorage-dependent animal cells, microcarrier cultures have several advantages such as high cell yields and simple medium/cell separation (van Wezel, 1967; Feder and Tolbert, 1983). Anchorage-dependent animal cells grown on microcarriers are, however, susceptible to mechanical damage due to agitation (Cherry and Papoutsakis, 1986). Croughan et al. (1987) found that a correlation exists between the specific growth rate and the size of the smallest turbulent eddy. By considering the hydrodynamic environment around microcarriers, the effects of microcarrier concentration and agitation rate on the specific growth rate were analyzed by Cherry et al. (1988) and Croughan et al. (1988).

A new type of microcarrier with porous internal structure was first developed by Nilsson et al. (1986). A variety of porous microcarriers are now available from several manufacturers. The advantages of porous microcarrier cultures are that the porous microcarrier can support a higher cell number than the conventional solid microcarrier because of an increase of surface for cell attachment and cell growth, and that the matrix protects fragile cells against mechanical damage due to agitation.

In this report, the effects of the agitation rate on cell growth in porous microccarrier cultures and the cell density in porous microcarrier cultures are discussed.

R. Sasaki and K. Ikura (eds.), Animal Cell Culture and Production of Biologicals, 121–126.
© 1991 *Kluwer Academic Publishers.*

Materials and methods

Cell line and cell maintenance

CHO-K1 cells and HeLa cells were used in the experiment. The growth media used were Ham's F-12 (Flow Laboratories) for CHO-K1 cells and DMEM (Flow Laboratories) for HeLa cells. The media were both supplemented with 10% fetal calf serum (FCS). Antibiotics used were penicillin (Gibco Laboratories) at 100 U/mL and streptomycin (Gibco Laboratories) at 100 μg/mL. Stock cultures were propagated in 150 cm^2 plastic tissue culture flasks (Costar Plastics) at 37°C in a humidified 5% CO_2 incubator. Under these conditions, CHO-K1 cells and HeLa cells grew exponentially with 17 h and 28 h doubling times, respectively.

Porous microcarrier cultures

The porous microcarriers used were CultiSpher-G (Percell Biolytica) and Asahi microcarrier (Asahi Chemical Industry Co., Ltd.) as shown in Fig. 1. The properties of these porous microcarriers are shown in Table 1 according to data from the manufacturers except the number of porous microcarriers for the Asahi microcarrier. The number of porous microcarriers for Asahi microcarrier was obtained by counting the porous microcarriers with a microscope. The surface areas of CultiSpher-G and Asahi microcarrier are 40-fold and 16-fold larger than that of a 200 μm diameter solid microcarrier.

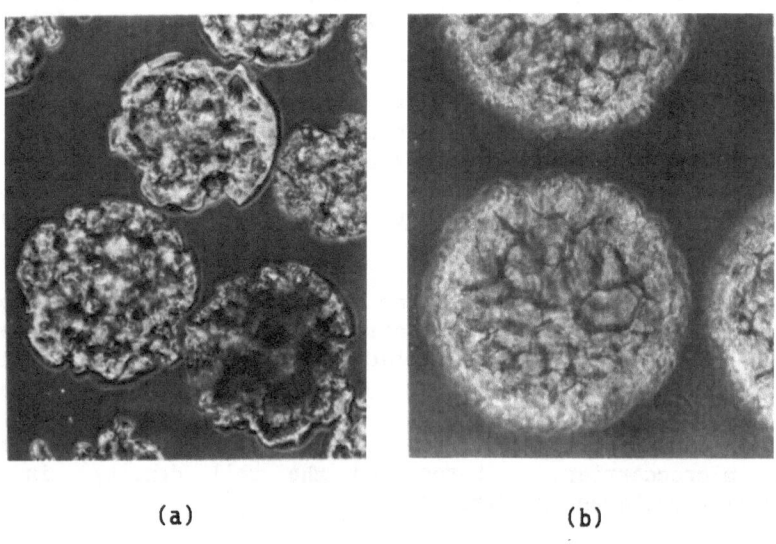

(a) (b)

Fig.1 Photographs of porous microcarriers
(a) CultiSpher-G (b) Asahi microcarrier

Table 1   Properties of porous microcarriers

	CultiSpher-G	Asahi microcarrier
matrix	gelatin	cellulose
density [g/cm^3]	1.04	1.03
mean diameter [μm]	200	220
number of porous microcarriers [carriers/g]	8.0×10^5	1.5×10^6
surface area [m^2/carrier]	5.0×10^{-6}	2.0×10^{-6}
pore size [μm]	10	30

The 100-mL microcarrier culture was done in a 250-mL spinner vessel (Bellco Glass). The rod impeller used was 4.2 cm in length and 0.9 cm in diameter. The spinner vessel was siliconized before use to prevent porous microcarriers from sticking to the vessel wall. Trypsinized cells from tissue culture flasks were inoculated in the spinner vessel containing 50 mL of growth medium and porous microcarriers. At the initial stage of the culture, an intermittent agitation was used. Stirring for 1 min and settling for 59 min were repeated for 6 h. After these periods, the culture volume was increased to 100 mL and continuously stirring was started. The inoculated spinner vessel was placed on a magnetic stirrer equipped with a water bath at 37℃. Half of the culture medium was replaced with 50 mL of fresh medium daily from the second day of the culture. CHO-K1 cells were cultured with CultiSpher-G at 5 g/L porous microcarrier concentration at 60, 90, and 120 rpm. HeLa cells were cultured with CultiSpher-G at 5 g/L and Asahi microcarrier at 2 g/L at 60 rpm. The total porous microcarrier volume of CultiSpher-G at 5 g/L and Asahi microcarrier at 2 g/L were both 1.67 mL.

Measurements

Cell growth was monitored by the following method. A 1-mL sample was withdrawn from a well-mixed spinner vessel. In the case of CultiSpher-G, the sample was mixed with 1 mL of 0.2% collagenase solution to dissolve the CultiSpher-G and incubated for 5 min at 37℃. The cells dispersed by 0.25% trypsin in phosphate buffered saline (PBS) solution with 0.5 mM ethylenediamine tetraacetic acid (EDTA) were counted with a hemocytometer. Viable cells were counted by a dye exclusion method using 0.2% trypan blue in PBS solution. In the case of the Asahi microcarrier, the sample was treated with trypsin-EDTA solution and then mixed with 1 mL of 0.1% crystal violet-0.1% Tween 20 in 0.1 M citric acid. After incubation for 24 h at 37℃, the suspension was sheared using a Pasteur pipette to dissociate the nuclei from the cells. The stained nuclei were counted with a hemocytometer.

Results and discussion

The effects of the agitation rate on cell growth for CHO-K1 cells in porous microcarrier cultures are shown in Fig. 2. The specific growth rate of 0.041 h^{-1} in exponential growth phase was independent of agitation rates within the range of the experiments and was equivalent to that in flask culture. From this result, cells grown on the interior surface of the porous microcarrier were considered to be insusceptible to mechanical damage due to agitation. The maximum cell density at 60 rpm was 2.2×10^6 cells/mL. The cell viability was 92 percent. The total surface area occupied by viable cells was estimated to be 0.11 m^2, considering CHO-K1 cells were propagated up to 5×10^6 cells in a 25 cm^2 flask. This surface area is only 6 percent of all the porous microcarriers' surface area, 2.0 m^2. Cell growth at the central region of the porous microcarrier is probably difficult because of the limitation of oxygen or nutrient supply into this region.

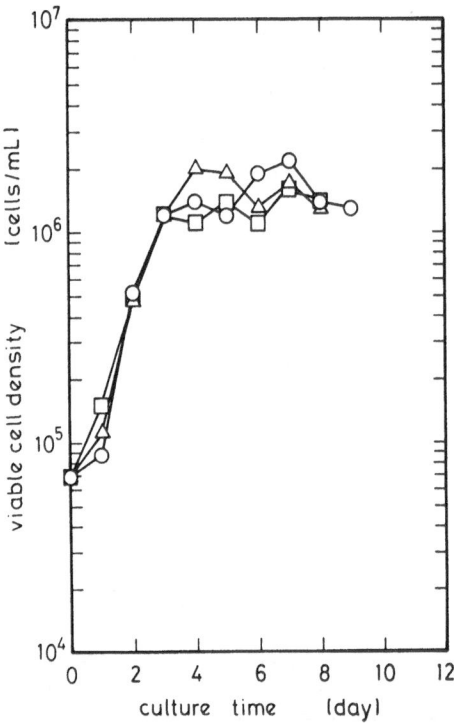

Fig.2 Effects of agitation rate on cell growth
for CHO-K1 cells in porous microcarrier culture.
CHO-K1 cells were cultured with CultiSpher-G
at 60 rpm(○), 90 rpm(△), and 120 rpm(□).

HeLa cells were cultured with CultiSpher-G and Asahi microcarrier at 60 rpm. The cell growth for HeLa cells in CultiSpher-G and Asahi microcarriers are shown in Fig. 3. The cell density is expressed by the cell number per unit volume of the porous microcarrier. The specific growth rate in exponential growth phase for CultiSpher-G was calculated to be 0.025 h^{-1}. This value was equivalent to that in flask culture. In the case of Asahi microcarreir, the specific growth rate decreased slightly to 0.020 h^{-1}. This bit of decrement of the specific growth rate is considered to be due to the difference of matrix. The maximum cell density was 7.2×10^7 cells/mL-carrier for CultiSpher-G and 1.1×10^8 cells/mL-carrier for Asahi microcarrier. The total surface area occupied by viable cells was estimated to be 0.10 m^2 for CultiSpher-G and 0.16 m^2 for Asahi microcarrier, considering HeLa cells were propagated up to 3×10^6 cells in a 25 cm^2 flask. Considering that all the porous microcarriers' surface area was 2.0 m^2 for CultiSpher-G and 0.6 m^2 for Asahi microcarrier, the surface usage is 5 percent for CultiSpher-G and 26 percent for Asahi microcarrier. The cell density

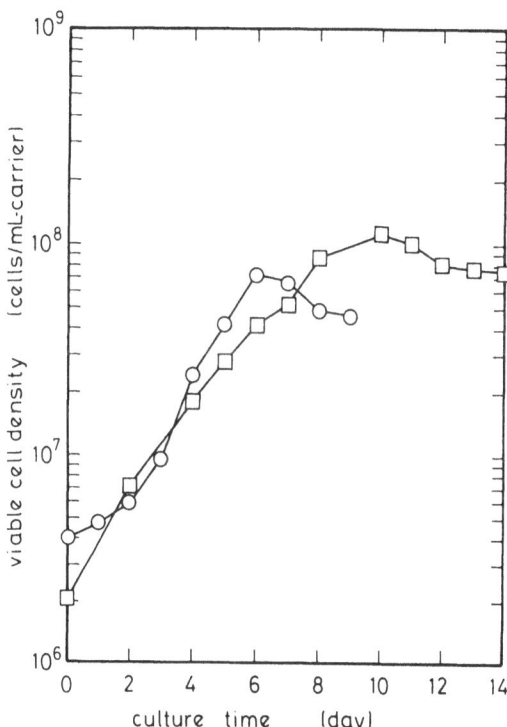

Fig.3 Cell density for HeLa cells in porous microcarrier culture. HeLa cells were cultured by CultiSpher-G(○) and Asahi microcarrier(□) at 60 rpm.

and the surface usage for Asahi microcarrier are both higer than these for CultiSpher-G. The supply of oxygen or nutrient into the central region of the porous microcarrier seems to be improved by the enlargement of the pore size.

The cell density in porous microcarrier cultures will be further enlarged by optimizing operational conditions such as microcarrier concentration and agitation rate with the perfusion of medium and the aeration. Sparged aeration is expected to be still more effective than the conventional surface aeration, because cells grown on the interior surface of the porous microcarrier are insusceptible to the hydrodynamic environment around the porous microcarrier.

Acknowledgments

The authors thank Mr. Naoyoshi Keshi for his experiments and Miss Fumie Takahata for her technical assistance in cell maintenance. The authors also thank Mr. Masami Yokota (Yamanouchi Pharmaceutical Co., Ltd.) for providing us with the CHO-K1 and the HeLa cell lines and Asahi Chemical Industry Co., Ltd. for giving us the Asahi microcarrier. The authors thank Dr. Hiroyuki Honda for his helpful suggestions.

References

1. Cherry, R.S. and Papoutsakis, E.T. (1986) 'Hydrodynamic effects on cells in agitated tissue culture reactors', Bioprocess Eng. 1, 29-41.
2. Cherry, R.S. and Papoutsakis, E.T. (1988) 'Physical mechanisms of cell damage in microcarrier cell culture bioreactors', Biotechnol. Bioeng. 32, 1001-1014.
3. Croughan, M.S., Hamel, J.F. and Wang, D.I.C. (1987) 'Hydrodynamic effects on animal cells grown in microcarrier cultures', Biotechnol. Bioeng. 29, 130-141.
4. Croughan, M.S., Hamel, J.F. and Wang, D.I.C. (1988) 'Effects of microcarrier concentration in animal cell culture', Biotechnol. Bioeng. 32, 975-982.
5. Feder, J. and Tolbert, W.R. (1983) 'The large-scale cultivation of mammalian cells', Sci. Am. 248, 24-31.
6. Nilsson, K., Buzsaky, F. and Mosbach, K. (1986) 'Growth of anchorage-dependent cells on macroporous microcarriers', Bio/Technol. 4, 989-990.
7. van Wezel, A.L. (1967) 'Growth of cell-strains and primary cells on micro-carriers in homogeneous culture', Nature. 216, 64-65.

FORMATION OF MULTICELLULAR AGGREGATES OF ADULT RAT HEPATOCYTES

YASUYUKI SAKAI AND MOTOYUKI SUZUKI
Institute of Industrial Science,
University of Tokyo
7-22-1 Roppongi, Minato-ku,
Tokyo 106
Japan

1. Introduction

Since hepatocytes in primary culture express various differentiated hepatic functions *in vitro* for a certain period of time, large scale culture of the cells with high density are expected to have many important applications. Particularly, a hybrid artificial liver using normal hepatocytes is expected for an effective treatment of acute hepatic failure.

Adult rat hepatocytes are usually cultured in monolayers on collagen-coated surfaces for 1~2 weeks, but they rapidly lose the ability for performing differentiated functions. Moreover, it may be difficult to attain a high culture density as long as hepatocytes are cultured in monolayers.

Recently, it was reported that adult rat hepatocytes inoculated on the proteoglycan fraction of rat liver homogenate or Falcon Primaria dishes formed multicellular spheroidal aggregates (spheroids) (Koide *et al.* (1989, 1990)). They secrete albumin at a higher rate for longer than monolayer-cultured cells. Furthermore, a spheroid has a high cell density in itself and the cells in it have a cubic shape like that *in vivo*. Large scale culture of spheroids may lead to a compact module of hepatocyte culture with high density.

We also found that a polylysine-coated surface with more than a certain amount of adsorbed polylysine improved the spheroid formation, and epidermal growth factor (EGF) improved the maintenance of the spheroids formed. Since polylysine solution is convenient to precoat various kinds of culturing support materials, our results would extend the feasibility of large scale formation and maintenance of hepatocytes spheroids.

This study focuses on the next two points. First is to establish the culture conditions for the formation and maintenance of the spheroids formed on polylysine-coated surfaces. Second is to confirm the expression of hepatic functions sufficient for a hybrid artificial liver in the established conditions.

R. Sasaki and K. Ikura (eds.), Animal Cell Culture and Production of Biologicals, 127–134.
© 1991 *Kluwer Academic Publishers.*

2. Materials and Methods

2.1. DISH COATING

Falcon 35-mm dishes (3001) were coated with 1.0 mL of 0.1 mg/mL polylysine (poly-D-lysine, M.W.= 30,000~70,000; Sigma, St.Louis, MO) in PBS for 1 h at a room temperature, washed twice with plenty of PBS, air-dried, and stored at 4°C. Collagen-coated dishes were also prepared. The dish was coated with 0.03% collagen (Nitta Gelatin, Osaka, Japan) in 1 mM hydrochloric acid, air-dried, washed, and stored in the same way.

2.2. PREPARATION AND CULTURE OF HEPATOCYTES

Adult rat hepatocytes were isolated by the collagenase-perfusion method (Seglen (1976)) from Wister male rats (5~7-weeks old, 150~250 g weight). About 77~82% viability was obtained. The basal medium used here was William's medium E (WE) supplemented with 20 mM HEPES, 20 ng/mL mouse-EGF (Takara, Kyoto, Japan) 10^{-7} M insulin (Takara), 10^{-7} M dexamethasone (Wako, Osaka, Japan), 100 units/mL penicillin 100 μg/mL streptomycin, and 0.25 μg/mL Fungizone. Cells were finally suspended at 3.5 X 10^5 cells/mL with the basal medium and inoculated 1.5 mL in each dish. The medium were changed first after 6 h of culture, second after 1 d of culture, and then at 2-d intervals. The medium were changed by centrifugation (50 X g, 1 min) in case of spheroid culture.

In the experiments for the morphological examination of the effects of various supplements, 10^{-7} M glucagon (Sigma) or 200 ng/mL aprotinin (Takara) were added to the medium at the 6-h medium exchange. In the experiment for measuring DNA and the expression of hepatic functions, the basal medium supplemented with 20 ng/mL EGF, 10^{-7} M copper ($CuSO_4 \cdot 5H_2O$), and 3 X 10^{-8} M selenium (H_2SeO_3) were used. This is the medium for the formation phase of the spheroids. Glucagon (10^{-7} M) and aprotinin (200 ng/mL) were added to the medium after the 5th-day medium exchange. This is the medium for the maintenance phase of the spheroids.

2.3. MEASUREMENT OF HEPATIC FUNCTIONS

The cell number was described by the DNA amount. The amount of DNA in cultured cells was measured fluorometrically by the method of Brunk et al. (1979) using calf thymus DNA as a standard. The cells were washed two times with PBS, digested by trypsin / EDTA and homogenized with an Ohtake (Tokyo, Japan) 5201 power sonicator. The spheroid-cultured cells were washed and resuspended by centrifugation (50 X g, 1 min). Samples (50 μL) were taken into a test tube and reacted with 3.0 mL of 0.1 μg/mL 4',6-diaminodino-2-phenylindol (DAPI). The fluorescence was measured with a Hitachi (Tokyo, Japan) 650-40 spectro-photofluorometer at excitation wavelength 350 nm and emission wavelength 450 nm.

The amount of rat albumin secreted into the medium was measured by enzyme-linked immunosorbent assay (ELISA). Standard rat-albumin, anti-rat-albumin antibody and peroxidase-conjugated anti-rat-albumin antibody were purchased from Cappel (West Chester, Pa.). The absorbance was measured at 490 nm with a Tosoh (Tokyo, Japan) MPR-A4i microplate reader.

The amount of urea synthesized was measured by the diacetyl monoxime method (Marsh et al. (1965)). The cells were washed two times with PBS by the procedure used in measuring DNA , and resuspended with Hank's balanced salt solution (Hank's BSS; 10 mM HEPES, pH 7.2) supplemented with 5 mM ammonium chloride as a substrate. Absorbance at 490 nm was measured with a Shimadzu (Tokyo, Japan) UV-160 spectrophotometer.

The glucose synthesized was measured by the glucoseoxydase /peroxydase method (Tomomura et al. (1980)). The cells were washed with PBS and resuspended with Hank's BSS supplemented with 20 mM fructose as a substrate and 10^{-7} M glucagon. The absorbance at 550 nm was measured with the same spectrophotometer.

3. Results and Discussion

3.1. EFFECTS OF MEDIUM SUPPLEMENTS ON SPHEROID MORPHOLOGY

Effects of glucagon and aprotinin supplemented in the culture medium on the spheroid formation were examined. Glucagon is a hormone improving glucose synthesis markedly. Also, it improves hepatocyte survival in the presence of insulin. Aprotinin is a protease inhibitor. It was reported to improve the survival of monolayer-cultured hepatocytes in serum-free medium by inhibiting the protease in hepatocyte membranes (Nakamura at al. (1984)).

Figure 1. shows microscopic photographs of the hepatocyte on the 5th day. Most of the spheroids floated on the 5th day in the glucagon·aprotinin-free media (A), while in the presence of 10^{-7} M Glucagon, the cells formed attached spheroids or hemispheroids. Few floating spheroids were observed (B). However, the addition of glucagon in the medium improved the spheroid survival markedly. In the presence of 200 ng/mL aprotinin, the cells aggregated, but formed only multicellular islands (C). They never changed to spheroids. On the other hand, when we transferred floating spheroids formed in aprotinin-free medium to aprotinin-containing medium from the 5th-day medium-exchange, the spheroid was maintained 2~3 days longer than aprotinin-free medium. It is interesting that hormones required for the survival of hepatocytes influenced the spheroid morphology.

Therefore, the media supplements optimized both for the formation and maintenance of spheroids were decided on as shown in Table 1. For the formation phase (0~5 d), neither glucagon nor aprotinin was added, and after the spheroid formation was completed, we changed the media to that for the maintenance phase (5 d~) supplemented with glucagon and aprotinin.

130

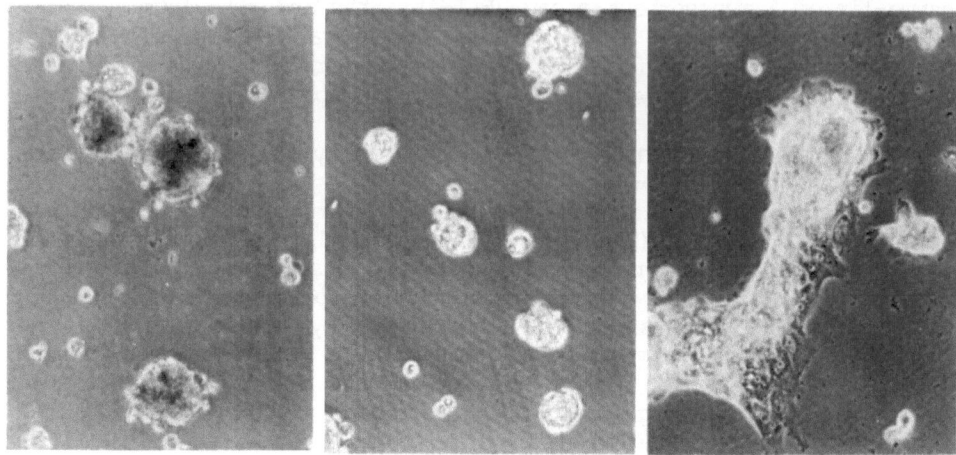

(A) 20 ng-EGF/mL   **(B) 20 ng-EGF/mL**   **(C) 20 ng-EGF/mL**
                   10^{-7} M Glucagon   **200 ng-Aprotinin/mL**

⊢———⊣
100μm

Figure 1. Effects of medium supplements on the hepatocyte morphology inoculated on polylysine-coated surfaces.

TABLE 1. Supplements in the optimized medium.

Supplement	Formation phase (0~5 d)	Maintenance phase (5 d~)
20 ng-EGF/mL	(+)	(+)
10^{-7} M Glucagon	(−)	(+)
200 ng-Aprotinin/mL	(−)	(+)

3.2. HEPATIC FUNCTIONS ON THE OPTIMIZED MEDIUM

In the optimized media as mentioned above, we investigated the expression of various hepatic functions of the spheroid formed on polylysine-coated surfaces.

On the 4th day, most of the spheroids floated similarly as shown in Figure 1.(A). After the addition of glucagon and aprotinin on the 5th day, however, the floating spheroids were attached onto the surface, partially spread, and they changed to hemispheroids or multicellular islands on the 7th day as shown in Figure 2. They were maintained morphologically up to at least 5 weeks. The results suggest that the morphology of the aggregates at last reaches a equilibrium state according to the supplements.

The albumin secretion, urea synthesis, and glucose synthesis of the spheroids formed on polylysine-coated surfaces were compared with those of monolayer-cultured cells on collagen-coated dishes. Total cell number was described as the DNA amount on each dish, and the expression of these hepatic functions were normalized by the DNA amount.

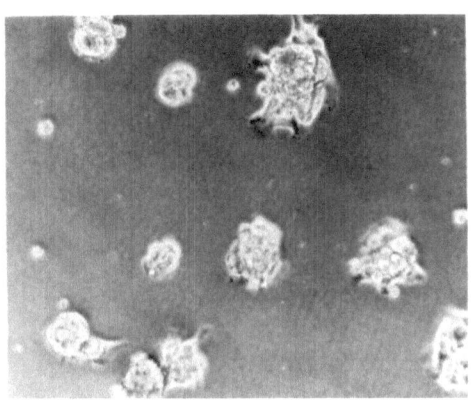

├────────┤
100 μm

Figure 2. Morphology of the spheroids in the optimized medium on the 7th day.

Cell maintenance described by the DNA amount is shown in Figure 3. The spheroids were maintained better than monolayers on collagen.

Figure 4 shows the maintenance of various hepatic functions.

Albumin secretion of the spheroids was much better than that of the monolayer. The secretion rate of the spheroids increased with the culture day (A).

Urea synthesis of the spheroids was maintained at an almost constant level, but monolayer-cultured cells rapidly lost the function (B).

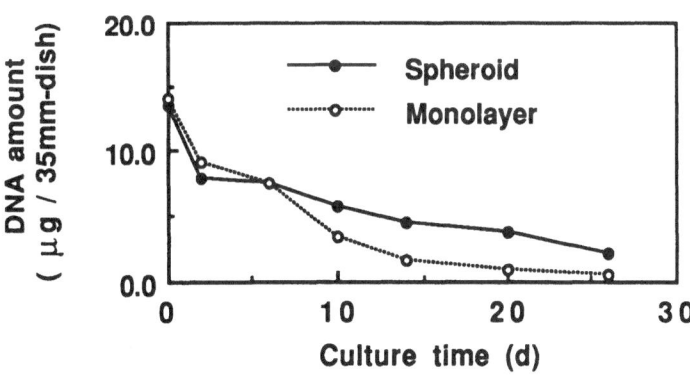

Figure 3. Cell maintenance described by DNA amount.

132

Glucose synthesis ability of the spheroids was markedly lost during the initial 6 days of culture. It may be due to the absence of glucagon. From the 6th day, the synthesis rate of the spheroids were maintained at almost the same level as that of the 6th day. However, monolayer-cultured cells rapidly lost the function (C).

Furthermore, we investigated the effects of initial presence of glucagon on the maintenance of glucose synthesis and spheroid morphology. Figure 5. shows the results. More than 10^{-8} M was needed

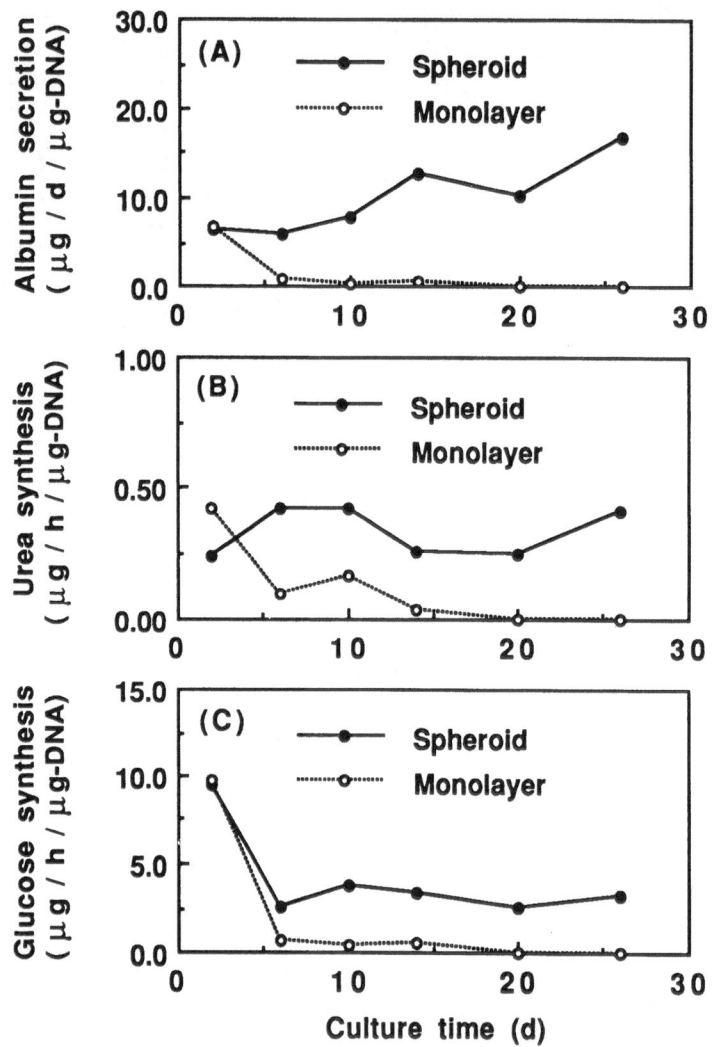

Figure 4. Various hepatic functions of the spheroids formed on polylysine-coated surfaces.

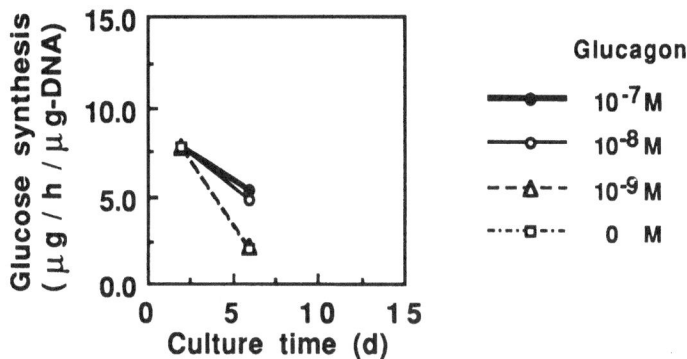

Figure 5. Effects of initial presence of glucagon on the maintenance of glucose synthesis ability.

for preventing the initial loss of the glucose synthesis ability. Most of the spheroids floated at this glucagon concentration. However, the supplement of 10^{-7} M glucagon improved the formation of attached spheroids or hemispheroids.

The spheroids formed on polylysine-coated surface were found to express not only albumin secretion but also the synthesis of urea and glucose by the optimization of medium supplements.

Since polylysine solution is convenient for precoating various kinds of culturing supports with a high surface-to-volume ratio, such as fibrous and macroporous materials, compact and high performance modules would be developed by the combination of spheroids and these supports.

4. Conclusions

To establish the basic culture conditions for the formation and maintenance of the spheroids formed on polylysine-coated surfaces, effects of medium supplements considered to be effective on the cell survival in monolayer culture were investigated. In the basal medium containing insulin, dexamethasone, and EGF, floating spheroids were stably formed within 4 days. Aprotinin inhibited the spheroid formation. More than 10^{-7} M glucagon improved the attached spheroids and hemispheroids. However, these supplements improved the spheroid survival.

According to the these results, the medium supplements were optimized both for the formation and maintenance of the spheroids. Namely, aprotinin and glucagon were added after the spheroid formation was completed (5 d~).

The spheroids formed were mainly floating ones first, but changed their morphology to attached spheroids or hemispheroids by the addition of aprotinin and glucagon from the 5th day. Spheroid morphology appeared to reach at last an equilibrium state decided by the

134

supplements in the medium. These attached spheroids and hemispheroids expressed high hepatic functions, that is, albumin synthesis, urea synthesis, glucose synthesis, for 26 days.

References

Brunk, C. F., Jones, K. C., and James, T. W. (1979) 'Assay for nanogram quantities of DNA in cellular homogenates', Anal. Biochem., 92, 497~500

Koide,N., Shinji, T., Tanabe, T., Asano, K., Kawaguchi, M., Sakaguchi, K., Koide, Y., Mori, M., and Tsuji, T. (1989) 'Continued high albumin production by multicellular spheroids of adult rat hepatocytes formed in the presence of liver-derived proteoglycans', Biochem.Biophys.Res.Commun., 161, 385~391

Koide,N., Sakaguchi, K., Koide, Y., Asano, K., Kawaguchi, M., Matsushima, H., Takenami, T., Shinji, T., Mori, M., and Tsuji, T. (1990) 'Formation of multicellular spheroids composed of adult rat hepatocytes in dishes with positively charged surfaces and under other nonadherent environments', Exp. Cell Res., 186, 227~235

Marsh, W.H., Fingerhut, B., and Miller, H.,T (1965) 'Automated and manual direct methods for the determination of blood urea', Clin. Chem., 11, 624~627

Nakamura, T., Asami, O.,Tanaka,K., and Ichihara, A.(1984) 'Increased survival of rat hepatocytes in serum-free medium by inhibition of a trypsin-like protease associated with their plasma membranes', Exp. Cell Res., 155, 81~91

Seglen,P.O., (1976) 'Preparation of isolated rat liver cells', Methods Cell Biol., 13, 29~83

Tomomura, A., Nakamura, T., and Ichihara, A. (1980) 'Role of the cytoskeleton in glycogenolysis stimulated by glucagon in primary cultures of adult rat hepatocytes', Biochem. Biophys. Res. Commun., 97, 1276~1282

OPERATION OF AN AIR LIFT REACTOR FOR PRODUCTION OF IMMUNOCHEMICALS BY IMMOBILIZED HYBRIDOMA CELLS

B.Bugarski[1] ,G.Vunjak[1] ,G.Jovanovic[1] ,K.Cuperlovic[2] ,M.F.Goosen[3]
1.Department of Chemical Engineering,University of Belgrade,Karnegijeva 4,11000 Belgrade,Yugoslavia . 2.Institute for Immunology,Endocrinology and Nutrition,INEP,Zemun,Yugoslavia. 3.Department of Chemical Engineerig,Queen's University,   Kingston,K7I 3N6,Canada

1. ABSTRACT

The feasibility of using external loop bioreactor system for the production of monoclonal antibodies by mouse/mouse hybridoma cells immobilized in Poly-l-lysine(PLL)-alginate microcapsule was investigated. Batch culture of immobilized cells followed logarithmic growth, reaching a concentration of 7 x 10^7 cells/mL of capsule after 20 days giving monoclonal antobody (mAb) concentrations of up to $600\,\mu g/ml$ of capsule. The main advantages of the bioreactor system used is control of process parameters, minimized shear stresses on immobilized cells and retention of the product (mAb) within the microcapsule.

2. INTRODUCTION

There is, at present, a growing market for animal cell derived products such as monoclonal antibodies,lymphokines, hormones,growth factors,tissue plasminogen activator and enzymes. For bulk production using microbial systems, high performance fermentation equipment and sophisticated mass culturing techniques with advanced computer control are currently available in pharmaceutical and chemical industries. It is very challenging to introduce these existing technologies and bioreactors designed for bacterial production also in large-scale mammalian cell cultivation. Unfortunately, in most cases, this only can be done by adapting or even drastically modifying such kind of bioreactors in order to overcome the difficulties which are arising from the fundamental differences in structure, properties and physiology between microbial and animal cells. For instance, animal cells are 10 to 100 times larger than microbial cells and thus have a much higher settling velocity which is a key parameter in the choice of a cell retention or cell recycling method. They also present a 10 to 100 times lower surface area-to-volume ratio from which results a lower metabolic activity and consequently a lower growth rate. Therefore at the same cell concentration, animal cell cultures will generally need a much lower mass

R. Sasaki and K. Ikura (eds.), Animal Cell Culture and Production of Biologicals, 135–140.
© 1991 *Kluwer Academic Publishers.*

transfer capacity to be assured in the bioreactor than microbial cultures. Mechanical agitation and aeration may cause damage to the fragile cells which are freely suspended in the media. Shear stress over 1 N/m^2 appears to be critical for most hybridoma and insect cells (Tramper et al.,(1986)). Therefore the combination of low sheer air lift reactor and cell immobilization is particularly appropriate for shear sensitive animal cells. The importance of liquid circulation has been recognized as a key factor in reactor design, since both mixing and volumetric mass transfer coefficient are influenced by the circulation rate. When considering design of bioreactor, one of the requirements is to provide an adequate supply of nutrients and oxygen to the cells through the permeable microcapsule. Specific respiration rate in the order of 0.3 μmol $O_2/10^6$ cells/h is typical of murine hybridomas. To maintain the required respiration rate, the total oxygen transfer rate (OTR) must exceed or at least match the total oxygen consumption rate by the cells under equilibrium conditions. Efficient. large scale production of immunochemicals by animal cell culture requires a good understanding of the way bioreactors, medium and cells interact.

3. EXPERIMENTAL

3.1. Cell culture. Mouse/mouse hybridoma cells,secreting an β-HCG (human chorionic gonadotropin) specific antibody, were grown in DMEM medium (Sigma Chemical) supplemented with 3.7 g/l NaHCO3 , 5 μg/ml geneticin disulfate (Sigma Cmemical) and 10% (V/V) heat inactivated fetal calf serum (FCS) (Gibco. Labs.) in 75 mL culture flasks. All other chemicals used were of reagent grade. All solutions were sterilized by passing through a 0.22 μm filter apparatus (Nalgene Sterilization Filter Unit Type S, Nalgene Company Rochester, N.Y.). Hybridoma cells were maintained in culture flasks and kept in humidified incubator at 37 ^{0} C under 5% CO and 95% air. For experimental purposes hybridoma cells were seeded from these stocks.Calcium Cloride (1.5%) and saline were sterilized by filtration . Sodium alginate 1.5 % (w/v) in 0.85 % of NaCl was sterilized by boiling in a water bath for 15 minutes.
3.2.Immobilization of hybridoma cells. The cell pellet resuspended in 1.2 % (w/v) sodium alginate (Kelco Gel LV) solution was extruded into a 1.0 % CaCl2 solution. Spherical droplets 600-800 μm in diameter were formed by an air-jet droplet generator (DG.1 Chem.Eng.Depar.,Belgrade). A semipermeable capsule was formed by reacting the alginate beads with aqueous PLL (Mw=22,000) for six minutes. The capsules were then reacted with 0.04% of sodium alginate for 5 minutes. The interior of the microcapsules was liquefied with the 0.05M sodium citrate solution for 4 minutes.
3.3.Assays. Cell viability was assessed by staining the cells with Trypan blue dye. An enzyme-linked immuno-sorbent assay-ELISA was used to determine the concentration of monoclonal antibody.
3.4.Experimental bioreactor. The gas-stirred external air loop reactor simple in design used in this work was made of glass (Fig 1a.) and had a working volume of 1L (Bioloop II, T.M.F. Belgrade). For aeration a sintered disk sparger with holes in the range of 0.1-0.3 mm in diameter was placed at the bottom of the reactor. It was found that the ratio of liquid level to the reactor diameter, the

slenderness grade , approximately equal 8 gave reasonably good circulation rate . That is, the taller and slimmer the reactor the higher the residence time of the bubbles, and therefore the better the mixing within the system. The gas mixture consisting initially of 95% air and 5% CO_2 was, adjusted to maintain a constant pH of 6.9 to 7 throughout the culture period. In addition, the pH of the medium was kept constant using an automatic PID regulator coupled to an electromagnetic valve on the CO_2 tank. The bioreactor system also included optical sensor for foam control (Af.d2 University of Belgrade, Depart. of Chem. Eng). The anti foam unit for detection and control of foam consists of one infra red emitter and one phototransistor. A signal from the phototransistor is amplified before entering the logical circuit which drives the anti-foam feeding pump.The bioreactor was placed in a water bath where the temperature was maintained constant at 37^0 C. Samples of the media and immobilized cells were

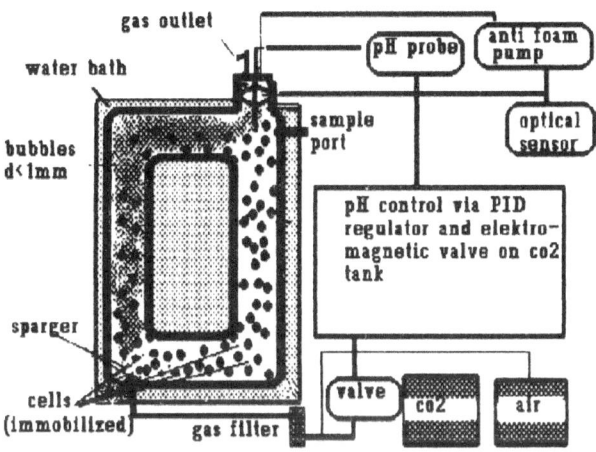

Figure 1. Shematic diagram of the air lift biorector system.

withdrawn every 24 h. The average liquid circulation velocity was measured by observing distance and time for a dye to travel around the riser-down corner circuit and this value was verified by recording the change in the pH profile when CO_2 was injected . The Kla value was determined experimentally by the dynamic gassing-out method (Atkinson and Mavituna 1983). The gas holdup was measured by comparing the change in volume of aerated media and gas free media. The immobilized cell volume was about 5 % that of the overall reactor volume .

4. RESULTS

Liquid velocity becomes critical if the liquid circulation is not sufficient enough to maintain the immobilized cells in suspension, in which case mass transfer becomes non-uniform within the reactor. The superficial gas velocity, was adjusted to just maintain microcapsules in suspension and also to minimize foaming. The average bubble size ranged from 1 to 2 mm diameter. When the superficial gas velocity was varied from 2 to 35 cm/min (Fig.2), the measured average liquid velocity in the reactor increased from 3 to 14 cm/s. Based on the observation of tracer and mirocapsule circulation, it did not appear that any stagnant regions were developing in the reactor at low gas superficial velocities used in our studies. Gas holdup is another parameter which is useful in characterizing the hydrodynamics of air sparged reactors. When the inlet gas velocity was varied from 2 to 36 cm/min ,the gas holdup increased from 0.2 to 3 %.For such a low range of gas inlet flows, an almost linear correlation was obtained between the superficial gas velocity and the gas holdup (Fig.2). Our results were in agreement with the findings of Hill (1974).

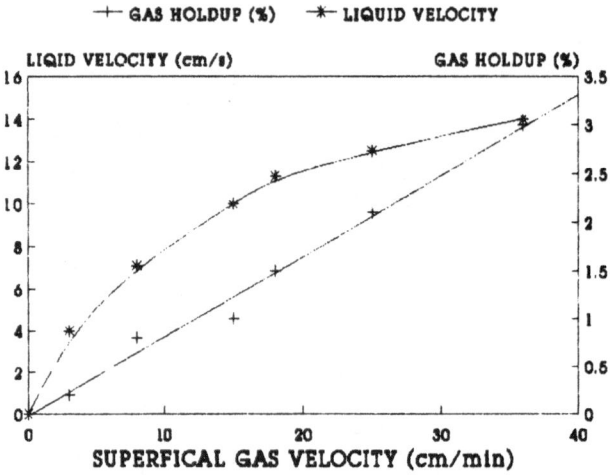

Figure.2.Effect of the superficial gas velocity in the air lift bioreactor, on the average liquid velocity and gas holdup.

Experimental findings of the volumetric mass transfer coefficient in the bioreactor indicates its strong dependence on the gas superficial velocity (Fig.3). In addition, we derived an empirical expression which predicts the overall volumetric oxygen mass transfer coefficient as a function of superficial gas velocity, V_g , for experimental air lift boireactor used in this experiment. $Kla = Vg^{0.82}$, where Kla is given in 1/h and Vg has units of cm/min. Transition from hydrodynamics, to the biological requirements was given by the Kla value, where for example, oxygen supply as an essential factor for cell growth was provided by the appropriate reactor performance.In order to test the suitability of our air lift reactor, for monoclonal antibody production, immobilized hybridoma cells were cultured in a batch process for up to tree weeks. The cell density in the capsules , followed a logarithmic growth phase (doubling time was about 20 h), reaching a density of 7×10^7 (cell/ml capsule) after 20 days giving antibody concentration of 600 μg/ml capsule (Fig.4).

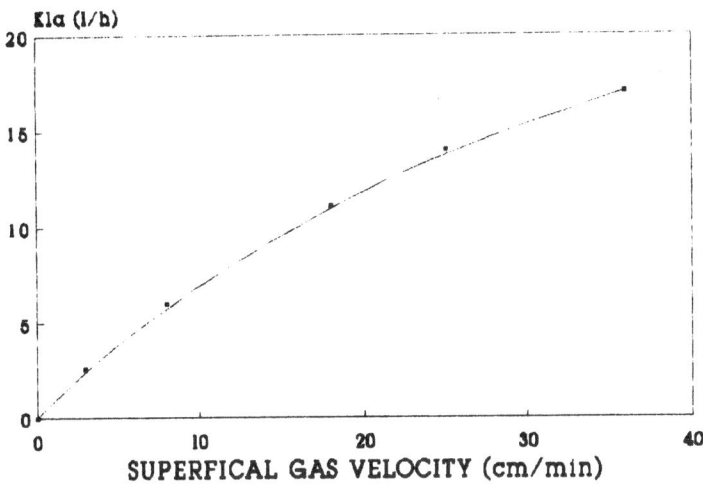

Figure.3. Overall volumetric mass transfer coefficient as a function of superficial gas velocity .

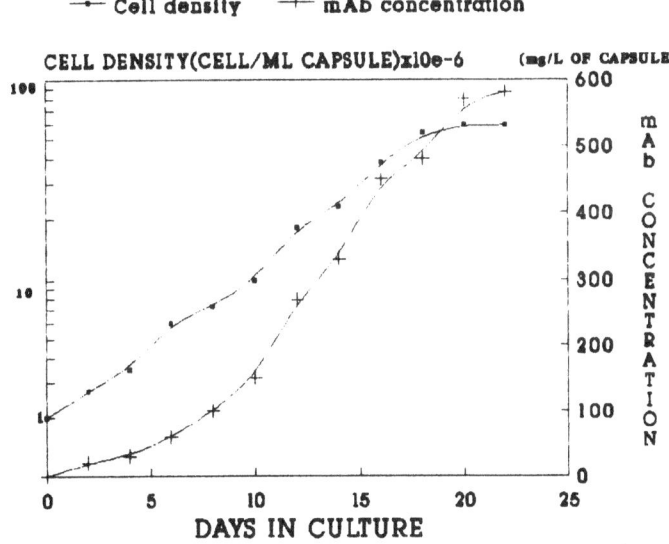

Figure.4. Culture of encapsulated hybridoma cells in air lift biorector.

5.DISCUSSION

140

In comparison to the conventional system of culturing cells in flasks, 80-100 times higher antibody concentration was obtained with more than 95 % of the mAb being retained within permeable alginate-PLL capsule. The successful use of the microcapsule system for the production and concentration of mAb depends on the retention of the product within the microcapsule. At present, products with a molecular weight of perhaps 60,000 and higher can be retained by varying the molecular weight of PLL, alginate/PLL reaction time and PLL initial concentration. The tissue culture studies with alginate-PLL encapsulated mouse hybridoma cells confirmed similar work reported by Posillico (1986) and Rupp (1985): ' Hybridoma cells preferentially grow near the interior surface of the capsules, reaching a maximum cell density of about 2-4 x 10^7 cells/ml of capsule after 10-12 days of growth.'

However, we found that while the encapsulated hybridioma cells produced active mAb , approximately 60 % of the capsule volume remained free of cells and was actually occupied by calcium alginate gel and not by liquid alginate. The difference in the physical state of the capsule core and cell density, may have been due to the fact that, in the present system, the capsule membrane molecular weight cut-off was lower (60,000) than that reported by Posillico (80,000).

In this work an attempt was made to built a simple system capable of controlling only the necessary bioreactor parameters which can provide an adequate environment for microencapsulated cell growth and product formation. Work is continuing on the interactions between mixing, fluid flow, mass transfer and cell processes in animal cell culture system. One of our objectives is to employ hydrodynamics and kinetic analysis techniques in conjunction with capsule permeability studies and capsule loading experiments to identify conditions for maximum volumetric productivity of bioreactor system.

6. REFERENCES

Atkinson, B., and Mavituna, F. (1983) Biochemical Engimeering and
    Biotechnology Handbook, The Nature Press, NY, Chapter 9, p 737.
Hill, J. (1974) ' Radial non-uniformity of velocity and voidage in a
    bubble.column ', Trans 1 Chem E 5, 1.
Posillico, E. (1986) 'Microencapsulation technology for large-scale
    antibody production ' , Bio/Technology 4(2), 114-117.
Rupp, R. (1985) Large scale mammalian cell culture, Feder, J. and
    Tolbert, W. (Eds), Academic Press.
Tramper, J. , Williams, J. and Joustra, D. (1986) 'Shear sensitivity
    of insect cells in suspension ', Enzyme Microb. Technol. 8, 33-36

SIMULATION OF GROWTH OF HYBRIDOMA CELLS IMMOBILIZED IN ALGINATE GEL
BEADS BASED ON AN OXYGEN LIMITED MODEL

Y. SHIRAI, K. HASHIMOTO and A. KUBO
Department of Chemical Engineering
Faculty of Engineering
Kyoto University
Sakyo-ku, Kyoto 606
Japan

ABSTRACT. A model is developed for predicting the growth of hybridoma
cells immobilized in alginate gel beads with several diameters under
different conditions of oxygen partial pressures. Hybridoma cells in
gel beads proliferate evenly at the early stage of culture, but only
the cells near the surface of the gel bead can survive in compensation
for culture death of the cells in the center of the gel owing to lack
of oxygen at the late stage. Moreover, the cells start to leak from
the gel bead in time. The cell growth in the gel would be inhibited
by cell compression caused by grown-up cells. These factors are
included in the model. Changes in cell concentration in alginate gel
beads with several diameters were investigated in the CSTR fermentor.
The results indicate that concentration of the immobilized cells
increased exponentially, reached a maximum point, and was stabilized
during culture. The calculated line shows a good agreement with the
experimental data points and predicts well a growth pattern of the
cells immobilized.

1. INTRODUCTION

Alginate immobilization methods are widely used for high density
cultivation of animal cells [1-5]. However, no animal cell growth
model for alginate gel beads has been developed yet.
    The following characteristics of animal cells should be considered
for modeling of their growth in gel beads: 1) Growth rate of animal
cells is much lower than that of microorganisms like E. coli [6]. 2)
The Oxygen consumption rate of animal cells is also much lower [6, 7].
    Animal cells proliferate slowly, and every cell in a gel bead can
survive at the early stage of cultivation because oxygen has not been
consumed by the cells. But the cells near the center of gel beads
would not survive with the cell growth near the surface of gel beads
owing to lack of oxygen. At last, only the cells very near the
surface can survive and the cell distribution in a bead would reach a
steady state. Therefore, a model including changes in cell distribu-

141

R. Sasaki and K. Ikura (eds.), Animal Cell Culture and Production of Biologicals, 141–149.

tion in a gel bead in an unsteady state should be developed.

Here, a model is proposed for predicting growth of hybridoma cells immobilized in alginate gel beads with several diameters under different conditions of oxygen partial pressure. The model also includes effects of cell leakage from the gel surface on changes in cell concentration in the gel beads.

2. THEORY

Oxygen is essential for animal cell growth, but too much oxygen is also toxic for their growth. Very a little oxygen can be dissolved in water (0.21 mol/m^3). An oxygen concentration level in a culture medium should be carefully set within a narrow range. Therefore, oxygen would be easily consumed by the cells, and is assumed to be a growth-limiting factor.

Figure 1 shows schematic diagrams of changes in animal cell distributions in a alginate gel bead at the early stage of culture as well as that at the late stage. At the early stage, every cell can survive in the bead, but the cells near the surface of the bead can survive at the late stage, while the others die owing to lack of oxygen.

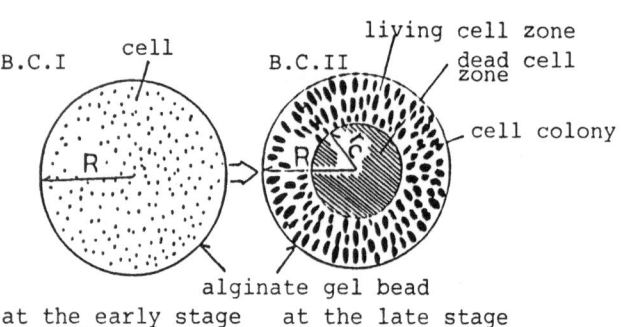

Fig. 1. Schematic diagrams of changes in animal cell concentration in an alginate gel bead at the early and late stage of cultivation.

The material balance equation for oxygen in a gel bead is derived with the boundary conditions, B. C. I for the early stage of culture and B. C. II for the late stage as

$$\frac{\partial C_A}{\partial t} = \frac{D_{eA}}{r^2}\frac{\partial}{\partial r}(r^2\frac{\partial C_A}{\partial r}) + r_{Am}C_x \tag{1}$$

$$\text{B. C. I} \quad r = R: C_A = C_{AS}, \quad r = 0: \frac{\partial C_A}{\partial r} = 0 \tag{2}$$

$$\text{B. C. II} \quad r = R: C_A = C_{AS}, \quad 0 \leq r \leq r_c: C_A = 0; C_x = 0$$

$$4\pi R^2 D_{eA}\frac{\partial C_A}{\partial r}\bigg|_{r=R} = 4\pi \int_{r_c}^{R}(-r_{Am})C_x r^2 dr \tag{3}$$

where D_{eA} is a diffusion coefficient in an alginate gel bead, C_A is concentration of oxygen, r_{Am} is a specific oxygen consumption rate per cell, C_x is cell concentration and r_c is a critical radius of boundary surface at which oxygen concentration reach zero.

Shirai et al. [8] found that the oxygen consumption rate was expressed by the zeroth order reaction kinetics for oxygen concentration of the 4H11 hybridoma cells immobilized in an alginate gel bead, which were adopted in the experiments for this paper, and increased with the cell growth in the gel. The following equation was obtained.

$$\frac{(-r_{Am})}{(-r_{Am})_0} = (\frac{C_x}{C_{x0}})^{0.23} \tag{4}$$

The growth rate of the 4H11 hybridoma cells was independent of the oxygen concentration levels. Cells in a gel particle cannot proliferate infinitely because there is no room to grow in the gel particle at the late stage of culture even when enough nutrients and oxygen are supplied and waste products remain low. The following relationship is assumed between the maximum growth rate and the cell concentration in the gel bead.

$$\mu_{max} = \mu_{max(free)} - kC_x^m \tag{5}$$

The constant, k and the power number, m should be measured experimentally.

A very few cell leakage models [9] have been proposed for predicting cell growth in immobilized gels. A concept of a cell leakage model presented here is explained in Fig. 2. Only cells staying in an outer shell of a bead with a certain depth are hypothesized to leak from the bead. The cell balance in this cell leakage zone and the cell staying zone are expressed as

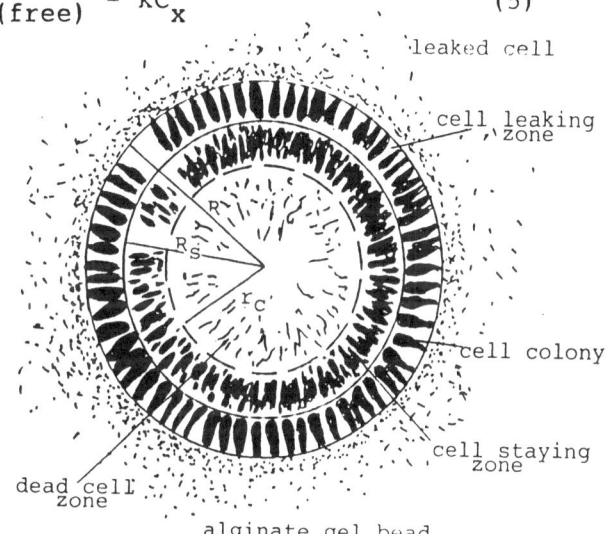

Fig. 2. Schematic diagram of the cell leakage model.

$$\frac{dC_{xs}}{dt} = \mu C_{xs} - r_{leak} \qquad : R_s \leq r \leq R \tag{6}$$

$$\frac{dC_x}{dt} = \mu C_x \qquad : 0 \leq r \leq R_s \tag{7}$$

where C_{xs} is a cell concentration in the cell leakage zone and r_{leak} is a leakage rate of the cells from the surface of the bead. The cell balance in a CSTR type fermentor is described with the boundary conditions as

$$V \frac{dC_{xL}}{dt} = V\mu C_{xL} + V_s r_{leak}; t = t_0 : C_{xL} = 0 \quad (8)$$

where t_0 is the time interval until the medium be changed and C_{xL} is the cell concentration in the medium.

Changes in cell distribution in a bead are calculated as follows: 1) Changes in oxygen consumption rates caused by cell proliferation are neglected in a short increment of time because the hybridoma cell growth rate is too low, and the left side of Eq. (1) could be equal to zero. 2) Equation (1) is solved with the boundary conditions, yielding an oxygen concentration distribution in a gel bead. 3) Changes in cell concentration in a gel bead as well as in a medium are calculated by solving Eqs.(6), (7) and (8), based on the oxygen concentration distribution calculated.

3. MATERIALS AND METHODS

3.1. Materials

The mouse-human hybridoma cells used, 4H11 produce human IgA mono-clonal antibody. A serum-free RDF medium, of which detailed composition was reported in our previous paper wsa used [1].

3.2. Production of the Immobilized Cells

The 4H11 cells were entrapped in calcium alginate gels using a special apparatus we designed. One percent of sodium alginate solution with cells was transferred to a inner syringe equipped in a reversed air bottle by a peristaltic pump. An air jet flowed in the space between the inner syringe and the bottle mouth. Liquid particles of alginate-cell suspension were dropped down into a 0.1 M calcium chloride solution by a shear stress caused by air jet. Uniform sized gel particles were formed. The particle size is strongly dependent

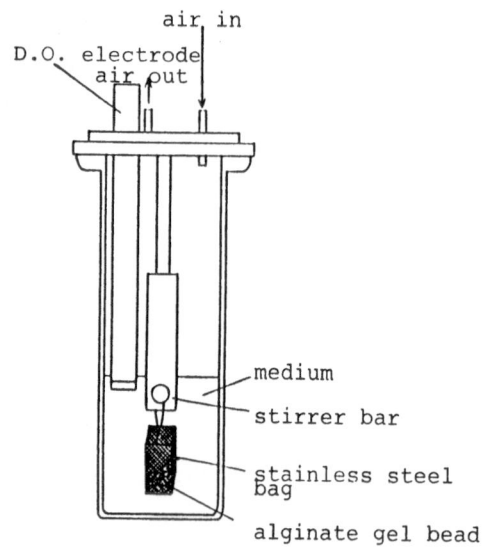

Fig. 3. A fermentor for alginate immobilized cells cultivated in a stainless steel net bag.

on the flow velocity of the jet.

3.3. Cultivation of the Immobilized Cells

The 4H11 cells immobilized in calcium alginate gel particles were incubated in a CSTR type fermentor shown in Fig. 3 under adjusted conditions. Immobilized cells remained in stainless steel bags equipped at an agitator bar which rotated at the rate of 50 rpm.

Oxygen concentration in the medium was monitored with a electrode. The concentration of oxygen were controlled by adjusting the feeding rate of the gas of which composition was 95% and 5% CO_2. The fermentor was contained in a thermostatic chamber at 37 °C.

Living cell number in gel particles was measured by a trypan blue dye exclusion method after the gels were dissolved by 1% trisodium citrate solution.

3.4. Leakage Experiment of Immobilized Cells

An experimental apparatus for checking cell leakage from the gel is shown in Fig. 4. It consists of two parts: Two stainless steel discs were fitted identically and a culture medium was circulated between the discs and a medium reservoir at a constant flow velocity. A very thin alginate gel sheet in which hybridoma cells were growing was set on the lower disc. A constant shear stress be forced on the alginate gel sheet.

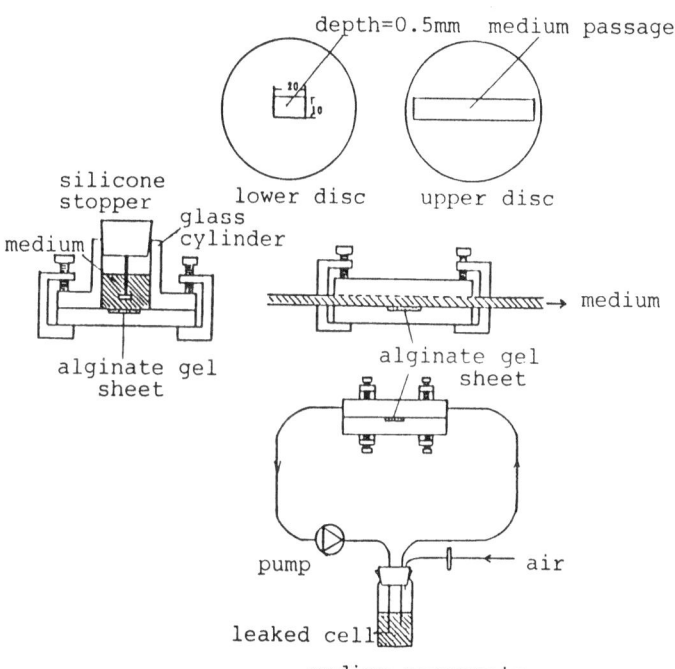

Fig. 4. Experimental apparatus for determining cell leakage rate.

In case of cultivation of immobilized cells in an alginate gel sheet, a glass cylinder capped with a silicone stopper equipped with a stirrer bar was adjusted to the lower disc, culture medium was poured and the immobilized cells were cultivated on a magnetic stirrer in a CO_2 incubator.

After 24 hours, the cell suspension in the reservoir was sampled and the cell number was counted. Either the initial cell concentration in the alginate gel sheet or the cultivation time of the immobilized cells was changed to clarify the relationships among the cell leakage rate, initial cell concentration, and the cultivation time.

4. RESULTS AND DISCUSSION

4.1. Hybridoma Growth Rate

The growth rate of hybridoma 4H11 cells was measured under various oxygen partial pressure. The oxygen concentration did not affect the cell growth rate much. The cell growth rate is assumed to be independent of oxygen concentration. However, the specific growth rate of the immobilized cells might be different from that of non-immobilized cells because of the effects of immobilization. The maximum specific growth rate of immobilized cells measured were almost the same as that of non-immobilized cells this time.

4.2. Cell Leakage Rate from the Gel

Figure 5 shows a relationship between the initial cell concentration in an alginate gel sheet and the cell leakage rate from the gel. It indicates that the cell leakage rate is linearly proportional to the initial cell concentration.

The cell concentration in the gel sheet was fixed at 1 x 10^7 cells/cm^3-gel by adjusting the initial cell concentration to clarify how the cell leakage rate was af-

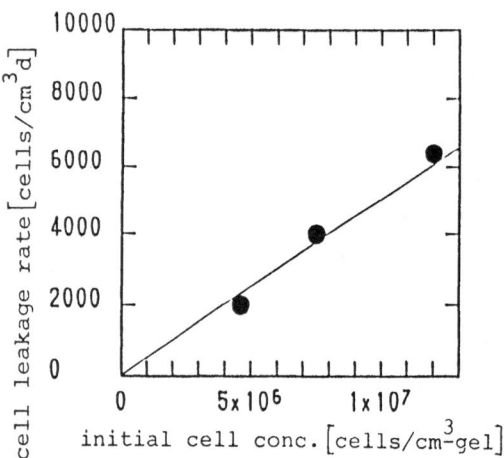

Fig. 5. Relationship between initial cell concentration in an alginate gel sheet and cell leakage rate.

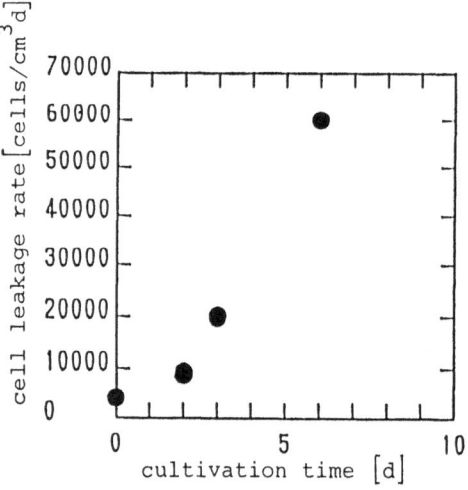

Fig. 6. Relationship between cultivation time of immobilized cells and cell leakage rate.

fected by the cultivation time. Figure 6 shows a relationship between the cultivation time and the cell leakage rate, indicating that the cell leakage rate is represented by a nonlinear equation. Now, the cell leakage rate is considered to be proportional to the cell density and the n-th power of cultivation time, t. Equation (6) is rearranged as Eq. (9). The rate constant M and the power number n are determined experimentally.

$$\frac{dC_{xs}}{dt} = \mu C_{xs} - Mt^n C_{xs} \qquad (9)$$

4.3. Simulation of Cell Proliferation in the Gel Bead

Now five parameters should be determined experimentally; the constants and the power numbers in Eqs. (5) and (9), and the depth of the zone in an alginate gel bead from which the cells would leak. It was found from the observation of alginate gel beads that the size of cell colonies near the surface of gel beads was around 0.15 mm. Then, the depth was assumed to be 0.15 mm regardless of gel bead sizes. The other parameters were set so that the calculated values might agree with the experimental ones.

Figure 7 shows the culture results of immobilized cells entrapped in a 3.2-mm diameter gel particle and the cells leaked into the medium in a oxygen concentration of 0.2 mol/m^3. The lines in Fig. 7 were obtained by calculation. The parameter values determined were listed in Fig. 7. Here, the medium in the reactor was exchanged every 24 hours to sustain favorable conditions for cell growth, except oxygen. The calculated results fitted the experimental ones, indicating the validity of the model.

Figure 8 shows the changes in cell concentration in the gel beads of 1.5 mm in diameter and in the concentration of cells leaked to the medium. The same constants and the same power numbers shown in Fig. 7 were used for the simulation in concentration change in the gel bead. The calculated results agree with the experimental ones for immobilized cells as well as the cells leaked, indicating that a good prediction can be achieved.

4.4. Discussion

The growth model based on a cell leakage model on detailed experimental data can cover any changes in immobilized cell concentration even under unsteady state conditions. There is no growth model for immobilized cells by which not only cell growth in gel beads but also cell leakage from the beads can be precisely predicted. This facilitates estimation of total cell number in a fermentor.

However, this model now has a serious weak point: In the cell leakage model, the cell leakage rate is proportional to the third power of the culture time, but this is not true. If so, the cell leakage rate would increase infinitely, and the cell leaking zone in a gel bead would be emptied soon. However, this is not observed at all with a microscope.

148

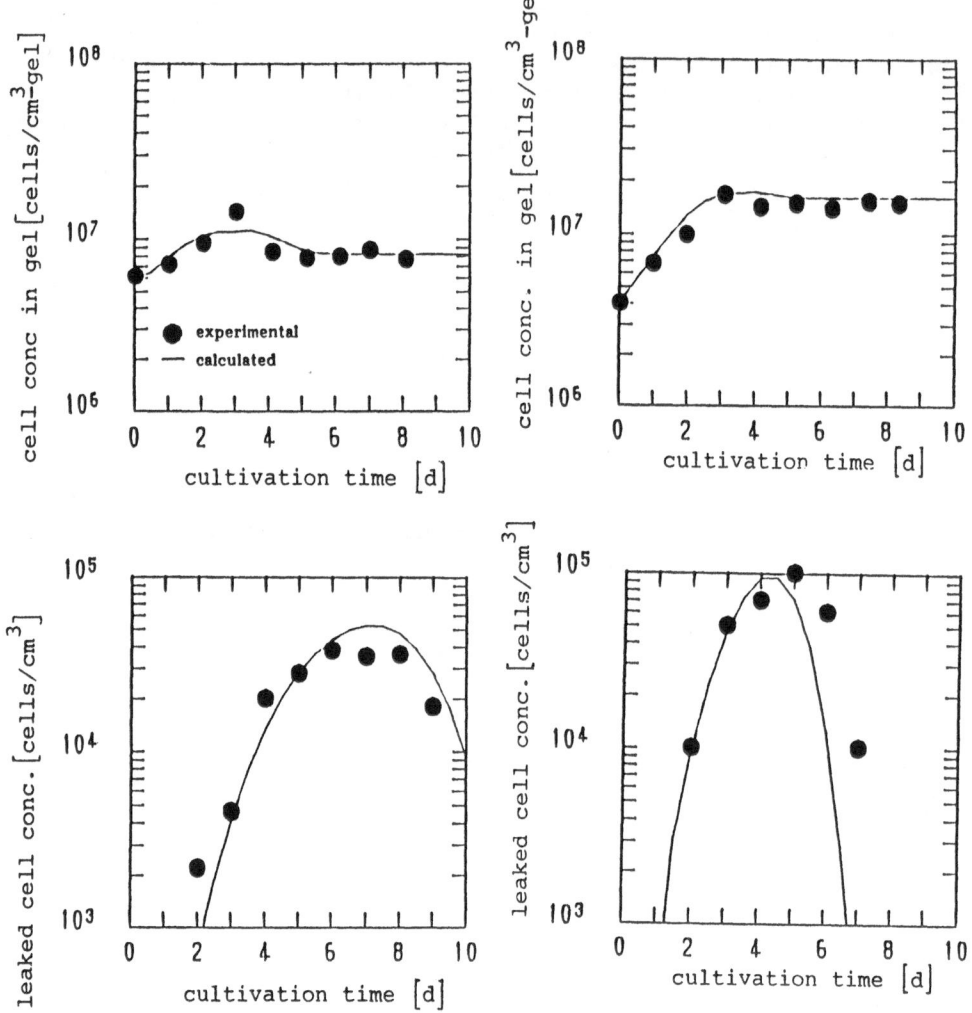

Fig. 7. Growth of hybridoma cells in a gel bead (a) and in a medium (b) and its computer simulation: oxygen concentration = 0.2 mol/m^3, bead size = 3.2 mm, k = 1.3 x 10^{-17}cm^6/h, m = 2, M = 3.9 x 10^{-9} h^{-4}, n = 3.

Fig. 8. Growth of hybridoma cells in a gel bead (a) and in a medium (b) and its computer simulation: oxygen concentration = 0.2 mol/m^3, bead size = 1.5 mm.

This weak point should be addressed and overcome as early as possible.

Acknowledgment

Appreciation is due to Teijin Limited for kindly supplying the hybridoma 4H11 cells.

Nomenclature

c_A = concentration of oxygen [mol/m^3]
c_{AS} = concentration of oxygen at the gel surface [mol/m^3]
c_x = concentration of cells in gel particles [cells/m^3]
c_{xL} = concentration of cells in medium [cells/m^3]
c_{xs} = concentration of cells in a cell leakage zone [cells/m^3]
D_{eA} = diffusion coefficient of oxygen in the gel [m^2/s]
k = constant in Eq. (5) [m^{3m}/s]
M = constant in Eq. (9) [1/s^{n+1}]
m = power number in Eq. (6) [-]
n = power number in Eq. (9) [-]
R = radius of a gel particle [m]
R_s = radius of a zone in a gel particle from which cells do not leak [m]
r_c = critical radius of a boundary sphere at which oxygen concentration reaches zero [m]
r_{Am} = specific oxygen consumption rate [mol/s/cell]
r_{leak} = cell leakage rate [cells/s/m^3]
t = time [s]
t_0 = time interval until a whole medium is changed [s]
V = volume of medium [m^3]
V_s = volume of whole gel particles [m^3]
μ = specific cell growth rate [1/s]
μ_{max} = maximum specific growth rate of cell [1/s]

References

1. Shirai, Y. et al., Appl. Microb. Biotechnol., 26, 495 (1987).
2. Shirai, Y. et al., Appl. Microb. Biotechnol., 29, 544 (1988).
3. Iijima, S. et al., Appl. Microb. Biotechnol., 28, 572 (1988).
4. Familletti, P. C. and Fredricks, J. E., Bio/Technol., 6, 41 (1988).
5. Bugarski, B., et al., Appl. Microb. Biotechnol., 30, 264 (1989).
6. Glacken, M. W., et al., Annal. NY. Acad. Sci., 413, 355 (1983).
7. Spier, R. E. and Griffiths, B., Develop. Biol. Standard., 55, 81 (1984).
8. Y. Shirai, et al., Appl. Microb. Biotechnol., 29, 113 (1988).
9. Mori, A. , et al., Biotechnol. Let., 11, 183 (1989).

HIGH DENSITY SUSPENSION CULTURE OF INSECT CELLS IN A STIRRED BIOREACTOR

S. DEUTSCHMANN and V. JÄGER
Gesellschaft für Biotechnologische Forschung mbH
Mascheroder Weg 1
D-3300 Braunschweig, Germany

ABSTRACT. In order to develop an efficient process for the production of recombinant proteins with the baculovirus expression system, various factors concerning the culture conditions of Sf-21 - a cell line derived from the "Fall Armyworm" *Spodoptera frugiperda*- were studied. These insect cells are very common in insect cell culture and were used as host cells for the infection with recombinant *Autographa californica Nuclear Polyhedrosis Virus*.
To optimize the growth of the Sf-21 cells, four different media supplemented with 5% fetal calf serum (FCS) were tested in 125 ml spinner flasks. The highest cell density was achieved in IPL-41 modified with 2.5 g/l tryptose phosphate broth + 5% FCS. This medium was used for further continuous and batch experiments of Sf-21 cells in a perfused 1.4 l bioreactor.
In these experiments the exponential growth phase could be maintained for more than 7 d and a maximal cell density of $2 \cdot 10^7$ viable cells/ml was achieved. A setpoint of 70% DO was found to be optimal for growth of these cells.

1. Introduction

During the last 10 years the baculovirus expression system has been established for the expression of a variety of heterologous proteins [1]. The virus normally used is *Autographa californica Nuclear Polyhedrosis Virus* which infects nearly 20 species of Lepidoptera in addition to *Spodoptera frugiperda*, its natural host. The system is based on the construction of recombinant baculoviruses by replacing the polyhedrin gene, which is under the control of a very strong promoter, with foreign DNA [13, 15]. The insect cells are infected with the recombinant baculoviruses, and during the replication of the virus the foreign gene of interest, instead of the polyhedrin gene will be expressed at relativly high levels of 1 - 500 mg/l [18, 19]. The factors which determine how well a foreign gene is expressed by the baculovirus expression system are not well characterized, but it is known that the culture conditions of the insect cells have great influence on the expression. To get a maximum product yield it is necessary to optimize culture conditions of the *Spodoptera frugiperda* cells and to establish high density suspension cultures in bioreactors.

R. Sasaki and K. Ikura (eds.), Animal Cell Culture and Production of Biologicals, 151–158.

The culture conditions of the insect cells have been the subject of many investigations concerning media [10, 21], the adaption to suspension culture [27], and scale-up strategies with Pluronic F-68 - a nonionic surfactant - to protect the cells against shear stress [14, 16]. Oxygen supply has not yet been optimized [17].

2. Material and Methods

2.1. CELL-LINE AND MEDIA

The cell line IPLB-SF-21AE (Sf-21), derived from ovaries of *Spodoptera frugiperda* insect cells and originally cloned 1977 by J. L. Vaughn [25], was kindly provided by Alberto J. Marcipar, Universidad National del Litoral, Santa Fé, Argentina.

The Sf-21 cells were cultured in three different media, each supplemented with 5% fetal calf serum (Gibco-BRL, Eggenstein, Germany). The cells were grown in TC100 [5](Gibco-BRL), IPL-41 [26](BIOCHROM, Berlin, Germany) or TNM-FH medium prepared from Grace medium [6](Gibco-BRL) modified by the addition of 3.3 g/l yeast autolysate (Ohly, Marl, Germany) and 3.3 g/l lactalbumin hydrolysate (SIGMA, Deisenhofen, Germany). In further experiments IPL-41 medium was modified with 2.5 g/l tryptose phosphate broth (SIGMA).
The yeast autolysate/lactalbumin hydrolysate mix was filter sterilized by 0.22 μm filters (MILLEX GV, Millipore Corporation, Bedford, Massachusetts, USA) [7, 22].

2.2. REACTOR SYSTEMS

2.2.1. Spinner flasks. The insect cells were kept in 125 ml spinner flasks (Techne, Cambridge, U.K.) and stirred at 50 rpm in a 27°C incubator (Heraeus, Hanau, Germany). Bioreactor inocula were grown under the same conditions in 500 ml spinner flasks.

2.2.2. Bioreactor. The reactor experiments were carried out in a 1.4 l (working volume 1.2 l) stirred tank bioreactor in batch or continuous perfusion mode. Both reactors were equipped with a hydrophobic polypropylene membrane for bubble-free aeration which was mounted on a stirrer [11]. The perfusion reactor had an additional hydrophilized polypropylene membrane for continuous medium exchange [12].

2.3. ANALYSIS OF SAMPLES
Viable and dead cells were estimated by trypan blue exclusion with a hemacytometer.

Glucose and lactate concentrations of supernatants were measured with YSI analysers (YSI, Yellow Springs, Ohio, USA).

18 different amino acids were quantified by reversed phase HPLC (Beckman, München, Germany) after precolumn derivatization with o-phthalaldehyde [2, 3].

3. Results and Discussion

The first step was to compare different media for the cultivation of Sf-21 insect cells. These experiments were carried out in 125 ml spinner flasks. The seeding density was $3 \cdot 10^5$ viable cells/ml. The best growth was achieved with 50 to 75 ml volume. The maximum cell density decreased, if the volume was increased to 125 ml medium. This effect was observed with TC 100 as well as with TNM-FH (see Fig 1). The reduction of the volume increased the maximum cell density from $2.08 \cdot 10^6$ to $2.77 \cdot 10^6$ viable cells/ml (TNM-FH) and from $2.51 \cdot 10^6$ to $3.75 \cdot 10^6$ viable cells/ml (TC 100), which could only be explained by a relativly high oxygen demand of the *Spodoptera frugiperda* cells.

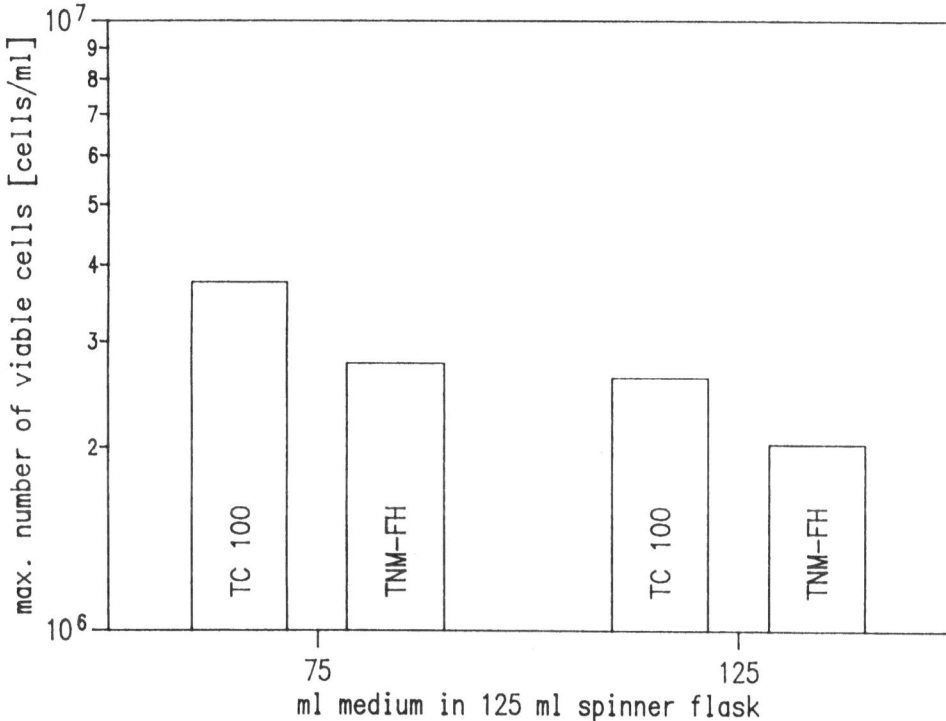

Fig 1:   Influence of medium volume on oxygen transfer and thereby on growth and maximum cell density of *Spodoptera frugiperda* insect cells in two different media in 125 ml spinner flasks

The following comparison of all four media was carried out with 125 ml starting volumes. Twice daily samples of 5 ml were taken, and the volumes decreased finally to 35 ml. This small volume provided a sufficient aeration. The results of these cultivations, the maximum cell densities and the averages of the population doubling times during the exponential growth phases are summarized in Fig. 2. Best growth was observed with TC 100. The maximum number of viable cells was $3.75 \cdot 10^6$/ml and the population doubling time had an average of 24.3 hours. In this experiment IPL-41 was inferior to TC 100 (a maximum cell

154

density of only $2.75 \cdot 10^6$ viable cells/ml could be achieved and the population doubling time increased to 31.2 h). However, we decided to improve IPL-41 by the addition of 2.5 g/l tryptose phosphate broth and use it rather than TC 100, because IPL-41 had been shown to be suitable as a basal medium for the development of a protein-free medium [8]. The use of serum-free medium might facilitate the isolation and purification of the produced recombinant proteins.

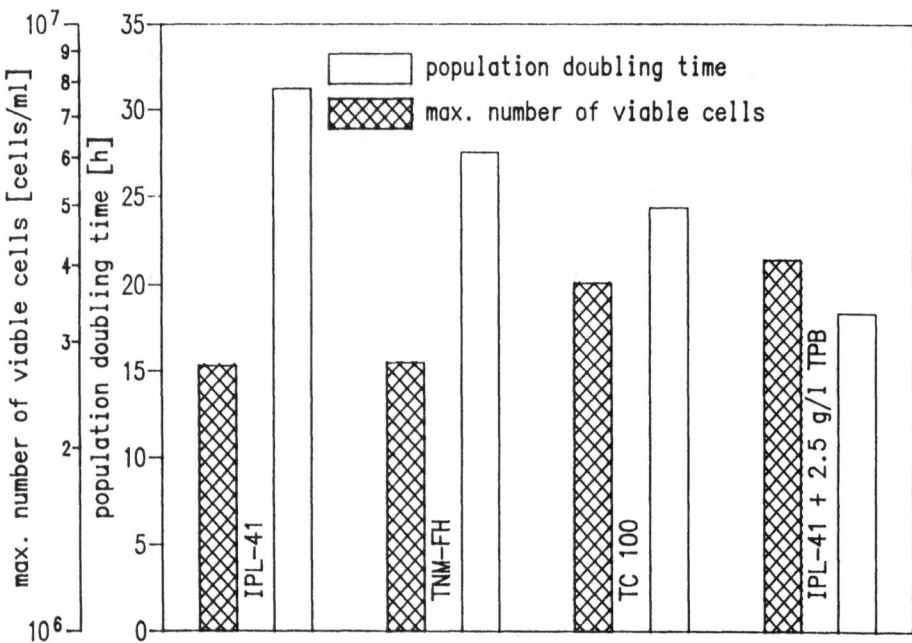

Fig 2: Test of four different media for their suitability for the suspension culture of *Spodoptera frugiperda* cells. Criteria for good growth were the maximum cell density and the population doubling time.

The next step, the continuous cultivation of Sf-21 cells in a perfused bioreactor in IPL-41 with 2.5 g/l tryptose phosphate broth supplemented with 5% FCS, is shown in Fig. 3a and b. After a lag-phase of about 20 hours the cells started to grow exponentially. This exponential phase could be maintained for more than 7 d, with the cells reaching a maximum concentration of $2.1 \cdot 10^7$ viable cells/ml with a viability of 92%. This long exponential growth phase during the continuous cultivation allows late infection with virus at high cell densities. The infection of the insect cells during the exponential growth phase is very important, because an infection during the stationary phase of growth reduces the product yield drastically [24]. On the other hand, the high cell densities of more than $2 \cdot 10^7$ viable cells/ml obtained here promise a higher product yield than with the maximum cell densities of $5 - 6 \cdot 10^6$ cells/ml which are normally reported [4, 9, 14]. Detailed quantitative analyses of samples showed that there were no limitations in C-sources and amino acids. Nevertheless growth stopped after 8 d of cultivation because of limited oxygen supply. The limitation could be prevented for several hours by increasing the stirrer speed from 40 to 60 rpm (see Fig 3b).

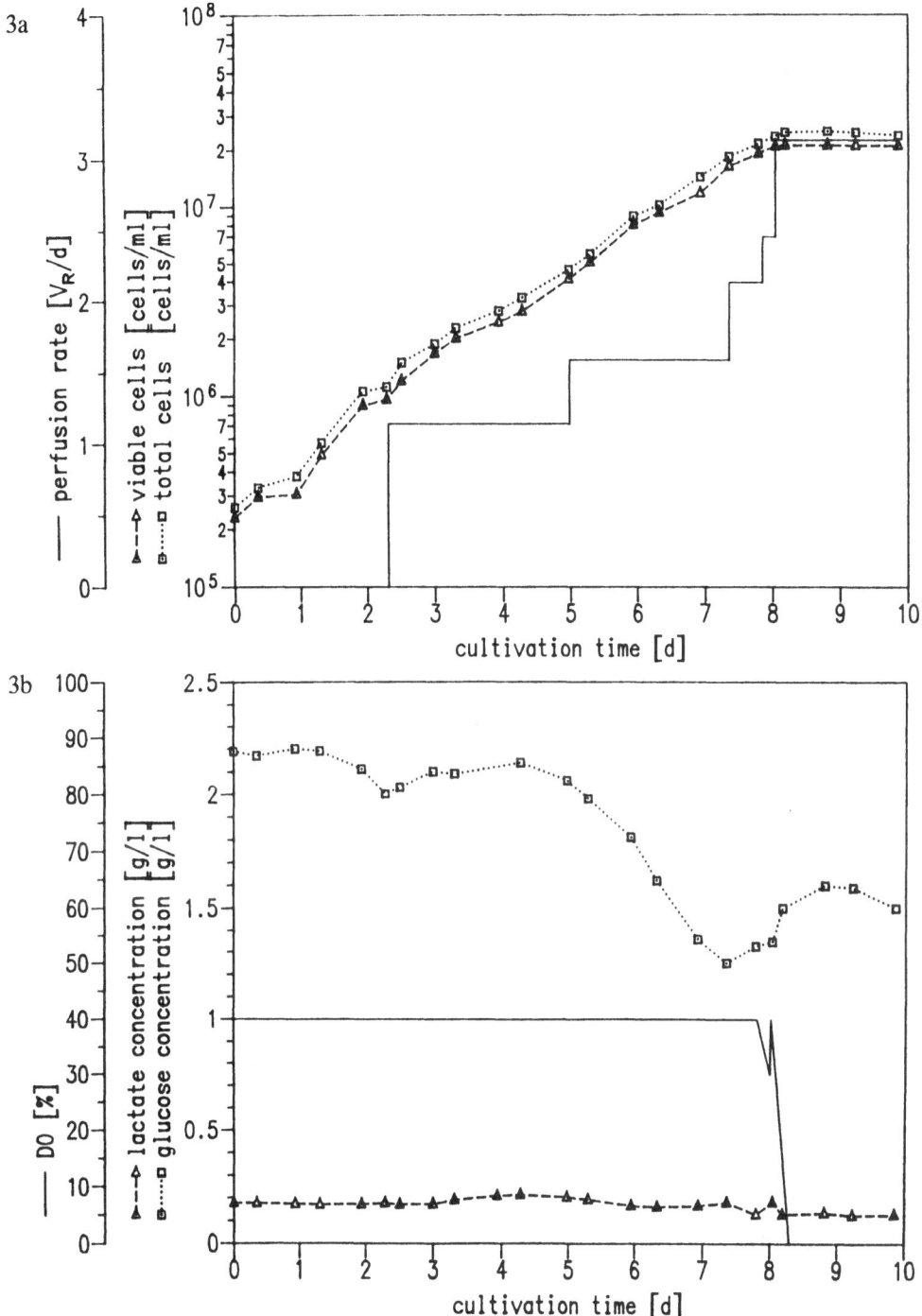

Fig 3: Continuous cultivation of *Spodoptera frugiperda* cells in a 1.4 l stirred tank perfusion bioreactor system.

This was another indication for a relativly high oxygen concentration requirement of the Sf-21 cells. Therefore sufficient aeration is the most important factor for high-density cultivation of these cells.

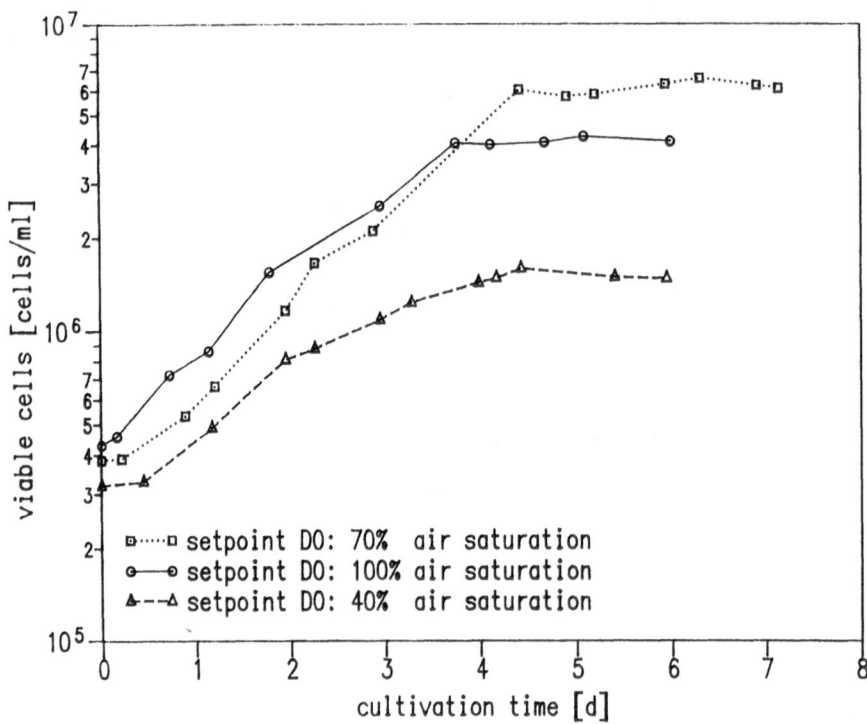

Fig 4: Growth curves of *Spodoptera frugiperda* cells of three batch cultivations at different aeration rates.

To optimize culture conditions with respect to their high oxygen demand, cells were cultivated in three batch experiments at different setpoints for DO, which were adjusted to 40, 70 and 100% air saturation, respectively. The results of these three cultivations are presented in Fig 4 and 5. With 40% DO the Sf-21 cells reached a maximum of only $1.65 \cdot 10^6$ viable cells/ml at an averaged population doubling time of 43.2 hours. The best growth could be achieved with a setpoint of 70% DO. The highest cell density was $6.6 \cdot 10^6$ viable cells/ml and the population doubling time decreased to 23.7 hours. When DO was increased to 100% air saturation, maximum concentration of viable cells dropped to $4.25 \cdot 10^6$/ml at an increased population doubling time of 29.3 hours.

The cell specific oxygen uptake rate (OUR) was determined during the exponential growth phase of the cells in all batch experiments. This rate was shown to correlate very well with cell growth. The highest OUR was monitored at a setpoint of 70% DO with 4.5 pg/cell·h. At 100% DO the specific OUR decreased to 2.2 pg and at 40% the rate was only 0.9 pg/cell·h. Compared with most mammalian cell lines [20], the specific oxygen uptake rate of *Spodoptera frugiperda* cells is in the same range, but optimum oxygen concentration for the cultivation of these cells is significantly higher (70% instead 40%).

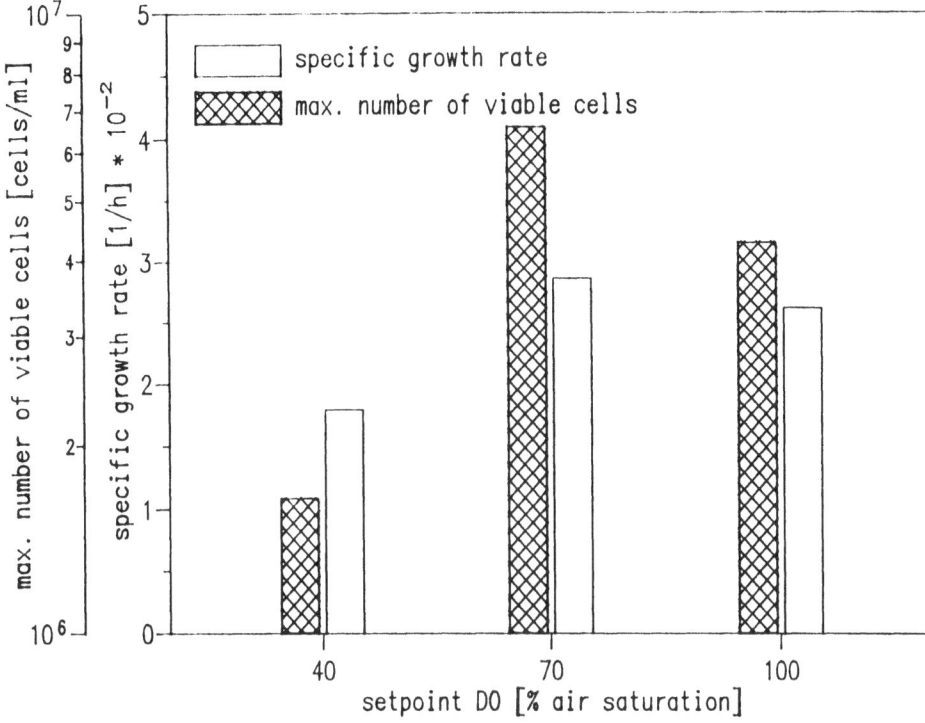

Fig 5: Influence of oxygen concentration on specific growth rate and maximum number of viable cells in batch cultivations.

4. Acknowledgement

We are grateful to Dr. D. Monner for linguistic advice.

5. References

[1]   Cameron, J.R., Possee, R.D., and Bishop, D.H.L. (1989) 'Insect Cell Culture Technology in Baculovirus Expression System', Trends Biotechnol. 7, 66-70

[2]   Cooper, J.D.H., Lewis, M.T., and Turnell, D.C. (1984) 'Pre-Column o-Phthalaldehyde Derivatization of Amino Acids and their Separation using Reversed-Phase High-Performance Liquid Chromatography I. Detection of the Imino Acids Hydroxyproline and Proline', J. Chromatogr. 285, 484-489

[3]   Cooper, J.D.L., Lewis, M.T., and Turnell, D.C. (1984) 'Pre-Column o-Phthalaldehyde Derivatization of Amino Acids and their Separation using Reversed-Phase High-Performance Liquid Chromatography II. Simultaneous Determination of Amino and Imino Acids in Protein Hydrolysates', J. Chromatogr. 285, 490-494

[4]   Fraser, M.J., (1988) 'Expression of Eucaryotic Genes in Insect Cell Cultures', In Vitro 25, 225-235

[5]   Gardiner, G.R., and Stockdale, H. (1975) 'Two Tissue Culture Media for Production of Lepidopteran Cells and Nuclear Polyhedrosis Viruses', J. Invertebr. Pathol. 25, 363-370

158

[6]   Grace, T.D.C. (1962) 'Establishment of Four Strains of Cells from Insect Tissues grown In Vitro', Nature 195, 788-789

[7]   Hink, W.F., (1970) 'Established Insect Cell Line from the Cabbage Looper Trichoplusia ni', Nature 226, 466-467

[8]   Inlow, D., Shauger, A., and Maiorella, B. (1989) 'Insect Cell Culture and Baculovirus Propagation in Protein-Free Medium', J. Tissue Cult. Methods 12, 13-16

[9]   Kompier, R., Tramper, J., and Vlak, J.M. (1988) 'A Continuous Process for the Production of Baculoviruses using Insect Cell Cultures', Biotechnol. Lett. 10, 849-854

[10] Kurtti, T.J., Chaudary, S.P.S., and Brooks, M.A. (1974) 'Influence of Physical Factors on the Growth of Insect Cells In Vitro I. Effects of Osmotic Pressure on Growth Rate of a Moth Cell Line', In Vitro 10, 149-156

[11] Lehmann, J., Vorlop, J., and Büntemeyer, H. (1988) 'Bubble-Free Reactors and their Development for Continuous Culture with Cell Recycle', in Spier, R.E., and Griffith, J.B. (eds) "Animal Cell Biotechnology Vol 3", Acad Press, London 1988, 221-237

[12] Lehmann, J., Piehl, G.W., and Schulz, R. (1987) 'Bubble-Free Cell Culture Aeration with Porous Moving Membranes', Develop. Biol. Standard. 66, 227-240

[13] Luckow, V.A., and Summers, M.D. (1988) 'Trends in the Development of Baculovirus Expression Vectors', Bio/Technology 6, 47-55

[14] Maiorella, B., Inlow, D., Shauger, A., and Harano, D. (1988) 'Large-Scale Insect Cell Culture for Recombinant Protein Production', Biotechnol. 6, 1406-1410

[15] Miller, D.W., Safer, P., and Miller, L.K. (1986) 'An Insect Baculovirus Host-Vector System for High-Level Expression of Foreign Genes', Genet. Eng. 8, 277-298

[16] Murhammer, D.W., and Goochee, C.F. (1988) 'Scaleup of Insect Cells: Protective Effects of Pluronic F-68', Biotechnol. 6, 1411-1418

[17] Schopf, B., Howaldt, M.W., and Bailey, J.E. (1990) 'DNA Distribution and Respiratory Activity of Spodoptera frugiperda Populations Infected with Wild-Type and Recombinant Autographa californica Nuclear Polyhedrosis Virus', J. Biotechnol. 15, 169-186

[18] Smith, G.E., Ju, G., Ericson, B.L., Moschera, J., Lahm, H., Chizzonite, R., and Summers, M.D. (1985) 'Modification and Secretion of Human Interleukin 2 Produced in Insect Cells by a Baculovirus Expression Vector', Proc. Natl. Acad. Sci. USA 82, 8404-8408

[19] Smith, G.E., Summers, M.D., and Fraser, M.J. (1983) 'Production of Human ß-Interferon in Insect Cells Infected with a Baculovirus Expression Vector', Mol. Cell. Biol. 3, 2156-2165

[20] Spier, R.E., and Griffith, B. (1984) 'An Examination of the Data and Concepts Germane to the Oxygenation of Cultured Animal Cells', Develop. Biol. Standard. 55, 81-92

[21] Stockdale, H., and Gardiner, G.R. (1977) 'The Influence of the Condition of Cells and Medium on Production of Polyhedra of Autographa californica Nuclear Polyhedrosis Virus In Vitro', J. Invertebr. Pathol. 30, 330-336

[22] Summers, M.D., and Smith, G.E. (1987) 'A Manual of Methods for Baculovirus Expression Vectors and Insect Cell Culture Procedures', Texas Agricultural Experiment Station, Bulletin No. 1555

[23] Vaughn, J.L. (1976) 'The Production of Nuclear Polyhedrosis Viruses in Large Volume Cell Cultures', J. Invertebr. Pathol. 28, 233-237

[24] Vaughn, J.L., Goodwin, R.H., Tompkins, G.J., and McCawley, P. (1977) 'The Establishment of two Cell Lines from the Insect Spodoptera frugiperda (Lepidoptera; Noctuidae)', In Vitro 13, 213-217

[25] Weiss, S.A., Smith, G.E., Kalter, S.S., and Vaughn, J.L. (1981) 'Improved Methods for the Expression of Insect Cell Cultures in Large Volume', In Vitro 17, 495-502

[26] Wu, J., King, G., Daugulis, A.J., Faulkner, P., Bone, D.H., and Goosen, M.F.A. (1990) 'Adaption of Insect Cells to Suspension Culture', J. Ferment. Bioeng. 70, 90-93

MEDIA FOR CULTIVATION OF ANIMAL CELLS: AN OVERVIEW

A. Mizrahi and A. Lazar
Department of Biotechnology,
Israel Institute for Biological Research,
P.O.Box 19,
Ness-Ziona 70450,
Israel.

ABSTRACT. The increasing interest in products from animal cells has caused an extensive research effort toward development of media for cell cultivation. The basic components in the media used for cultivation of animal cells vary depending upon the characters of the cells and the cultivation method. Basic components consist of an energy source, nitrogen source, vitamins, fats and fatty soluble components, inorganic salts, nucleic acid precursors, antibiotics, oxygen, pH buffering systems, hormones, growth factors and serum. Extensive efforts are directed towards developing serum-free media: Low-protein media or chemically defined media. Several approaches in developing serum-fre media were reported. These are described in detail in this review.

1.   Introduction

Almost 50% of the biologicals produced to-day or planned to be produced in the future are from animal cells origin. Therefore, there is an increasing interest in developing technologies for cultivation of animal cells and production of a wide spectrum of biologicals (Mizrahi, 1986, 1986a). The world-wide activities and markets of biologicals from mammalian cells were recently reviewed (Ratafia, 1987).
     Although the major achievements in the field of animal cell cultivation have been accomplished in the last three decades, it has a long history of about one hundred years. Apart from the development of various types and sizes of culturing vessels (Hoare, 1987), research and development (R & D) of optimal media for cell cultivation is also carried out among most groups involved in the field of animal cell cultivation and production of biologicals.
     Media used for animal cell cultivation is considered to include two major parts:

159

R. Sasaki and K. Ikura (eds.), Animal Cell Culture and Production of Biologicals, 159–180.
© 1991 Kluwer Academic Publishers.

1.1. Essential basal ingredients that fulfill all cellular requirements for nutrients and known as the basal growth medium. The wide range of basal media used today are mainly based on formulations determined by Eagle more than 30 years ago (Eagle, 1955). Eagle's formulation was developed for laboratory scale experiments, but not for large scale production of animal cells.

1.2. A set of supplements that satisfy other types of cellular growth requirements and make it possible to grow cells in the basal growth medium.

From a cell culture point of view, a nutrient is defined as a chemical that enters the cell and is used as a structural component: as a substrate for biosynthesis: as a substrate for energy metabolism or in a catalytic role in such metabolism (Bettger and Ham, 1982). A supplement is defined as anything else needed for cellular proliferation. This group includes all undefined additives (such as serum and animal or plant extracts) and defined additives (such as growth factors, attachment factors, extracellular matrix, hormones, etc.). The exclusion of these supplements from being termed 'nutrient' is due to the fact that some of them often enter the cells, promote cellular multiplication, or undergo catabolism.

There are many comprehensive reviews which summarize the field of media for animal cells each from its own point of interest (Morton, 1970; Higuchi, 1973; Higuchi, 1977, Katsuta and Takaoka, 1977; Rizzino et al., 1979; Ham, 1981; Bettger and Ham, 1982, Ham, 1984; Waymouth, 1984; Jayme and Blackman, 1985; Lambert and Birch, 1985; Mizrahi and Lazar, 1988). This review will cover only cell growth media composition and will not deal with techniques such as media preparation and sterilization. Detailed procedures, methods and recommendations of these points are comprehensively described in a book by Freshney (Freshney, 1984).

1.3. WATER FOR ANIMAL CELL MEDIA

One of the most important points for consideration when preparing media is the required high quality of the water. Water used for culture media should be pyrogen-free (especially if the product is for human or animal use). It should have a resistance of 1.5 - 2.0 mOhms, indicating a low salt content.

It is highly recommended to use fresh ultra-pure water and not stored-water, since in some storage tanks, organic material or ions from plastic or glass may dissolve in the water. Coppola reviewed ultrapure water from various purification processes (Coppola, 1984).

1.4. PURITY OF CHEMICALS, STABILITY AND SHELF LIFE

Chemicals of the highest purity are required for preparation of media. Commercial chemicals, although being pure, inevitably contain traces of contaminants (Aronowitz, 1981), and in several cases, maximum permissible concentrations are listed on the labels. Some of the traces may be toxic (like Hg). On the other hand, there are several reports that some of the trace contaminants (like As, Cr, Se, Sn, V), have beneficial effects on some of the cells (McKeehan et al., 1976; Nielsen, 1981; Bettger and Ham, 1982).

With regards to stability of media ingrediants, inorganic chemicals are indefinitely stable. Vitamins are the least. Hormones, several antibiotics and growth factors are recommended to be stored frozen (-20°C) or refrigerated (0-4°C).

Several ingredients used in animal cell culture media are known for their instability, e.g. ascorbic acid (which is oxidized in natural and basic pH) and glutamine.

Many factors affect the shelf-life of media. Among them are the following: natural decay rate of unstable compounds, pH, storage temperature, moisture, access of oxygen and exposure to near-ultraviolet, daylight or fluorescent light. Most media should be stored at $4 \pm 1°C$ and in a dark place. Storage of media by freezing may cause loss of some poorly soluble ingredients, such as tyrosine, which may not redissolve completely when the medium is thawed. It is highly recommended not to store liquid media longer than 3-4 weeks. Powdered media may be stored for several years.

1.5. BASIC COMPONENTS IN MEDIA

1.5.1. Energy Sources. Glucose is one of the most important energy and carbon sources for cultivated cells. Its concentration varies in the range 5-20 mM. When added at high concentration, it is often rapidly converted to lactate which lowers the pH, sometimes to an inhibitory level. Glutamine (0.7-5 mM) is another major carbon and energy source in most media. This may be due to the fact that the glucose conversion in cultured cells does not use the citric acid cycle, as in vivo. Thus, much of the required carbon is derived from the glutamine. The degree of conversion of glucose to lactate was found to be lower when glutamine is the principle energy source (Reitzer et al., 1979; Zielke et al., 1980).

Fructose (the most abundant carbohydrate in mammalian fetal blood), galactose, mannose and maltose were also reported to be used instead of glucose as another means of controlling the lactate production (Reitzer et al., 1979; Cristofalo and Kritchevsky, 1965; Leibovitz, 1963; Immanura et al., 1980).

Several amino acids and α-keto acids (e.g. pyruvate, α-ketoglutarate, oxaloacetate) were also reported as contributors of energy to cultivated cells (McKeehan and McKeehan, 1979; Groelke, et al., 1979). These ingredients cannot totally substitute for carbohydrates which, in most media, are the main energy source. Uridine or cytidine may also serve as a carbon source in several cell growth media (Wice et al., 1981).

1.5.2. Nitrogen Sources. Amino acids are the nitrogen source in all media, but their number an concentration vary from one medium to another. Eagle's MEM contains a mixture of 13 amino acids, Dulbecco's modified medium contains 15 amino acids. M-199 medium contains 19 amino acids, while RPMI #1640 medium and Iscove medium each contain 20 amino acids.

Apart from being the basic units for protein synthesis, some of the amino acids also serve as an energy source (McKeehan and McKeehan, 1979; Groelke et al., 1979). The classical classification of the amino acids into 'essential' and 'nonessential' is done according to the basic requirements of the whole organism. In many culturing systems, this classification has no value. But 'tradition' in media formulation of animal cell cultures is a 'very stable' phenomenon. It is also known that media having a high concentration of amino acids, provide insurance against depletion and also act as an energy source. It should be mentioned that too high a concentration of amino acids, may be toxic to cultivated cells. Hydrolyzates of proteins (enzymatic digest) from animal tissue: Primatone RL - a product of Sheffield, Norwich, N.Y. (Mizrahi, 1977), or Primatone HS (Sheffield, Unpublished data, and plant protein hydrolyzates (Mizrahi and Shahar, 1977), were reported as a cheap and satisfactory nitrogen source in cultivation of several animal cells in production of biologicals: like human lymohoblastoid interferon (Mizrahi et al., 1980) and monoclonal antibodies (Velez et al., 1986). A comprehensive discussion of the role of the amino acids in media is given by Waymouth (Waymouth, 1984).

1.5.3. <u>Vitamins</u>. All media contain vitamins, mainly water-soluble ones. Most of them act as cofactors in enzyme in enzyme reactions. The use and the effect of fat-soluble vitamins (e.g.: A,D,E,K) on cultured cells is limited (Smith, 1981; Lotan, 1980). Vitamin A and Vitamin E (also know as α-tocopherol) are included in medium M-199 (Morgan et al., 1950). Regarding the water-soluble vitamins, members of the vitamin B group and vitamin C are included in most media formulation at different variations and concentrations. More detailed information on vitamins in media is given by Lambert and Birch (1985) and Waymouth (1984).

1.5.4 <u>Fats and Fat-Soluble Components.</u> Fatty acids were reported to be essential nutrients for certain cultured cells, among them are linoleic acid (Ham, 1963; Dubin et al., 1965) and oleic acid (Jenkins and Anderson, 1970). Cholesterol was also found to be an essential medium component for several animal cell cultures (Sato et al., 1957; Higuchi, 1970). Cholesterol and phospholipids are found in the basal medium of several serum-free media formulations (Chen & Kandutsch, 1981) as are ethanolamine or phosphoethanolamine (Kano-Sueska and Errick, 1981) which have been shown to be required for both growth and monoclonal antibody secretion by hybridoma cells (Murakami et al., 1982).

Rothblatt and Kritchevsky (1967) in their excellent book comprehensively cover the subject of the role of the lipids in cultured mammalian cells. Later King and Spector (1981) comprehensively reviewed this subject.

1.5.5. <u>Inorganic Salts.</u> The inorganic salts perform a variety of essential functions in sustaining cell proliferation in vitro (Ham and McKeehan, 1979; Freshney, 1984; Jayme and Blackman, 1985). The majority of the inorganic ions (e.g. Na^+ K^+ CA^{+2}, Mg^{+2}, Cl^-, HPO_4^{-2}) in principle are involved in several functions, amongst them: maintaining the electrolyte balance, the extracellular environment iso-osmotic to the native environment (in the range of 285-300 mOsm), mediation of transmembrane signals and buffering the external medium to maintain the required pH (in the range 7.0 - 7.4).

In most media, one of two principle inorganic salt bases are used:

1.5.5.1. Earle's base salts: a bicarbonate base salt solution, which relies on the physiologically relevant pKa for carbonic acid dissociation and the equilibration of gaseous and liquified carbon dioxide to maintain constant pH (Earle et al., 1943).

1.5.5.2. Hank's base salts: are designed for atmospheric equilibration based upon the physiologically relevant pKa of phosphoric acid and contain only sufficient bicarbonate ions for carboxylation and similar biological processes (Hanks and Wallace, 1949).

Most convential media contain trace elements in their formulations. Their importance was established by Shooter and Gey (1952). Fe^{+2}, Cu^{+2}, Zn^{+2}, CO^{+2}. Mn^{+2}. are the most used elements. Comprehensive reviews on their functions are given by Higuchi (1973) and Waymouth (1984).

1.5.6. <u>Nucleic Acid Precursors.</u> Nucleic acid precursors are not an essential element in animal cell growth medium, since cells are capable of synthesizing them (Kelley, 1972). However, they are included in several rich media formulations.

1.5.7. <u>Antibiotics.</u> Although antibiotics are used commonly in most media, their addition has several disadvantages: resistant microorganisms are developed in several cases, antibiotics may often have adverse effects on cell growth and function, antibiotics can reduce cell yield, growth rate and longevity. Cytotoxicity of antibiotics may be increased in low protein media. Therefore, if possible, it is recommended to avoid antibiotics in medium.

However, antibiotics are added almost routinely in most laboratories and large scale animal cell production systems. The use of antibiotics in cell culture media was reviewed by Perlman (1979), penicillin, streptomycin, neomycin and gentamycin are the most used antibacterial antibiotics. Amphotericin B and nystatin are the most used antifungal antibiotics.

The relative stability of the antibiotics should be kept in mind. Stability is usually pH-dependent. Gentamycin is the most stable antibiotic being stable in wide pH range (6.5-8.0) and has a mycoplasmacidic effect (not on intracellular mycoplasma) as well as a bacteriostatic effect. The main disadvantage is the high cost of this antibiotic.

1.5.8.   pH and Buffering Systems. Most animal cells grow well
around pH 7.2 - 7.4. Some normal fibroblasts grow at pH 7.4 - 7.7,
but there are reports on epidermal cells that are maintained at pH 5.5
(Eisinger et al., 1979). In most media, phenol red is used as the pH
indicator. Buffering is done traditionally using the CO_2/HCO_3 system:
5 - 10% CO_2 in compressed air and sodium bicarbonate (usually added in
a concentration of 26 mM which is comparable to that available in the
circulating blood). The pKa of the sodium bicarbonate (6.1), results
in suboptimal buffering throughout the physiological pH range.
    During the last decade, the zwitterionic (hydrogen) buffers first
described by Good et al. (1966), have become popular and usuable.
HEPES (N-2-hydroxyethyl-piperazine-n'-2-ethane sulfonic acid) is the
most commonly used hydrogen buffer. Its pKa is 7.3, which covers the
optimal pH of most cell cultures. Waymouth (1984) in her review,
discussed various points regarding HEPES, other zwitterionic buffers,
and their corelation to the CO_2/HCO_3 buffering system.

1.5.9.   Oxygen. Oxygen is a very essential gas for the growth of
all animal cells and its supply in submerged cultures is complicated
due to the following factors:
a. Solubility of oxygen in culture media is very poor (approximately
7 ppm at saturation of air).
b.   Aeration of submerged cultures with sparged air, as done in
bacterial culture, may cause damage to animal cells.
    This has resulted in the development of various types of culture
vessel, (e.g. airlift fermenters, fluidized-bed reactors) and reliable
oxygen controlling and monitoring devices that ensure the supply of
the required amount of oxygen in submerged large scale cultures of
animal cells. Lambert and Birch, (1958) and Waymouth (1984) also
discussed this subject in their reviews, while McLimans and his group
(McLimans et al., 1968), reviewed the requirements for gases by
mammalian cells in culture.

1.5.10.   Hormones and Growth Factors. The field of hormones and
growth factors as important ingredients in a wide spectrum of media
used for cultivation of various animal cells, is one of the most
discussed and reported subjects. This has become an important subject
due to the development of serum-free media (as will be reviewed
later). Due to its importance, this subject will be reviewed and
discussed more comprehensively than other points in this article.
    The proliferation and survival of normal cells is controlled by a
variety of substances, among them, a group of agents called growth
factors (GF's). GF's are mitogenic polypeptides capable of promoting
DNA synthesis in resting cells and keeping growing cells from entering
the resting phase (G_0). They are insufficient to stimulate sustained
proliferation of a cell culture without the appropriate nutritional
factors. The growth of cultured cells in serum-free defined medium
has demonstrated that cellular proliferation is controlled by multiple
factors (Barnes and Sato, 1980).

Growth factors are coded by cellular oncogenes (C-onc) that participate in molecular pathways essential for normal cell growth. The GF's exert their biological message by interacting with specific cell membrane receptors. Binding of a GF to its receptor activates a kinase activity in the cell, leading to the phosphorylation of various cellular proteins which activate a program for cell proliferation (Kris et al., 1985). Lesions in the expression and regulation of the GF's might release the cells from their normal physiological control, leading to cell transformation. Cancer, seems to be initiated when a cell begins to over-produce a normal GF for which it has receptors (Huang and Huang, 1985; Hunter, 1984). The normal C-onc delivers its signal only on appropriate occasions, such as tissue growth to repair damage with a limited number of cell divisions. However, the tumor encoded protein may do it continually causing unlimited replication of the cells, resulting in tumor formation.

In addition to the GF's that directly affect DNA synthesis and cell growth, there are other factors that indirectly affect cell growth, these are cell attachment factors and transport proteins. All the above mentioned factors exist in serum and are the main reason why serum is so important for cell growth in culture.

Serum is a heterogenous mixture of a variety of hormones and GF's, the most studied ones are listed in Table 1. These factors are capable of promoting attachment, spreading and growth of a variety of cells in vivo, as well as in vitro. No single GF is sufficient to stimulate maximal DNA synthesis in a normal G_0-arrested cell population. A combination of factors are needed. Many reports have described combinations of GF's that stimulated DNA synthesis and, in some studies, cell proliferation in low serum or even in serum-free media (Gospodarowicz and Moran, 1976; Scher et al., 1976; Cherington, 1984).

GF's are present in serum in very small amounts (nanograms per ml). Platelet-derived growth factor (PDGF) which is the major GF in human serum, is present in clotted blood at a concentration of 15-20 ng/ml (Pierson, Jr. and Temin, 1972). Many cell types, like fibroblasts, epithelial cells, lymphocytes and endothelial cells, require 1-2 ng/ml of PDGF for optimum growth in vitro. This requirement can be satisfied by supplementing cell culture medium with 10%-20% serum.

Insulin in combination with PDGF and other factors has been used in a variety of serum-free media (Barnes and Sato, 1980). Insulin initiates DNA synthesis, stimulates protein and lipid synthesis and promotes transport of nutrients across the cell membrane. The somatomedins can replace insulin for growth stimulation of some cell lines of one hundredth the concentration required for insulin (Stiles et al., 1979). A wide range of factors act synergistically with insulin including fibroblast growth factor (FGF), epidermal growth factor (EGF) and glucocorticoids.

EGF is a mitogenic factor for all endodermal and ectodermal cells and some mesodermal cells (for review, see Cohen, 1987; Carpenter and Cohen, 1979). In serum-free condition, EGF stimulates DNA synthesis synergistically with somatomedin, PDGF and insulin (Leof et al., 1980;

Dicker and Rozengurt, 1978). FGF (basic) is mitogenic for all mesodermally derived cells and some endodermal and ectodermal cells (Gospodarowicz, 1975; Gospodorowicz, et al., 1978). Transforming growth factor (TGF) activity may be composed of more than one component and has growth activity in serum-free media. The growth of a rat cell line in a defined medium containing transferrin, insulin and TGF was nearly as rapid as growth in 10% serum (Kaplan et al., 1982).

A major group of factors essential for cell growth are the attachment factors (for review, see Barnes, 1984). Collagen is the most abundant protein in stroma or basement membrane (Kleinman et al., 1981). Collagen mediates cell attachment in culture and is an essential factor for the growth of anchorage dependent cells. Fibronectin which is found in vivo in many sera or plasma, is capable of binding to both collagen and cells and acts as a substratum for many cell types plated in vitro on collagen matrices. The fibronectin may come from serum added to the culture medium or may be synthesized by the cells themselves.

Another important GF in serum is serum-spreading factor (SF) (Barnes et al., 1983). It is capable of stimulating the attachment and spreading of both fibroblastic and epithelial cell types in serum-free media. SF also influences cell migration in vitro. Fetuin is the major protein in fetal calf serum. It promotes attachment, spreading and growth of cells in culture (Fisher et al., 1958; Fliorini and Roberts, 1979). Fetuin has been used in serum-free cell cultures of a variety of anchorage-dependent cells cultivated in culture (Rizzino and Sato, 1978).

Several proteins in blood serve as carriers for compounds essential for cell growth and survival. Hemoglobin transports oxygen, albumin binds a wide range of essential compounds such as steroid hormones or vitamins, and transferrin binds iron and facilitates its transport into cells. Albumin has successfully been used in serum-free media for human lymphoblastoid cells used to produce interferon (Lazar et al., 1982; Zoon et al., 1979). Hemoglobin has been used in cell culture to increase oxygen uptake (Adlercreutz amd mattiasson, 1982). Transferrin and iron are essential for growth of cells in culture and most serum-free media utilize transferrin (Barnes and Sato, 1980).

Several cells in culture, especially transformed cells, produce and secrete into the culture medium GF's that 'auto-stimulate' their growth. TGF was found to be produced by several human carcinoma cells (Todaro et al., 1980). HeLa cells secrete into serum-free culture media, a factor (HDGF) that can reduce the serum requirement of culture normal cells and, in combination with other defined componements, replace serum completely (Lazar et al., 1986). These cell-derived GF's may serve as a low-cost source for serum substitutes, since the current sources for serum replacements exist in limited amounts and are expensive if used on a large scale.

1.6. SERA IN ANIMAL CELL MEDIA

Sera is the most important, most discussed and most problematic component in animal cells media. During more than three decades, sera has been an essential medium component with the following functions:

1. Provides nutrients (eg. sugars: glucose and fructose; amino acids, trace elements, vitamins, minerals, nucleosides and lipids).
2. Provides proteins that solubilize essential nutrients that do not dissolve readily.
3. Provides enzymes needed to convert components in the medium to a utilizable form onto a non-toxic state.
4. Provides carrier proteins for low molecular weight substances.
5. Provides protease inhibitors that neutralize proteases introduced during trypsinization, or produced by the cells.
6. Contains 'bulk' proteins that prevent non-specific adsorption of critical factors to culture vessel walls or filters.
7. Binds essential nutrients that are toxic when present in excessive amounts, and releasing them slowly in a controlled manner.
8. Provides hormones and growth factors, binding and protecting labile essential nutrients, releasing them slowly when needed.
9. Modulates physical and chemical properties of the medium (eg. viscosity, colloid osmolality and rate of diffusion) - protects cells in agitated culture.
10. Has a pH buffering function.
    Despite these advantages, there are several problems associated with the use of serum for cell cultures:
1. Serum is the most expensive component (sometimes 90% of the medium cost). When purchased in large lot sizes, it should be stored frozen (-20°C or -70°C), a process which increases its cost (refrigeration and space).
2. Being highly viscous, sera slows down the sterilization by filtration of the media.
3. From time to time, there is a shortage in world supply of sera.
4. Availability of serum in media increase the complexity of the downstream processing of the desired biological media.
5. Possible availability of contaminants: mycoplasmas and OR/and viruses in the used sera.
6. Variation in sera component concentration may occur due to one or more of the following reasons, a disadvantage which may negatively affect the production of the desired biological:
a) Effect of soil conditions on content of animal feed.
b) Kind of feed given to the animal donating serum.
c) Contaminants found in the food, such as teratogens from poisonous plants, herbicides and pesticides, heavy metals, alkaloids, aflatoxins.
d) Season of the year.
e) Stress on animal at time of slaughter.
f) Time of day of blood collection.
g) Size of production lot.
h) Substances released as a result of fetal infection (eg.: adventitious microbial contaminants, antibodies, interferons).
i) Geographic location of the animals.

Various types of sera are used as one of the main components in media: human, calf, polyethylene glycol (PEG)-treated calf, dialyzed calf, newborn calf, fetal bovine (FBS) and horse sera are all used. FBS (also called fetal calf serum, FCS), is the most expensive and the most used serum in animal cell cultures. The fact that fetal sera (mainly FBS) are the most usable ones, is due to the fact that it contains much higher concentrations of some components than maternal or other adult sera contain. Examples are given in Waymouth's review (Weymouth, 1984).

1.7. SERUM SUBSTITUTION IN MEDIA

The attempts to substitute the serum from the media, partially or completely, have targeted into one or two of the following four approaches:

1.7.1. Sera Processing Types. The basic idea is to process bovine serum so that the treated serum will substitute the FBS successfully. The reported method of processing bovine sera are the following:
a)   Pretreatment of the serum with PEG 6000 (Inglet et al., 1975). After an overnight storage of the serum (at 4°C) with PEG, the formed precipitate (contains the lipoprotein and the immunoglobulin fractions) is removed by centrifugation. PEG-treated bovine serum successfully substituted the FBS in the lymphoblastoid interferon production systems (Mizrahi, 1981).
b)   Processing bovine serum by filtration through depth filters and microporous filters. The processed serum was successfully used as a substitute to FBS in mouse myeloma and hybridoma cell cultures (Maldonado and fulbright, 1984).
c)   Dialysed serum was also found to be able to support growth of animal cells (Klinman and McKean, 1981).

1.7.2. Partial substitution of serum in non-defined or defined medium.

a)   Partial substitution of the serum with low cost nitrogen sources (eg. proteins, peptones): enzymatic or acid digests of low-cost peptones, such as: Primatone RL (Mizrahi, 1977) (a product of Sheffield, Norwich, N.Y.), Primatone HS (unpublished data): Tryptose Phosphate Broth (TPB) (Reuveny et al., 1980); lactalbumin hydrolyzate (Reuveny et al., 1982); dialyzable bacto peptone (Anborski and Moskowitz, 1968); milk products (Steiner and Klagsbrun, 1981; Fassolitis et al., 1981); colostrum (Klagsbrun, 1980), bovine serum albumin (Zoon et al., 1979), yolk emulsion (Roder, 1982), and vegetable proteins (Mizrahi and Shahar, 1977; Murakami et al., 1982a).
The use of Primatone RL or TPB were successful in the medium used for growth of Namalva cells in the production of human lymphoblastoid interferon (Reuveni et al., 1980a).
b)   Partial substitution of the serum with synthetic polymers: methylcellulose (Bryant, 1969); carboxymethylcellulose; hydroxyethyl-

starch, dextran; polyvinyl pyrrolidine (Mizrahi and Moore, 1970); α-cyclodextrin (Yamana et al., 1982); β-cyclodextrin (Ohmori and Yamatoma, 1987) and Pluronic polyols (Mizrahi, 1975), have been used as protective agents in low serum media. An interesting review on serum and its substitutes was written by Taylor (1974).

1.7.3. "Low-Protein" Serum-Free Media. A media where the serum was substituted with purified proteins, hormones and GFS is called "low-protein" serum-free medium.

An enormous number of papers and reviews are published on this subject, mainly for cultivation of hybridomas, human diploid fibroblasts and lymphoblastoid cell lines. The variation in the reported media are mainly due to the differences in the nutritional requirements of the cultivated cells and also on the researchers theoretical and practical approach to develop media. Albumin, insulin, transferrin and selenite are employed in most reported media.

1.7.4. Chemically Defined Media. The ingredients in these media are of known molecular structure and contain no proteins. The publication of White (1946), should be recognized as the earliest detailed account of attempts to develop defined culture medium. Since White, there has been a vast number of publications on chemically defined media. Although published, in practice chemically defined media are limited in application in large scale cultivation of animal cells and production of biologicals.

The main problem in using serum-free media in a large scale system, is their high cost. The price of serum-free media was estimated to be £20 - 30 per liter (Griffiths, 1986). In many cases, cells that grow at laboratory scale in developed serum-free chemically defined media lose their ability to grow in the developed media when cultured in agitated cultures at large scale after several passages. Therefore, "low-protein serum-free media" are the most recomended media for production of biologicals.

Today, a long list of commercially available serum-free media are available. Murakami in his comprehensive review on serum free media for cultivation of hybridomas summarized the list of these media (Murakami, 1989).

Selected reviews and articles which deal with low protein serum-free media, or chemically defined media are the following: (Morgan et al., 1950; Higuchi, 1973; Pirt and Lambert, 1976; Higuchi, 1977; Katsuta and Takaoka, 1977; Kay, 1978; Guibert and Isacove, 1979; Honma et al., 1979; Chang et al., 1980; McHugh et al., 1983; Weinstein, 1983; Barnes et al., 1984; Waymouth, 1984; Kovar and Franek, 1984; Walpe, 1984; Hajiwara et al., 1985; Lambert and Birch, 1985; Kawamoto et al., 1986; Bartal et al., 1986; Tharakan et al., 1986; Shive et al., 1986; Lipson, 1986; Jou et al., 1986, Barnes, 1987; Cole et al., 1987; Kovar and Franek, 1987; Silbermann et al., 1987; Murakami, 1989; Clark and Chick, 1990; Hoover and Martin, 1990; Miyazaki et al., 1990).

Detailed composition of selected and commonly used media are given by Jayme and Blackman (1985) and Freshney (1984).

TABLE 1. Hormones and growth factors present in serum('(0

Name	Source	Structure	Effects	References
Insulin	Pancreas	Protein (2 chains) M.W.=5734	-Stimulates protein and lipid synthesis -Potentiates other GF's -Promotes transport of nutrients across cell membrane	B.V. Howard et al (1979)
Somatomedins A,B,C(insulin-like growth factor)	Liver	Protein M.W.of A=7600 M.W.of B=5000 M.W.of C=7600	-Stimulates cell division -Insulin-like activities	R.W. Pierson and H.M. Temin (1972)
Platelet-derived growth factor (PDGF)	Platelets	Glycoprotein (2 chains) M.W.=28.000-31.000	-Stimulates cell division -Elicits a wide range cellular responses	C.D.Stiles (1983) R.Ross et al. (1985)
Epidermal growth factor (EGF)	Mouse submaxil-glands and human urine	Protein(1 chain) M.W.=6100	-Stimulates cell division -Elicits a wide range of cellular responses	R.M.Kris et al. (1985); S.Cohen (1987)
Fibroblast growth factor (FGF) basic and acidic	Hypothalamus retina,kidney pituitary brain	Protein(1 chain) basic M.W.=16.400 acidic M.W.=15.800	-Stimulates mesoderm cell division	D.Gospodarowicz et al.(1976); D.Gospodarowicz et al.(1978)
Liver cell growth factor (GHL)	Plasma	Tripeptide	-Stimulates growth and maintains cell viability -Acts synergistically with Cu^{++} and Fe^{++}	L.Pickart and M.M.Thaler (1973)
Transforming growth factor-B (B-TGF)	Kidney platelets	Protein(2 chains) M.W.=25.000	-Stimulates cell division -Inhibits cell division	G.J. Todaro et al (1980)
Cortisol (hydrocortison)	Adrenals	Steroids M.W.=362	-Stimulates protein, carbohydrate and lipid synthesis	R.P.Bunge et al. (1982)
Fibronectin	Plasma	Glycoprotein (2 subunits) M.W.=440.000	-Stimulates cellular adhesion	R.O.Hynes (1986)

TABLE 1 (continued)

Name	Source	Structure	Effects	References
Fetuin	Serum	Protein mix	-Promotes attachment, spreading and growth of cells	H.W.Fisher et al. (1958);J.R.Florini and S.B. Roberts (1979)
Transferrin	Serum	Protein M.W.=80.000	-Facilitates iron transport through cell membrane	J. Kovar and F.Franek (1984)
Serum spreading factor (SF)	Serum	Glycoprotein M.W.=65.000-78.000	-Promotes attachment and spreading of cells	D.W.Barnes et al. (1983)

1.9. References

Adlercreutz, P. and Mattiasson, B. (1982) 'Oxygen supply to immobilized cells', Eur. J. Appl. Microbiol. Biotechnol. 16, 165-170.

Allegra, J.C. and Lippman, M.E. (1978) 'Growth of a human breast cancer cell line in serum-free hormone-supplemented medium'. Cancer Res. 38, 3823-3828.

Anborski, R.L. and Moskowitz, M. (1968) 'The effects of low molecular weight materials derived from animal tissues on the growth of animal cells in vitro'. Exp. Cell Res. 53, 117-128.

Aronowitz, J.L. (1981) 'How pure is 'pure'? in C. Waymouth, R.G. Ham and P.J. Chapple (eds.), The Growth Requirements of Vertebrate Cells. Cambridge University Press, N.Y. pp. 82-93.

Barnes, D. and Sato, G. (1980) 'Methods for growth of cultured cells in serum-free medium', Anal. Biochem. 102, 255-270.

Barnes, D.W., Silnutzer, J., See, C. and Shaffer, M. (1983) 'Characterization of human serum spreading factor with monoclonal antibody. Proc. Natl. Acad. Sci. USA 80, 1362-1366.

Barnes, D. (1984) 'Attachment factors in cell culture', in J.P. Mather (ed.), Mammalian Cell Culture. The Use of Serum-Free Hormone Supplemented Media, Plenum Press, N.Y. and London, pp. 195-237.

Barnes, D.W., Sirbasku, D.A. and Sato, G.H. (1984) 'Cell culture methods for molecular and cell Biology, Alan Liss, N.Y. Volumes 1-4.

Barnes, D. (1987) 'Serum-Free Animal Cell Culture', Biotechniques 5, 534-542.

Bartal, A.H., Fert, C. and Hirshaut, Y. (1986) 'The use of insulin-enriched culture medium and human-human hybridoma formation: A preliminary study', Acta Haematol. 76, 50-53.

Bettger, W.J. and Ham, R.G. (1982) 'The Nutritional Requirement of Cultured Mammalian Cells', in H.H. Draper (ed.), Advances in Nutritional Research. Plenum Press, N.Y. pp. 249-286.

Bryant, J.C. (1969) 'Methylcellulose effect on cell proliferation and glucose utilization in chemically defined medium in large stationary culture', Biotech. Bioeng. 11, 155-179.

Bunge, R.P., Bunge, M.B., Carey, D.J., Conbrooks, C., Higgins, D. Johnson, M.I., Kleinschmidt, D.C., Wood, P.M., Iacovitti, L. and Moya, F. (1982) 'Functional expression in primary nerve tissue cultures maintained in defined medium', in G.H. Sato, A.B. Pardee and D.A. Sirbasku (eds.), Growth of Cells in Hormonally Defined Media, Book B. Cold Spring Harbor, N.Y. pp. 1017-1031.

Carpenter, G. and Cohen, S. (1979) 'Epidermal growth factor', Ann. Rev. Biochem. 48, 193-216.

Chang, T.H., Steplewski, Z. and Koprowski, H. (1980) 'Production of monoclonal antibodies in serum free medium', J. Immunol. Methods 39, 369-375.

Chen, H.W. and Kandustsch, A.A. (1981) 'Cholesterol requirements for cell growth: Endogenous synthesis versus exgenous sources', in C. Waymouth, R.G. Ham and R.J. Chapple (eds.), The Growth Requirements of Vertebrate Cells In Vitro, Cambridge University Press, N.Y., pp. 327-342.

Cherington, P.V. (1984) 'Regulation of fibroblast growth by multiple growth factors in serum-free media', in J.P. Mather (ed.), Mammalian Cell Cultures. The Use of Serum-Free Hormone-Supplemental Media, Plenum Press, N.Y. and London, pp. 17-58.

Clark, S.A. and Chick, W.L. (1990) 'Islet cell culture in defined serum-free medium', Endocrynol. 126, 1895-1903.

Cohen, S. (1987) 'Epidermal growth factor. In Vitro Cellular & Developmental Biology 23, 239-246.

Cole, S.P.C., Vreeken, E.H., Mirski, S.E.L. and Campling, B.G. (1987) 'Growth of Human X human hybridomas in protein-free medium supplemented with ethanolamine', J. Immunol. Methods 97, 29-35.

Coppola, R.J. (1984) 'Ultrapure water for tissue culture. International Biotechnol. Lab. 2, 40-43.

Cristofalo, V.J. and Kritchevsky, D. (1965) 'Growth and glycolysis in human diploid cell strain W1-38', Proc. Soc. Exp. Biol. Med. 118, 1109-1113.

Dicker, P. and Rozengurt, E. (1978) 'Stimulation of DNA synthesis by tumor promoter and pure mitogenic factors, Nature 276, 723-726.

Dubin, I.N., Czernobilsky, B. and Herbst, B. (1965) 'Effect of albumin fraction and linoleic acid on growth of macrophages in tissue culture', J. Nat. Cancer Inst. 34, 43-51.

Eagle, H. (1955) 'Nutrition needs of mammalian cells in tissue culture. Science 122, 501-504.

Earle, W.R., Schilling, E.L., Stark, T.H., Straus, N.P., Brown, M.F. and Shelton, E. (1943) 'Production of maignancy in vitro. IV. The mouse fibroblast cultures and changes seen in the living cells. J. Natl. Cancer Inst. 4, 165-212.

Eisinger, M., Lee, J.S., Hefton, J.M., Danzykiewicz, A., Chiao, J.W. and Deharven, E. (1979) 'Human epidermal cell cultures-growth and differentiation in the absence of demand components or medium supplements', Proc. Nat. Acad. Sci. USA, 76, 5340-5342.

Fassolitis, A.C., Novelli, R.M. and Larkin, E.P. (1981) 'Serum substitute in epithelial cell culture media: Nonfat dry milk filtrate', Appl. Environ. Microbiol. 42, 200-203.

Fisher, H.W., Puck, T.T. and Sato, G. (1958) 'Molecular growth requirements of single mammalian cells: The action of fetuin in promoting cell attachment to glass', Proc. Natl. Acad. Sci. USA 44, 4-10.

Florini, J.R. and Roberts, S.B. (1979) 'Serum-free medium for the growth of muscle cells in culture', In Vitro 15, 983-992.

Freshney, R.I. (1984) 'Culture of Animal Cells'. Alan R. Liss, N.Y.

174

Good, N.E., Winget, G.D., Winter, W., Connolly, T.N., Izawa, S. and Singh, R.M/M/ (1966) 'Hydrogen ion buffers for biological research', Biochemistry 5, 467-477.

Gospodarowicz, D. (1975) 'Purification of a fibroblast growth factor from bovine pituitary', J. Biol. Chem. 250, 2515-2520.

Gospodarowicz, D. and Moran, J.S. (1976) 'Growth factors in mammalian cell culture, Ann. Rev. Biochem. 45, 531-558.

Gospodarowicz, D., Bialecki, M. and Greenburg, G. (1978) Purification of the fibroblast growth factor activity from bovine brain', J. Biol. Chem. 253, 3736-3743.

Griffiths, B. (1986) 'Can cell culture medium costs be reduced? Strategies and Possibilities'. Trends Biotechnol. 4, 268.

Groelke, J.W., Baseman, J.B. and Amos, H. (1979) 'Regulation of the G_1 S phase transition in chick embryo fibroblasts with α-keto and L-alanine. J. Cell Physiol. 101, 391-398.

Guilbert, L.J. and Isacove, N.N. (1976) 'Partial replacement of serum by selenite, transferrin, albumin and lecithin in haemopoietic cell cultures', Nature 263, 594-595.

Hagiwara, H., Ohtake, H., Yuasa, H., Nagao, J., Nonaka, S., Chigiri, E. and Aotsuka, Y. (1985) 'Proliferation and antibody production of human X human hybridoma in serum-free media, in H, Murakami, I, Yamame, D.W. Barnes, J.P. Mather, I, Haysashi and G.H. Sato (eds.), Growth and Differentiation of Cells in Defined Environment, Springer-Verlag, Berlin and N.Y., pp. 1-16.

Ham, R.G. (1963) 'Albumin replacement by fatty acids in clonal growth of mammalian cells', Science 140, 802-803.

Ham, R.G. and McKeehan, W.L. (1979) 'Media and growth require-ments', Methods in Enzymol. 58, 44-93.

Ham, R.G. (1981) 'Survival and growth requirements of non-transformed cells', Handbook Exp. Pharmacol. 57, 13-88.

Ham, R.G. (1984) 'Formulation of basal nutrient media', in D.W. Barnes, Sirbasku, D.A. and Sato, G.H. (eds.), Cell Culture Methods for Molecular cell Biology 1, Alan Liss, N.Y. pp. 3-21.

Hanks, J.H. and Wallace, R.E. (1949) 'Relation of oxygen and temperature in the preservation of tissues of refrigeration', Proc. Soc. Exp. Biol. Med. USA 71, 196-199.

Higuchi, K. (1970) 'Requirements for cholesterol, hematin and lecithin for optimal growth of a porine kidney cell line', In vitro 6, 239.

Higuchi, K. (1978) 'Cultivation of animal cells in chemically defined media, A review', Appl. Microbiol. 16, 111-136.

Higuchi, K. (1977) 'Cultivation of mammalian cell lines in serum-free chemically defined medium. Methods in Cell Biol. 14, 131-143.

Hoare, SW (1987) 'Equipment for culturing mammalian and plant cells', Intern. Labmate, (June Issue), 13-15.

Honma, Y., Kasukabe, T., Okabe, J. and Hozumi, M. (1979) 'Replacement of serum by insulin, transferrin, albumin, phosphatidyl choline, cholesterol and some trace elements in cultures of mouse myeloid leukemia cells sensitive to inducers of differentiations', Exp. Cell Res. 124, 421-428.

Hoover, C.S. and Martin, R.L. (1990) 'Antibody production and growth of mouse hybridoma cells in Nutridoma media supplements', Biotechniques 8, 76-80.

Howard, B.V., Mott, D.M., Field, R.M. and Bennet, P.H. (1979) 'Insulin stimulation of glucose entering in cultured human fibroblasts', J. Cell. Physiol. 101, 129-138.

Huang, J.S. and Huang S.H. (1985) 'Role of growth factors in oncogenesis: Growth factor-protein oncogene pathways of metogenesis', in Ciba Foundation Symposium 116, Growth Factors in Biology and Medicine, Pitman, London, pp. 46-65.

Hunter, T. (1984) 'The proteins of oncogenics', Sci. Am. 251, 60-69.

Hynes, R.O. (1986) 'Fibronectins', Sci. Am. 254, 32-41.

Imamura, T, Crespl, C.L. and Brumengraber, H. (1980) 'Utilization of carbohydrate by Madian Darby canine kidney (MDCK) cells in high density suspension of dextran microcarriers', Fed. Proc. Fed. Am. Soc. Exp. Biol. 39, 2145-2149.

Inglot, A.D., Godzinska, H. and Chudzio, T. (1975) 'Use of polyethylene glycol-treated calf serum for cell cultures in virus and interferon studies', Acta Virol. 19, 250-254.

Jayme, D.W. and Blackman, K.E. (1985) 'Culture media in propagation of mammalian cells, viruses and other biologicals, in A. Mizrahi and A.L. van Wezel (eds.), Advances in Biotechnological Processes 5, Alan R. Liss, N.Y. pp. 1-30.

Jenkin, H.M. and Anderson, L.E. (1970) 'The effect of oleic acid on the growth of monkey kidney cells (LLC-MK$_2$), Exp. Cell Res. 59, 6-10.

Jou, T.C., Liu, S.M., Chou, M.Y., Li, S.Y. and Jou, M.J. (1986) 'Growth response of rat brain astrocytes cell line in a serum-free, chemically defined medium', J. Formasan Med. Ass. 85, 657-666.

Kans-Sueko, T. and Errick, J.E. (1981) 'Effects of phospho-ethanolamine and ethanolamine on growth of mammary carcinoma cells in culture', Exp. Cell Res. 136, 137-145.

Kaplan, P.L., Anderson, M. and Ozanne, B. (1982) 'Transforming growth factor's production enables cells to grow in the absence of serum. An autocrine system', Proc. Natl. Acad. Sci. USA 79, 485-489.

Katsuta, H. and Takaoka, T. (1977) 'Improved synthetic media suitable for tissue culture of various mammalian cells', Methods in Cell Biol. 14, 145-158.

Kawamoto, T., Sato, D., McClure, D.B. and Sato, G.H. (1986) 'Serum-free medium for the growth of NS-1 hybridomas', Methods in Enzymol. 121, 268-277.

Keay, L. (1978) 'The growth of L-cells and Vero cells on an autoclavable MEM-peptone medium', Biotechnol. Bioeng. 19, 399-411.

Kelley, W.N. (1972) 'Purine and pyrimidine metabolism of cells in culture', in G.H. Rothblat and V.J. Cristofalo (eds.), Growth Nutrition and Metabolism of Cells in Culture, 1, Academic Press, N.Y. pp. 211-255.

King, M.E. and Spector, A.A. (1981) 'Lipid metabolism in cultured cells', in C. Waymouth, R.B. Ham, and P.J. Chapple (eds,). Growth Requirements of Vertebrate Cells, Cambridge University Press, N.Y. pp. 293-312.

Klagsburn, M. (1980) 'Bovine colostrum supports the serum-free proliferation of epithelial cells but not the fibroblasts in long-term culture', J. Cell Biol. 84, 808-814.

Kleinman, H.K., Klebe, R.J. and Martin, G.R. (1981) 'Role of collagenous matrices in the adhesion and growth of cells', J. Cell Biol. 88, 473-485.

Klinman, D.M. and McKearn, T.J. (1981) 'Dialyzable serum components can support the growth of hybridoma cell lines in vitro', J. Immunol. Methods 42, 1-9.

Kovar, J. and Franek, F. (1984) 'Serum-free medium for hybridoma and parental myeloma cell cultivation: A novel composition of growth supporting substances', Immunol. Letters 7, 339-345.

Kovar, J. and Franek, F. (1987) 'Iron compounds at high concentrations enable hybridoma growth in a protein-free medium', Biotechnol. Letters 9, 259-264.

Kris, R.M., Libermann, T.A., Avivi, A. and Schlessinger, J. (1985) 'Growth factors, growth-factor receptors and oncogenes', Biotechnology 3, 135-140.

Lambert, K.J. and Birch, J.R. (1985) 'Cell growth media', in R.E. Spier and J.B. Griffiths (eds.), Animal Cell Biotechnol. 1, Academic Press, N.Y. pp. 85-121.

Lazar, A., Reuveny, S., Traub, A., Minai, M., Grosfeld, H., Feinstein, S., Gez, M. and Mizrahi, A. (1982) 'Factors affecting the large scale production of human lymphoblastoid interferon', Develop. biol. Stand. 50, 167-171.

Lazar, A., Wilson, R. and Spier, R.E. (1986) 'Growth promoting materials derived from HeLa cell culture supernatants', Enzyme Microb. Technol. 9, 295-299.

Leibovitz, A. (1963) 'The growth and maintenance of tissue cell cultures of free gas exchange with the atmosphere', Am. J. Hyg. 78, 173-180.

Leof, E.G., Wharton, W., van Wyk, J.J. and Pledger, W.J. 1980) 'Epidermal growth factor and somatomedin C control G_1 progression of competent BALB/C3T3 cells: Somatomedin C regulates commitment to DNA synthesis', J. Cell Biol. 87, 5a.

Lipson, S.M. (1986) 'Applications of a serum-free medium in the growth and differentiation of human blood lymphocytes', Diagn. Microbiol. Infect. Dis. 4, 203-214.

Lotan, R. (1980) 'Effect of vitamin A and its analogs (retinoids) on normal and neoplastic cells', Biochem. Biophys. Acta 605, 33-91.

Maldonado, R.L. and Fulbright, J.G. (1984) 'Processed serum: A consistent growth support for hybridomas', American Biotechnol. Lab. March Issue 5-7.

McHugh, Y.E., Walthall, B.J. and Steiner, K.S. (1983) 'Serum-free growth of murine and human lymphoid and hybridoma cell lines', Biotechniques June/July Issue 72-77.

McKeehan, W.L. and McKeehan, K.A. (1979) 'Oxocarboxylic acids, pyridine nucleotide-linked oxidoreductase and serum factors in regulation of cell proliferation', J. Cell Physiol. 101, 9-16.

McKeehan, W.L., Hamilton, W.G. and Ham, R.G. (1976) 'Selenium is an essential trace nutrient for growth of WI-38 diploid human fibroblasts', Proc. Natl. Acad. Sci. USA 73, 2023-2027.

McLimans, W.F., Crouse, E.J., Tunnah, K.V. and Moore, G.E. (1968) 'Kinetic of gas diffusion in mammalian cell culture systems', Biotechnol. Bioeng. 10, 725-740.

Miyazaki, M., Bai, L. and Sato, J. (1990) 'Selection of medium for serum-free primary culture of adult rat hepatocytes', Acta Med. Okayama 44, 9-12.

Mizrahi, A. and Moore, G.E. (1970) 'Partial substitution of serum in hematopeitic cell line media by synthetic polymers', Appl. Microbiol. 19, 906-910.

Mizrahi, A. (1975) 'Pluronic polyols in human lymphoblastoid cell line cultures', J. Clin. Microbiol. 2, 11-13.

Mizrahi, A. (1977) 'Primatone, R.L. in mammalian cell culture media', Biotech. Bioeng. 19, 1557-1561.

Mizrahi, A. and Shahar, A. (1977) 'Partial replacement of serum by vegetable proteins in BHK culture medium', J. biol. Stand. 5, 327-332.

Mizrahi, A., Reuveny, S. Traub and Minai, M. (1980) 'Large scale production of human lymphoblastoid (Namalva) interferon. I. Production of crude interferon', Biotechnol. Letters 2, 267-271.

Mizrahi, A. (1981) 'Production of human lymohoblastoid (Namalva) interferon', Methods in Enzymol. 78, 54-68.

Mizrahi, A. (1986) 'Production of biologicals from animal cells - An overview', Proc. Biochem. (August Issue) 108-112.

Mizrahi, A. (1986a) 'Biologicals from animal cells in culture', Biotechnology 4, 123-127.

Mizrahi, A. and Lazar, A. (1988) 'Media for cultivation of animal cells: An overview', Cytotechnol. 1, 199-214.

Morgan, J.F., Morton, J.H. and Parker, R.C. (1950) 'Nutrition of animal cells in tissue culture I. Initial studies on a synthetic medium', Proc. Soc. Exp. Biol. Med. USA 73, 1-9.

Morton, H.C. (1970) 'A survey of commercially available tissue culture media', In Vitro 6, 89-108.

Murakami, H., Masui, H., Sato, G.H., Sneska, N., Chow, T.P. and Kano-Sueska, T. (1982) 'Growth of hybridoma cells in serum-free medium: Ethanolamine is an essential component', Proc. Natl. Acad. Sci. USA 79, 1158-1162.

Murakami, H., Masui, H. and Sato, G.H. (1982a) 'Suspension culture of hybridoma cells in serum-free medium: Soybean phospholipids as the essential components', Cold Spring Harbor Conferences on Cell Proliferation, Growth of Cells in Hormonally Defined Media, Vol. 9, 711-715.

Murakami, A. (1989) 'Serum-free media used for cultivation of hybridomas', in A. Mizrahi (ed.), Advances in Biotechnological Processes, Alan R. Liss, N.Y. 11, pp. 107-141.

Nielsen, F.H. (1981) 'Consideration of trace elements requirements for preparation of chemically defined media', in C. Waymouth, R.G. Ham and R.J. Chapple (eds.), The Growth Requirements of Vertebrate Cells In Vitro, Cambridge University Press, N.Y. pp. 68-81.

Ohmori, H. and Yamatoma, I. (1987) 'β-Cyclodextrin as a substitute for fetal calf serum in the primary antibody response in vitro', Eur. J. Immunol. 17, 79-83.

Perlman, D. (1979) 'Use of antibiotics in cell culture media', Methods in Enzymol. 58, 110-116.

Pickart, L. and Thaler, M.M. (1973) 'Tripeptide in human serum which prolongs survival of normal liver cells and stimulates growth in neoplastic liver', Nature, New Biol. 243, 85-87.

Pierson Jr. R.W. and Temin, H.M. (1972) 'The partial purification from calf serum of a fraction with multiplication - stimulating activity for chicken fibroblasts in cell culture and with non-suppressible insulin-like activity', J. Cell Physiol. 79, 319-330.

Pirt, S.J. and Lambert, K. (1976) 'Towards a chemically defined medium for the growth of normal human diploid cells', Develop. biol. Stand. 37, 63-66.

Ratafia, M. (1987) 'Mammalian cell culture: Worldwide activities and market', Biotechnology 5, 692-694.

Reitzer, L.Z., Wice, M.B. and Kennell, D. (1979) 'Evidence that glutamine, not sugar, is the major energy source for cultured HeLa cells', J. Biol. Chem. 256, 2669-2676.

Reuveny, S., Bino, T., Rosenberg, H., Traub, A. and Mizrahi, A. (1980) 'Pilot plant production of human lymphoblastoid interferon', Develop. biol. Stand. 46, 281-288.

Reuveny, S., Lazar, A., Minai, M., Feinstein, S., Grosfeld, H, Traub, A. and Mizrahi, A. (1982) 'Large-scale production of human (Namalva) interferon', Ann. Virol. 133E, 191-199.

Rizzino, A. and Sato, G. (1978) 'Growth of embryonal carcinoma cells in serum-free medium', Proc. Natl. Acad. Sci. USA 76, 1844-1848.

Rizzino, A., Rizzino, H. and Sato, G. (1979) 'Defined media and the determination of nutritional and hormonal requirements of mammalian cells in culture', Nut. Rev. 37, 369-378.

Roder, A. (1982) 'Development of a serum-free medium for cultivation of mixed cells', Naturwissenschaften 69, 92-93.

Ross, R., Bowen-Pope, D.F. and Raines, E.W. (1985) 'Platelet-derived growth factor: Its potential roles in wound healing, atherosclerosis, neoplasia and growth and development', in Ciba Foundation Symposium, Growth Factors in Biology and Medicine, Pitman, London, 116, pp.98-112.

Rothblatt, G.H. and Kritchersky, D. (1987) 'Lipid metabolism in tissue culture cells', Wistar Inst. Symp. Monogr. #6, Wistar Inst. Anat. Biol. Philadelphia, PA.

Sato, G., Fisher, H.W. and Puck, T.T. (1957) 'Molecular growth requirements of single mammalian cells', Science 126, 961-964.

Scher, C.D., Shepard, R.O., Antoniades, H.N. and Stiles, C.D. (1976) 'Platelet-derived growth factor and the regulation of the mammalian fibroblast cell cycle', Biochem. Biophys. Acta 560, 217-241.

Shive, W., Pinkerton, F., Humphreys, J., Johnson, M.M., Hamilton, G.W. and Matthews, K.S. (1986) 'Development of a chemically defined serum and protein-free medium for growth of human peripheral lymphocytes', Proc. Nat. Acad. Sci. USA 83, 9-13.

Shooter, R.A. and Gey, G.O. (1952) 'Studies of the mineral requirements of mammalian cells', Br. J. Exp. Pathol. 33, 98-103.

Silberman, M., Tenenbaum, H., Livne, E., Leapman, R., von der Mark, K. and Reddi, H.H. (1987) 'The in vitro behavior of fetal condylar cartilage in serum free hormone-supplemented medium', Bone 8, 117-126.

Smith, J.R. (1981) 'The fat-soluble vitamins, in C. Waymouth, R.G. Ham and P.J. Chapple (eds.), The Growth Requirements of Vertebrate Cells In Vitro', Cambridge University Press, N.Y. pp. 343-352.

Steiner, K.S. and Klagsbrun, M. (1981) 'Serum-free growth of normal and transformed fibroblasts in milk: Differential requirements of fibronectin', J. Cell Biol. 88, 294-300.

Stiles, C.D., Capone, G.T., Scher, C.D., Antoniades, H.N., van Wyk, J.J. and Pledger, W.J. (1979) 'Dual control of cell growth by somatomedins and platelet-derived growth factor.', Proc. Natl. Acad. Sci. USA 76, 1279-1283.

Stiles, C.D. (1983) 'The molecular biology of platelet derived growth factor'.

Todaro, G.J., Fryling, C. and Dellarco, J.E. (1980) 'Transforming growth factors produced by certain human tumor cells: Polypeptides that interact with epidermal growth factor receptor', Proc. Natl. Acad. Sci. USA 77, 5258-5262.

Taylor, W.O. (1974) 'Feeding the Baby' - Serum and other suppliments to chemically defined medium', J. Natl. Cancer Inst. 53, 1449-1457.

Tharakan, J.P., Lucas, A. and Chan, P.C. (1986) 'Hybridoma growth and antibody secretion in serum-supplemented and low protein serum-free media', J. Immunol. Methods 94, 225-236.

Velez, D., Reuveny, S., Miller, L. and MacMillan, J.D. (1986) Kinetics of monoclonal antibody production in low serum growth medium', J. Immunol. 86, 45-52.

180

Walpe, S.D. (1984) 'In vitro immunization and growth of hybridomas in serum-free medium', in J.P. Mather (ed.), Mammalian Cell Culture, Plenum Press, N.Y. pp. 103-127.

Weinstein, R. (1983) 'Serum-free culture of animal mammalian cells', Biotechniques June/July Issue, 61-64.

Waymouth, C. (1984) 'Preparation and use of serum-free culture media, in D.W. Barnes, D.A. Sirbasku and G.H. Sato (eds.), Cell Culture Methods for Molecular and Cell Biology, Alan Liss, N.Y., 1, pp. 23-68.

White, P.R. (1946) 'Cultivation of animal tissues in vitro in nutrients of known constitution', Growth 10, 281-289.

Wice, B.M., Reitzer, L.J. and Kennell, D. (1981) 'The continuous growth of vertebrate cells in the absence of sugar', J. Biol. Chem. 256, 7812-7819.

Yamane, I., Kam, M., Minamoto, Y. and Amatsuji, Y. (1982) 'Alpha-cyclodextrin: A partial substitute for bovine serum albumin in serum-free culture of mammalian cells', in G.H. Sato, A.B. Pardee and D.A. Sirbasku (eds.), 'Growth of Cells in Hormonally Defined Media, Cold Spring Harbor, N.Y. pp. 87-92.

Zielke, H.R., Sumbilla, C.M., Sevdalian, D.A., Hawkins, R.L. and Ozard, P.T. (1980) 'Lactate: A major product of glutamine metabolism by human diploid fibroblasts', J. Cell Physiol. 104, 433-441.

Zoon, K.C., Bridgen, P.J. and Smith, M.E. (1979) 'Production of human lymphoblastoid interferon by Namalva cells cultured in serum-free media', J. gen Virol. 44, 227-229.

CHARACTERIZATION OF A HUMAN DERIVED MACROPHAGE-LIKE CELL LINE, U-M

K. Shinohara,(1) Z-L. Kong,(2) and H. Murakami(3)
(1) National Food Research Institute, The Ministry of Agriculture, Forestry, and Fisheries, Kannondai, Tsukuba, Ibaraki 305, Japan
(2) Institute of Food Chemistry, Faculty of Agriculture, Kyushu University, Hakozaki, Fukuoka 812, Japan
(3) Graduate School of Genetic Resources Technology, Faculty of Agriculture, Kyushu University, Hakozaki, Fukuoka 812, Japan

Macrophage cells are immunologically important, producing various kinds of cytokines such as interleukin-1 (IL-1), interferon, tumor necrosis factor (TNF-alpha), prostaglandins, and active oxygens, and play a important functions in the cell-mediated immunity and the biodefensive functions such as tumorcidal and microcidal functions. Studies concerning the functions of macrophage cells have been exclusively done using in vivo macrophage cells in mouse and rabbit or established cell lines derived from mice such as J774A.1, RAW264.1 and P388D1 cells. However, no human-derived cell line has been established. Establishment of a human-derived macrophage that can grow in the serum-free medium is useful not only for study on the in vitro functions of human macrophage cells but also for production of such useful substances. In the preceding study, we established a human-drived macrophage-like cell line that proliferates well in serum-free medium, from a human histiocytic lymphoma cell line (U-937) by treatment with 12-o-tetradecanoyl phorbol 13-acetate (TPA)(1). In this study, we examined some properties of the established macrophage-like cells, compared with those of the original U-937 cells.

Materials and Methods

Materials
As monocyte/macrophage stimulants, retinoic acid(2), cyclo-heximide, sodium butyrate, a Vitamin D_3 derivative ($1,25(OH)_2D_3$), lipopolysaccharide (LPS), TPA, zymozan A, calcium-ionophore A23187, concanavalin A (Con A), phytohemaglutinine (PHA-L type) and DMSO were used. As the

R. Sasaki and K. Ikura (eds.), Animal Cell Culture and Production of Biologicals, 181–186.
© 1991 Kluwer Academic Publishers.

inhibitors, H-7, H-8, tetracaine, W-7, ethyleneglycol-bis-N,N,N,N'-tetraacetic acid (EGTA), and ethylenediamine-tetraacetic acid (EDTA) were used.

Cells and medium

The macrophage-like cells (U-M) were obtained by the treatment of U-937 with TPA in enriched eRDF medium containing FCS. These cells were cultured in serum-free enriched RDF medium supplemented with insulin (I: 5 µg/ml), transferrin (T: 10 µg/ml), ethanolamine (E: 1.5 pg/ml), and selenite (S: 4.3 µg/ml)(eRDF-ITES medium).

Results

Effects of ITES on the proliferation of U-M cells

The established U-M cells proliferate well in eRDF medium supplemented with ITES. The effects of ITES on the proliferation of U-M cells were examined by comparison with that of U-937 cells. As shown in Table 1, transferrin (T) promoted the proliferation of U-M and U-937 cells in eRDF medium most markedly among these growth factors. T also produced a higher viability of both cells. The promoting activity of T was more marked in the U-M cells than in the U-937 cells. Every supplementation of growth factors including T also promoted the proliferation of both cells. The U-M cells proliferated more than the U-937 cells in every combination of growth factors.

Table 1. Effects of ITES on Growth of U-M and U-937 Cells

ITES	Cell density (x 10^5 cells/ml)		Viability (%)	
	U-M	U-937	U-M	U-937
I (5µg/ml)	1.6	1.0	70	90
T (17.5 µg/ml)	2.6	1.8	90	93
E (10 µg/ml)	1.1	1.0	65	86
S (2.5 nm)	1.1	1.0	63	84
IT	2.9	1.9	94	93
IE	1.7	1.0	90	94
IS	1.6	1.0	91	94
TE	2.6	1.8	90	94
TS	2.6	1.8	90	94
ES	1.1	1.0	65	87
ITE	2.9	1.9	94	94
ITS	3.1	2.2	96	94
IES	1.7	1.1	92	94
TES	2.7	1.9	92	94
ITES	3.2	2.2	96	94
ITES(x2 fold)	3.2	2.2	96	95

Effects of some monocyte/macrophage stimulants or
inhibitors on U-M, U-937, HL-60 and P388D1 cells
    To characterize the properties of U-M cells, the modes
of action of some known monocyte/macrophage stimulants or
inhibitors on the cells were examined by the assays of NBT
reduction activity, morphological alteration and TNF production, compared with those on the U-937 or HL-60 cells. As
the stimulants, retinoic acid, cycloheximide, sodium
butyrate, $1,25(OH)_2D_3$), LPS, TPA, zymozan A, calcium-
ionophore A23187, Con A, PHA-L and DMSO were used, while as
the inhibitors, H-7 (protein kinase C inhibitor), H-8
(cyclo-GMP-dependent protein kinase inhibitor), W-7
(calmodulin antagonist), tetracaine HCl (protein kinase
inhibitor), EGTA and EDTA used. In these experiments, the
U-M or U-937 cells (1×10^5 cells/ml) in eRDF-ITES medium
were cultured with the stimulants or inhibitors for 2 days
in a CO_2 incubator, and NBT reducing activity, amount of
TNF-alpha, and numbers of elongated cells were measured or
counted. These results are summarized in Tables 2-4.

Table 2. Effects of Stimulators on the NBT Reduction Ability
of U-M or U-937 cells.

Stimulators	Dose	NBT reduction activity (%)	
		U-M	U-937
PBS	5 %	100	100
Retinoic acid	5 μM	120	150
Cycloheximide	0.5 μg/ml	70	80
Sodium butyrate	10 mM	80	200
$1,25(OH)_2D_3$	100 nM	350	750
LPS	0.5 μg/ml	150	300
TPA	10 nM	150	150
Zymosan A	5 mg/ml	100	350
Ionophore A23187	5 μg/ml	20	10
Con A	5 μg/ml	1700	70
PHA-L	10 μg/ml	100	100
DMSO	1 %	100	80

    The behaviors of stimulants on on the NBT reduction
activity of U-M and U-937 cells were found to vary with the
kinds of stimulants.   Con A induced the highest NBT
reduction activity toward U-M cells among the stimulants,
while it reduced the activity of U-937 cells.   Toward U-937
cells, $1,25(OH)_2D_3$ produced the NBT reduction activity most
markedly among the activators.   It also induced the
activity of U-M cells.

LPS and zymozan A induced the higher activity of U-937 cells than that of U-M cells. The NBT reducing activity of U-M cells was slightly stimulated by retinoic acid, while that of U-937 cells by retinoic acid, sodium butyrate and TPA. On the other hand, cycloheximide and Ca-ionophore A23187 evidently reduced the NBT reduction activity of U-M and U-937 cells.

The profiles of morphological alteration of U-M cells by the stimulators varied with those of U-937 (Table 3). Sodium butyrate, $1,25(OH)_2D_3$, TPA, Con A, and PHA-L caused the morphological alteration of U-M cells, while toward U-937 cells, retinoic acid, sodium butyrate, $1,25(OH)_2D_3$, and zymozan A did. U-M cells were subject to morphological alteration by Con A and their morphology was changed from the round shape to elongated and spread shape. The cells were aggregated by PHA-L. These results indicate the existence of mannose and N-acetylgluosamine residues on the surface membrane of the cells. On the other hand,

Table 3. Effects of Various Stimulators on the Morphological Alteration and TNF Production of U-M or U-937 cells.

Stimulators	Dose	Morphological alteration		TNF production	
		U-M	U-937	U-M	U-937
PBS	5 %	N	N	-	-
Retinoic acid	5 μM/ml	N	Y	+	+
Cyclohexamide	0.5 μg/ml	N	N	-	-
Sodium butyrate	10 mM	Y	Y	-	+
$1,25(OH)_2D_3$	100 nM	Y	Y	-	+
LPS	0.5 μg/ml	N	N	++	+
TPA	10 nM	Y	N	-	-
Zymosan A	5 mg/ml	N	Y	-	-
Ionophore A23187	5 μg/ml	N	N	-	-
Con A	5 μg/ml	Y	N	+	-
PHA-L	10 μg/ml	Y	N	-	-
DMSO	1 %	N	N	-	-

N : no morphological change Y : morphological change
- : no change, + : 2 fold, ++ : 3 fold, compared with the control

Table 4. Effects of Various Inhibitors on NBT Reduction Ability, Morphological Alteration and TNF Production of U-M or U-937 cells.

Dose		NBT reduction		Morphological activity (%)		TNF production alteration	
		U-M	U-937	U-M	U-937	U-M	U-937
PBS	5 %	100	100	N	N	-	-
H-7	10 μM	100	10	N	N	-	-
H-8	10 μM	100	100	N	N	-	-
W-7	80 μM	100	650	N	N	-	-
TC	20 μg/ml	100	20	N	N	-	-
EGTA	5 mM	100	100	N	N	-	-
EDTA	2 mM	100	100	N	N	-	-

N : no morphological change, - : no change, compared with the control

U-937 cells were not subject to morphological alteration by Con A and PHA-L, indicating the absence of mannose and N-acetylgluosamine residues on the surface membrane of the cells. The binding of Con A to the surface membrane of U-M cells was also confirmed. Con A did not bind to U-937 cells.

TNF-alpha production of U-M cells was increased by retinoic acid , LPS and Con A, and that of U-937 cells by retinoic acid, sodium butyrate, $1,25(OH)_2D_3$, and LPS (Table 3).

IL-1 production of U-M cells was also observed in the presence of Con A. The conditioned medium of U-M cells stimulated with Con A promoted the growth of a IL-1 dependent mouse helper T cell line (D10G4.1) in eRDF-ITES medium.

Some inhibitors had the different behavior on the induction of NBT reduction activity of U-M and U-937 cells (Table 4). H-7 and tetracaine inhibited the induction of NBT reduction activity of U-937 cells, while toward the U-M cells, they did not. On the contrary, a calmodulin antagonist (W-7) did not change any variation of the NBT reduction activity of U-M cells, but induced the activity of U-937 cells significantly. Other inhibitors such as H-8, EGTA, and EDTA caused no variation on either cell type. It was also found that these inhibitors caused no morphological alteration and TNF-alpha production of U-M and U-937 cells.

Discussion

The U-M cells established from U-937 cells proliferate well in eRDF medium supplemented with ITES. T(transferrin) promoted the proliferation of U-M and U-937 cells in eRDF medium most markedly among these growth factors. It also produced a higher viability of both cells. It was suggested that there are some differences in the glycoprotein compositions on the surface membranes between U-M and U-937 cells. Con A, is a lectin which recognizes mannose and glucose residue in the glycoprotein on the membrane of cells increased the NBT reduction activity of U-M cells markedly. The existence of binding sites on the surface membrane of U-M cells was confirmed. Con A did not bind to U-937 cells. The U-M cells were aggregated by PHA-L, which is a lymphocyte-activating agent and is known mainly to recognizes N-acetylgalactosamine residues in the glycoprotein on the membrane. These results indicate the existence of mannose and N-acetyl gluosamine residues on the surface membrane of the cells. On the other hand, U-937 cells were not subject to morphological alteration by Con A and PHA-L, indicating the absence of mannose and N-acetyl gluosamine residues on the surface membrane of the cells(3). No morphological alteration and TNF-alpha of U-M and U-937 cells occurred in the presence of inhibitors such as protein kinase inhibitors.

In the stage of differentiation processes, the protein and RNA synthesis in cells and the mitosis are generally inhibited. Cycloheximide which is a RNA synthesis inhibitor, decreased the NBT reduction activity of U-937 and U-M cells. It also caused no morphological alteration and TNF-alpha production in both cells. Sodium butyrate, which is reported to cause differentiation of U-937 cells into macrophage-like cells increased the NBT reduction activity and TNF production of U-937 cells, but did not those of U-M cells. It caused the morphological alteration of both cells. On the other hand, Ca-ionophore A23187, which is a calcium-transporting agent and stimulates the membranes of cells, decreased the NBT reduction activity and TNF production of both cells, and also caused no morphological alteration.

The established U-M cells are useful not only for the production of the cytokinines but also for the study of cell differentiation and stimulation factors of macrophages.

References
1)Kong, Z., Murakami, H. and Shinohara, K.(1990)
  'Establishment of a macrophage-like cell line derived from
  U-937, human histiocytic lymphoma, grown serum-free',
  In Vitro Cellular & Developmental Biology, 26, 949-954.
2)Olsson, I.L., Breitman, T.R.(1982)'Induction of
  differentiation of the human histiocytic lymphoma cell
  line U-937 by retinoic acid and cyclic AMP-inducing
  agents', Cancer Res., 42, 3924-3927.
3)Nilsson, K., Aderson, L.C. and Gahmberg, C.G.(1980)'Cell
  surface characteristics of human histiocytic lymphoma
  lines', Leukemia Res., 4, 271-279.

MITOGENIC ACTIVITY FROM FISH EMBRYOS AND THE GROWTH OF FISH BLASTOCYST CELLS IN CULTURE

PAUL COLLODI[1], YUTO KAMEI[2] and DAVID W. BARNES[1]
[1]Department of Biochemistry and Biophysics, Environmental Health Sciences Center, Oregon State University, Corvallis, Oregon, 97331 USA; [2]Agribusiness Department, Sapporo Breweries LTD., Tokyo 104, Japan

ABSTRACT. Advantages of fish over mammalian models in certain types of studies have led to increasing interest in the use of fish as an alternative experimental system for vertebrate developmental, neurological, and toxicological studies. Furthermore, with increased dependence on farmed fish as a food source a biochemical appreciation of the factors influencing fish growth would be of substantial practical importance. In this paper we describe a mitogenic activity obtained in an extract of 21-day old trout embryos that was identified by its ability to stimulate the serum-free growth of several established fish cell lines and early passage cultures. The level of mitogenic response induced by the extract could not be duplicated with purified mammalian growth factors added individually or in combination and the extract did not stimulate DNA synthesis in quiescent mouse fibroblasts. These results suggest that trout embryo extract may contain a novel growth-promoting activity for fish cells. The embryo extract was used to establish long-term cultures derived from blastula-stage zebrafish embryos.

1. INTRODUCTION

Over the past decade increasing interest has been focused on the use of fish as an alternative experimental system for vertebrate developmental [1-3], neurological [4] and toxicological studies [5-7]. Fish possess many characteristics which make them an excellent experimental system for these types of studies. The availability of a large amount of embryonic material is conducive for biochemical studies of proteins and nucleic acids at any stage of early development. Also, many species such as zebrafish and trout produce large embryos which develop ex utero making it convenient to perform manipulations such as microinjection and microdissection as part of an in vivo experimental approach. Furthermore, methods for generating clonal strains of fish make available a virtually unlimited supply of genetically identical individuals [3,8]. For toxicology studies fish are sensitive indicators of toxicity to environmental carcinogens such as aflatoxin and aromatic hydrocarbons and the relative inexpense of maintaining fish makes it feasible to carry out exposures using a large number of embryonic or juvenile individuals [7].

Traditionally, with few exceptions, experimental utilization of fish has involved an in vivo approach [9]. Therefore a need exists to develop piscine cell culture models for use in in

187

R. Sasaki and K. Ikura (eds.), Animal Cell Culture and Production of Biologicals, 187–195.

vitro studies designed to complement the ongoing in vivo work. Cultured fish cells have been used primarily as a host for the propagation of pathogenic fish viruses. Therefore a considerable potential remains in the broad application to fish cells of the in vitro experimental approaches commonly employed with mammalian cells [10]. A first step toward this goal is an increased understanding of the extracellular factors influencing fish cell proliferation and differentiation in culture. The present study describes a mitogenic activity obtained from trout embryos that stimulates the serum-free growth of several established fish cell lines and early-passage cultures. The mitogen was used to establish long-term cell cultures derived from blastula-stage zebrafish embryos.

2. MATERIALS AND METHODS

Fish cell lines were grown in a nutrient medium consisting of a 3:1 mixture of Dulbecco's modified Eagle's medium and Ham's F-12 supplemented with 10 nM sodium selenite, sodium bicarbonate (0.3 mg/ml), 15 mM 4-(2-hydroxyethyl)-1-piperazineethanesulfonic acid (Hepes) buffer (pH 7.4), penicillin (200 international units/ml), streptomycin sulfate (200 ug/ml), and ampicillin (25 ug/ml) (DF medium). Stock cultures were supplemented with 10% fetal bovine serum (FBS). Early passage trout embryo cell lines were derived as previously described [11] from newly hatched (28-day) Shasta strain rainbow trout. Zebrafish embryo cell lines were derived from blastula-stage embryos (2 hrs post-fertilization) and grown in nutrient medium consisting of a 1:1 mixture of DF medium and Leibovitz's L-15 supplemented with FBS (1%), insulin (10 ug/ml), trout serum (0.2%) and trout embryo extract (25 ug/ml). Primary cultures were initiated from zebrafish embryos as follows. Blastula-stage embryos produced by in vitro fertilization were rinsed three times in nutrient medium and once in a 0.5% bleach solution. Following three additional rinses in nutrient medium the embryos were dechorionated in trypsin [0.2% (wt/vol) trypsin/1mM EDTA in phosphate-buffered saline for approximately 10 minutes] and the cells were dissociated by pipeting the embryos gently until a suspension of small cell aggregates was obtained. The cell aggregates were then collected by centrifugation and plated in 96-well tissue culture dishes.

For serum-free experiments cells were plated in 6-well (35-mm diameter) tissue culture plates precoated with human fibronectin (10 ug/well). For each experiment assays were conducted in duplicate. The medium was supplemented with bovine insulin (10 ug/ml), human transferrin (25 ug/ml) and mouse epidermal growth factor (EGF) (50 ng/ml). All cell lines were grown at 20°C and the plates were wrapped in Parafilm to prevent evaporation. Medium was changed every 7 days. The cells were cultured in ambient air. Serum-free cells were passaged with trypsin and trypsin inhibitor and suspensions of trypsinized cells in phosphate-buffered saline were counted with a Coulter particle counter.

Preparation of fish serum from the blood of Shasta strain rainbow trout and embryo extract from 21-day old trout embryos has been previously described [11].

3. RESULTS

3.1 Growth-Promoting Activity from Trout Embryos

Initial attempts to formulate a serum-free medium for CHSE-214 chinook salmon embryo cells using combinations of known peptide growth factors, hormones, attachment factors and nutritional supplements were unsuccessful. As a result we began experimenting with other supplements in an attempt to identify mitogens capable of promoting sustained growth of piscine cells. During the process we discovered that an extract made from 21-day old trout embryos was mitogenic for CHSE-214 cells in the absence of serum (Fig. 1).

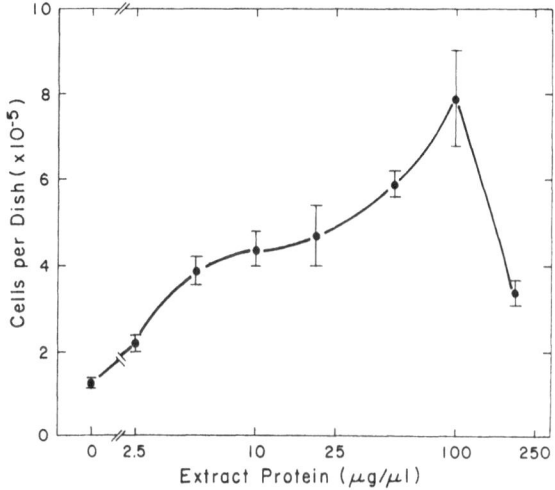

Fig. 1. Stimulation of CHSE-214 cell growth by trout embryo extract. Cells were plated at 10^5 per well in 6-well tissue culture plates as described in Materials and Methods. A suspension of trypsinized cells in phosphate-buffered saline was counted on day 10. Bars denote standard deviation. Data taken from Collodi and Barnes [11].

A mitogenic effect could be detected with a concentration of extract protein as low as 2.5 ug/ml in nutrient medium supplemented with insulin, transferrin and EGF. However optimal growth was achieved with a concentration of 100 ug/ml producing a 6-fold increase in cell number over the control after 14 days (Fig. 2). Concentrations greater than 100 ug/ml did not elicit a greater mitogenic response and in some cases were toxic to the cells. The embryo extract was also sufficient for long-term maintenance of CHSE-214 cultures. These cells were grown continuously through 25 passages and approximately 60 population doublings in basal nutrient medium supplemented with insulin (10 ug/ml), transferrin (25 ug/ml), EGF (10 ng/ml) and extract (40 ug/ml).

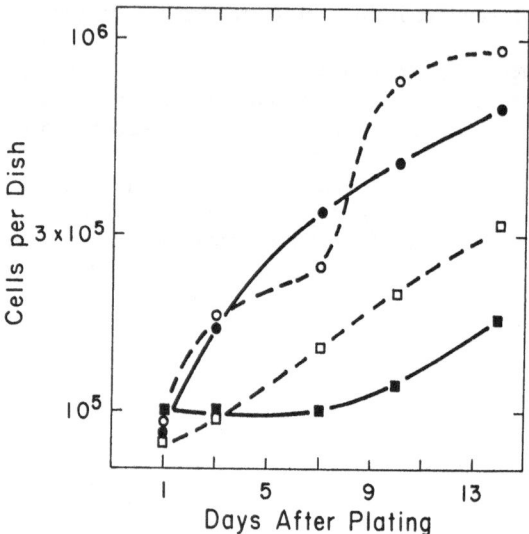

Fig. 2. Time course of CHSE-214 cell growth in the absence (■) or the presence of trout embryo extract protein at 2.5 ug/ml (□), 20 ug/ml (●), or 100 ug/ml (O). Cells were plated as described in Fig. 1 and were counted on the days indicated. Data taken from Collodi and Barnes [11].

3.2 Characteristics of the Growth-Promoting Activity.

Mitogenic activity was obtained in extracts made from both 21- and 28-day old embryos. The activity found in extracts prepared from unfertilized trout eggs was about 30% of that in extracts from 21-day old embryos (Table 1). The activity was nondialyzable (M_r 3500 cutoff membrane) against either 10 mM Hepes, pH 7.4/130 mM NaCl or 200 mM acetic acid. Proteolysis (24 hr at 37°C in the presence of 20 units of insoluble trypsin) or boiling (100°C for 5 min) reduced the activity by 48% and 35% respectively, whereas treatment with 1 mM dithiothreitol caused no decrease in activity. The activity was also stable to incubation at 60°C and was not removed by passage through a heparin-agarose column. When embryo extract was chromatographed on a phenyl-Sepharose column (loaded in 3 M NaCl, eluted in 20 mM sodium phosphate, pH 7.0) followed by chromatography on a

Sephadex G-100 column in 0.2 M acetic acid, activity was obtained from unretarded material and a peak of M_r approximately 14,000.

Purified peptide growth factors added at optimal concentrations for mitogenic effect on CHSE-214 were not able to mimic the growth-promoting effect of embryo extract on the cells (Fig.3). The optimal concentration of each growth factor was determined by dose-response experiments. TGF-B, IGF-I and IGF-II were moderately mitogenic but did not produce effects comparable to the embryo extract, even when added in various combinations. Conversely, the embryo extract, unlike the purified mammalian growth factors, did not stimulate DNA synthesis in quiescent mouse 3T3 fibroblasts (Fig.4).

Table 1. Stability of embryo extract activity.

Addition	Cells per plate no. x 10^{-5}
No extract	1.49 ± 0.08
21-day embryo extract	6.70 ± 0.15
28-day embryo extract	5.54 ± 0.67
Unfertilized egg extract	3.02 ± 0.00
21-day embryo extract dialyzed against 10 mM Hepes, pH 7.4/130 mM NaCl	5.78 ± 1.49
21-day embryo extract dialyzed against 200 mM acetic acid	5.27 ± 0.59

CHSE-214 cells were plated at 10^5 per well and counts were taken after 14 days. Data shown are the average cell count of duplicate plates and the standard deviation. Embryo extract protein was assayed at 20 ug/ml. Data taken from Collodi and Barnes [11].

3.3 Growth of Zebrafish Blastocysts

Trout embryo extract was used to derive cell cultures initiated from blastula-stage zebrafish embryos. These cells have been grown continuously for at least 7 passages and approximately 20 population doublings (3 months) in a nutrient medium consisting of a 1:1 mixture of Leibovitz's L-15 medium and DF. This medium was further supplemented with trout embryo extract (25 ug/ml), insulin (10 ug/ml), trout serum (0.2%) and FBS (1%). All of these factors were necessary to establish the cells in culture, however once established the cells could be grown in medium supplemented with only trout serum (0.2%) and FBS (1%). The mixture of L-15 and DF was important for optimal growth of the embryo cells. DF medium without L-15, which supported the serum-free growth of CHSE-214 cells supported, suboptimal growth of the zebrafish embryo cells. However, L-15 or DMEM alone, basal nutrient media traditionally used to culture fish cells (13), supported very poor growth of the zebrafish cells.

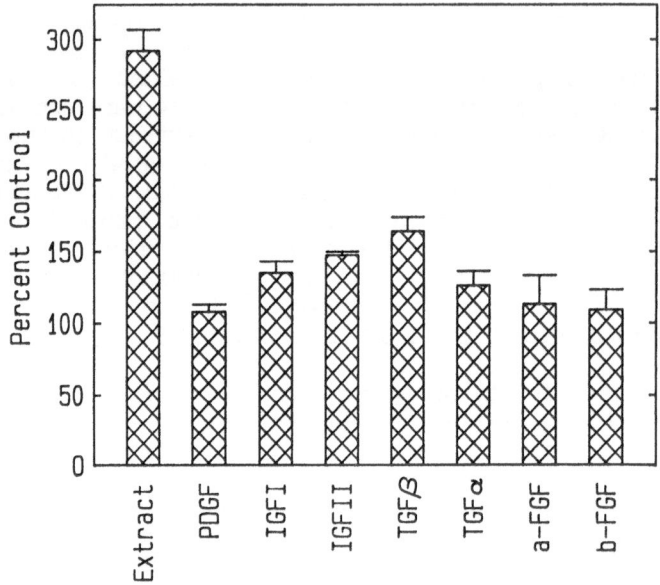

Fig. 3. Mitogenic activity of mammalian growth factors on CHSE-214 cells. Cells were plated as in Fig 1 and cultured in medium containing insulin, transferrin, EGF and each growth factor as indicated. Data shown are for the following concentrations: embryo extract, 20 ug/ml; PDGF, 10 ng/ml; IGF-I, 100 ng/ml; IGF-II, 1.0 ug/ml; TGF-B, 10 ng/ml; TGF-alpha, 50 ng/ml; a-FGF, 10 ng/ml; b-FGF, 10 ng/ml. Cells were counted on day 14 and the results are presented as percentages of the control (cells grown in medium with insulin, transferrin and EGF only). Data taken from Collodi and Barnes [11].

4. DISCUSSION

The results of this study demonstrate the presence of mitogenic activity in extracts made from 21-day old trout embryos which is able to stimulate the growth of established fish cell lines from several species. The activity is nondialyzable, stable in acetic acid and resistent to reduction by dithiothreitol. When chromatographed on Sephadex G-100 part of the extract-associated activity behaves as a molecule of $Mr \leq 14000$ whereas the remaining activity is not retarded by the column. The activity

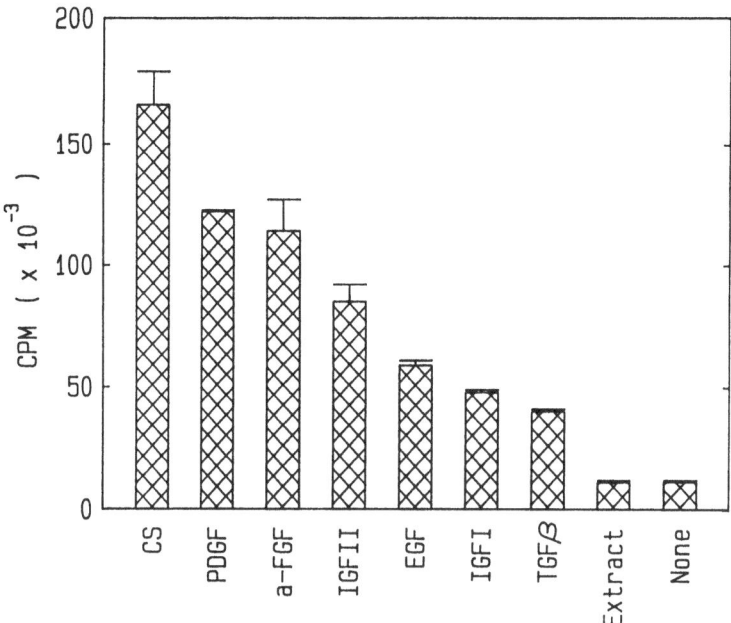

Fig. 4. Stimulation of DNA synthesis in quiescent mouse 3T3 cells. BALB/c 3T3 were plated in a 1:1 mixture of Ham's F-12 and Dulbecco's modified Eagle's medium supplemented with 10% calf serum (CS) at 2 x 10^5 cells per 35-mm-diameter tissue culture dish and the assays were conducted as described [13]. The following concentrations of growth factors were assayed: PDGF, 10 ng/ml; a-FGF, 10 ng/ml; IGF-II, 1 ug/ml; EGF, 20 ng/ml; IGF-I, 100 ng/ml; TGF-B, 10 ng/ml; dialyzed trout embryo extract, 50 ug/ml. Data taken from Collodi and Barnes [11].

was partially destroyed by protease treatment or boiling indicating that multiple activities may be present with different sensitivities to these treatments. The level of mitogenic activity elicited by the embryo extract could not be mimicked by purified mammalian growth factors added individually or in combination and the extract could not duplicate the stimulatory effect of the mammalian growth factors on DNA synthesis in quiescent mouse fibroblasts. These results indicate that the extract-associated activity may be a fish-specific version of a known mammalian growth factor or a factor not previously described.

The embryo extract was used to establish long-term cultures of blastula-derived zebrafish embryo cells. Growth of these cells was also dependent on the presence of small amounts of trout and fetal bovine serum. The ability to establish long-term piscine cell cultures initiated from early-stage embryos provides the potential to conduct in vitro studies of the parameters influencing fish embryonal cell growth and differentiation thus enhancing the attractiveness of the fish model for studies of vertebrate development. Furthermore, this work provides a first step in the establishment of a pluripotent piscine embryonal stem cell line. Such a line may be used to generate transgenic chimeric fish expressing exogenous genes introduced into the cells by transfection [12].

REFERENCES

1. Streisinger, G., Coale, F., Taggart, C., Walker, C. and Grunwald, D.J. (1989) Clonal origin of cells in the pigmented retina of the zebrafish eye. Dev. Biol. **131**, 60-69.

2. Chourrout, D., Guyomard, R. and Houdebine, L. (1986) High efficiency gene transfer in rainbow trout (Salmo gairdneri) by microinjection into egg cytoplasm. Aquaculture **51**, 143-150.

3. Streisinger, G., Walker, C., Dower, N., Knauber, D. and Singer, F. (1981) Production of clones of homozygous diploid zebra fish (Brachydanio rerio). Nature **291**, 293-296.

4. Kuwada, J.Y. (1986) Cell Recognition by neuronal growth cones in a simple vertebrate embryo. Science **233**, 740-746.

5. Streisinger, G. (1984) Attainment of minimal biological variability and measurements of genotoxicity: Production of homozygous diploid zebra fish. Natl. Cancer Inst. Monogr. **65**, 53-58.

6. Anders, F., Schartl, M. and Barnekow, A. (1984) Xiphophorus as an in vivo model for studies on oncogenes. Natl. Cancer Inst. Monogr. **65**, 97-109.

7. Hendricks, J.D., Meyers, T.R., Casteel, J.L., Nixon, J.E., Loveland, P.M. and Bailey, G.S. (1984) Rainbow trout embryos: Advantages and limitations for carcinogenesis research. Natl. Cancer Inst. Monogr. **65**, 129-137.

8. Parsons, J.E. and Thorgaard, G.H. (1985) Production of androgenetic diploid rainbow trout. J.Hered. **76**, 177-181.

9. Powers, D.A. (1989) Fish as model systems. Science **246**, 352-357.

10. Barnes, D.W. (1987) Serum-free animal cell culture. BioTechniques 5, 534-542.

195

11. Collodi, P. and Barnes, D.W. (1990) Mitogenic activity from trout embryos. Proc. Nat. Acad. Sci. USA **87**, 3498-3502.

12. Helmrich, A., Bailey, G.S. and Barnes, D.W. (1988) Transfection of cultured fish cells with exogenous DNA. Cytotechnology 1, 215-222.

13. Barnes, D.W. and Colowick, S.P. (1977) Stimulation of sugar uptake and thymidine incorporation in mouse 3T3 cells by calcium phosphate and other extracellular particles. Proc. Nat. Acad. Sci. USA. **74**, 5593-5597.

We thank Drs. J. Hendricks and G. Bailey for trout embryos. Portions of this work were supported by Postdoctoral Grant 1F32ES05445 (P.C.) and Center Grants ES00210 and ES03850 from the National Institute of Environmental Health Sciences. D.W.B. is a recipient of a Research Center Development Award from the National Cancer Institute.

PROTEIN FACTOR OBTAINED FROM RAT ADIPOSE TISSUE SPECIFICALLY PERMITS THE
PROLIFERATION OF 3T3-L1 AND OB1771 PREADIPOCYTE CELL LINES IN A COM-
PLETELY DEFINED SERUM-FREE MEDIUM

Teruo KAWADA, Naohito AOKI, and Etsuro SUGIMOTO
Laboratory of Nutritional Chemistry,
Department of Food Science and Technology,
Faculty of Agriculture, Kyoto University,
Kyoto 606, Japan

ABSTRACT.   The proliferation of preadipocytes is essential for the
formation of new adipocytes and important to the differentiation of
preadipocytes into adipocytes.   In this study, we have first developed
a completely defined serum-free medium that supports the growth of
preadipocytes, and then found the a novel protein factor, referred to as
PAGF (preadipocyte growth factor), in rat adipose tissue using the
serum-free medium.   PAGF specifically permitted the proliferation of
preadipocytes, acting like a competent factor, and its apparent molecular weight was about 20,000. PAGF activity existed in various mammalian
adipose tissues and should function in response to the energy intake.
These results strongly suggest that PAGF may play a key role in the
formation of new adipocytes by specifically and locally triggering the
proliferation of preadipocytes, and then contribute to the development
of adipose tissue.   PAGF may provide a useful tool for further understanding of obesity.

1. INTRODUCTION

Obesity, namely, abnormal adipose tissue development, is a highly complex phenomenon regulated by many factors.   However, several recent
studies suggested that the number of adipocytes may increase even in
adults.   Hirsh and Klyde demonstrated that the increase in the adipocyte
number was the result of the proliferation of preadipocytes rather than
lipid filling of differentiated cells normally too small to detect (1).
Furthermore, Björntorp has hypothesized that the de novo formation of
adipocytes is governed by the size of existing fat cells. When the mass
of the stored triglycerides exceeds a value of about 0.7 to 0.8 µg/cell,
specific precursor cells, namely preadipocytes, are caused, through an
unknown mechanism, to differentiate into adipocytes (2).   From these
results, it has been speculated that unknown factors that regulate adipose tissue development are present in adipose tissue, but this has not
been confirmed.   Therefore, we investigated the mechanism controlling
the proliferation of preadipocytes using 3T3-L1, established from Swiss

197

R. Sasaki and K. Ikura (eds.), Animal Cell Culture and Production of Biologicals, 197–204.
© 1991 *Kluwer Academic Publishers.*

mouse embryos by Green et al. (3), and Ob1771, established from the epididymal fat pads of a C57BL/6J ob/ob mouse by Ailhaud et al. (4), cell lines as models of preadipocytes (Figure 1).

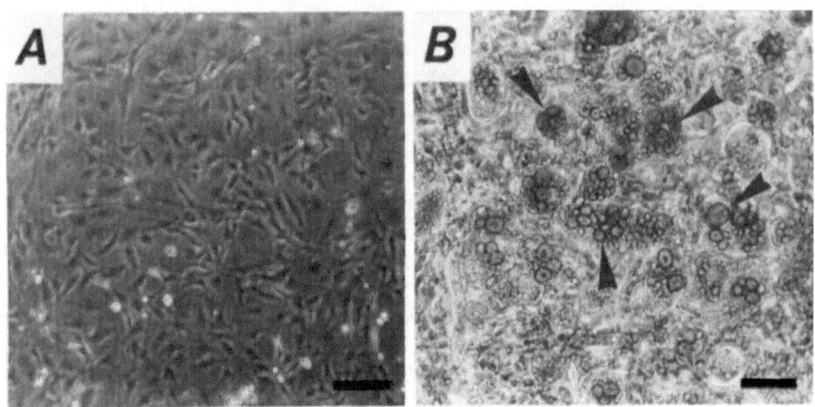

Figure 1. Photomicrographs of 3T3-L1 cells grown in Dulbecco's modified Eagle's medium containing 10% FCS. (A); Photomicrographs were taken 6 day after inoculation of cells grown in the medium. (B); Photomicrographs of Oil red O staining of culture wells 15 days after confluency reached. The arrows indicate triglyceride droplets stained with Oil red O. Bar=30 µm.

2. ADIPOSE TISSUE CELLULARITY

The de novo formation of adipocytes has three main phases (Figure 2): [1] the proliferation of preadipocytes; [2] the differentiation of preadipocytes into adipocytes; and [3] the maturation of adipocytes. There have been many studies focused on the differentiation and maturation phases, involving biochemical and cell culture techniques. For the maturation process, one of the most important hormone is insulin. In the differentiation process, many kinds of unknown factors are involved in the conversion to adipocytes. Recently, IGF-I has been considered to be important as a circulation factor (5,6). Furthermore, in our most recent experiment, it was found that IGF-I mRNA, independent of GH, is expressed in cultured preadipocytes in the arrest phase, using a completely defined serum-free medium. On the other hand, very little is known about the proliferation phase, especially its regulation, from the standpoint of its physiological and biochemical significance.

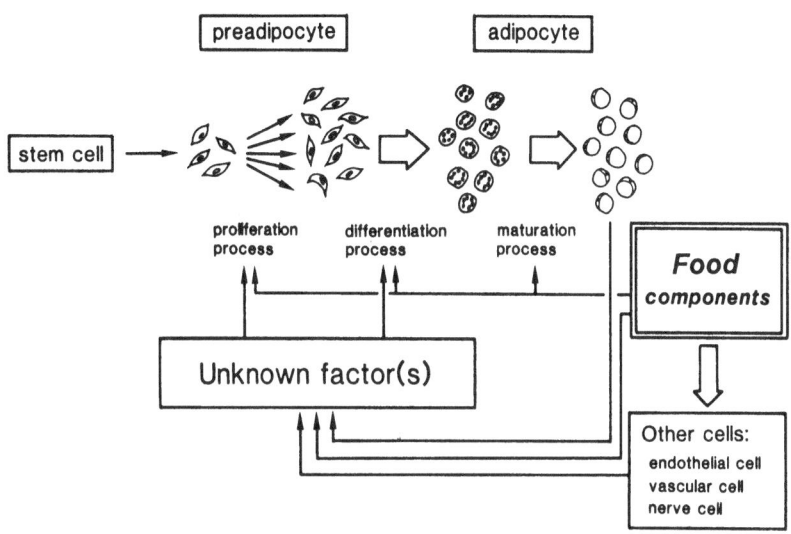

Figure 2.   Outlines of hypothesis concerning adipose tissue cellularity.

3. PROLIFERATION OF 3T3-L1 PREADIPOCYTES IN A COMPLETELY DEFINED SERUM-FREE MEDIUM

We have developed a completely defined serum-free medium that supports the growth of Swiss 3T3-L1 fibroblasts to nearly the same extent as DME medium supplemented with 10% fetal calf serum.   With ASF301 medium [former name, RITC 80-7] (7), most of the 3T3-L1 cells survived for at least 10 days, but did not grow.   ASF301 medium contains insulin and epidermal growth factor (EGF) as growth factors, which are termed "progression factors".   So we examined the effects of "competent factors" on the proliferation of 3T3-L1 preadipocytes.   The dose-related effects of competent factors [platelet-derived growth factor (PDGF) and bovine brain fibroblast growth factor (FGF) composed of acidic and basic FGF] on DNA synthesis were examined for incorporation of [^{3}H]thymidine into the DNA fraction (Figure 3).   PDGF and FGF were growth factors that significantly stimulated DNA synthesis in our culture system (at doses of 10-100 ng/ml).   PDGF and FGF dose-dependently showed mitogenic activity, the highest activity being observed at 20 and 50 ng/ml, respectively.   In each case, the highest DNA-synthesis activity was comparable to that in the case of the addition of 10% FCS.   In our other experiment, the mitogenic activity of acidic FGF was nearly equal to that of basic FGF in preadipocytes, in contrast to that in general fibroblasts. Acidic FGF was detected in the cerebrospinal fluid of rats after feeding and may participate in the regulation of feeding at the level of the central nervous system (CNS) (8).   So it is interesting that an increase of acidic FGF in CNS after feeding might contribute to the proliferation of preadipocytes and the development of adipose tissue.
      Furthermore, it was confirmed that 3T3-L1 cells grown in the

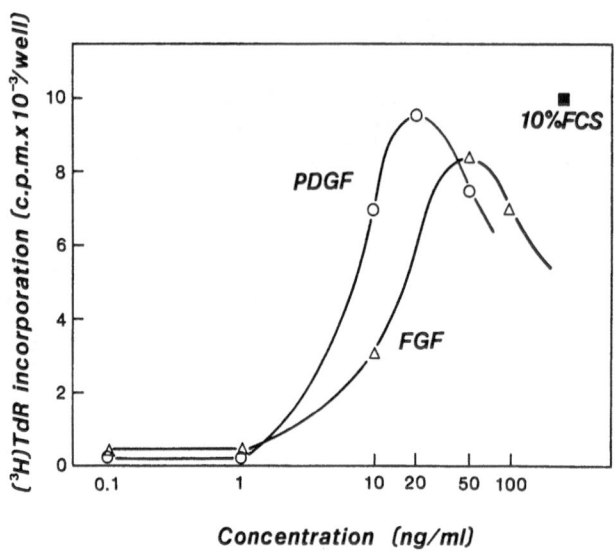

Figure 3. Dose-dependent effects of PDGF and brain FGF on [³H]thymidine incorporation into 3T3-L1 cells in ASF301 serum-free medium. Fetal calf serum (FCS) was used as a positive control for mitogenic activity at the concentration of 10%. TdR, thymidine.

serum-free medium retained the properties of differentiation into adipocytes. Our serum-free medium should be a useful tool for research on the growth and differentiation of 3T3-L1 preadipocytes.

4. CELL CYCLE AND GROWTH FACTORS IN PREADIPOCYTES

Cells reproduce by duplicating their contents and then dividing in two. Generally, many cells in vivo are in the arrest state (G_0 phase in the cell cycle), with unduplicated DNA (Figure 4). Cells can be activated to reenter the cycle during the G_1 phase and then prepare for DNA synthesis (S phase). This switching in and out of the G_1 phase mainly sets the post-embryonic proliferation rate and is defectively controlled in cancer cells. Recently, it was found that cell division in multicellular organisms was governed by different combinations of protein growth factors. They act at very low concentrations. We have demonstrated, using a completely defined serum-free medium, that in preadipocytes, proliferation is strongly dependent on "competent factors" and "progression factors" as protein growth factors (Figure 4). Preadipocytes require a competent factor such as PDGF or FGF during the G_0 to G_1 period. Also, during the G_1 to S period, preadipocytes require a progression factor such as epidermal growth factor (EGF) or insulin (9). Since insulin is present at a high enough concentration for preadipocyte growth, progression factors are probably not rate-limiting. However,

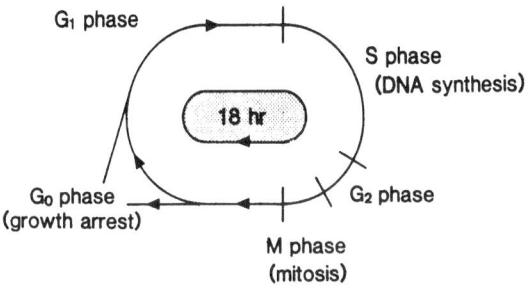

competent factor	progression factor
($G_0 \rightarrow G_1$)	($G_1 \rightarrow S$)
PDGF	EGF
FGF	insulin

Figure 4. Cell cycle and growth factors in preadipocytes. PDGF; platelet-derived growth factor, FGF; fibroblast growth factor, EGF; epidermal growth factor.

competent factors act as local chemical mediators and thus could be the most important growth factors that regulate preadipocyte proliferation.

5. PROTEIN FACTOR FROM RAT ADIPOSE TISSUE FOR THE PROLIFERATION OF CULTURED PREADIPOCYTES

As mentioned above, it has been speculated that unknown factors that regulate adipose tissue development exist in the tissue. So we searched for a proliferation factor for preadipocytes in rat adipose tissue. One hundred g of epididymal adipose tissue was excised from Wistar rats (male, 8-10 weeks), finely chopped in 10 mM sodium phosphate buffer (pH 7.0), frozen and thawed three times, and then centrifuged. The supernatant was lyophilized and 500 mg of crude extract was obtained. The extract was further purified by HPLC gel filtration (TSK G2000SW$_{XL}$ column). Mitogenic activity was measured as stimulation of the proliferation of preadipocytes, assessed by [^{3}H]thymidine incorporation into DNA and direct cell counting (9). After HPLC gel filtration, mitogenic activity was eluted as a peak corresponding to a molecular weight of 20,000, as calculated by comparison with standard proteins (10). Since this activity was not detected with a serum-free medium not containing a progression factor, it should be due to a competent factor. Furthermore, we confirmed the mitogenic activity of the active fraction by direct cell counting. After the replacement of serum-supplemented medium with serum-free medium containing the active fraction, the cell number increased gradually, confluency being reached on day 6 (Figure 5). After the medium had been replaced with a differentiation-inducing medium, the confluent monolayer differentiated into adipocytes loaded with multilocular droplets. As shown in Figure 6, the proliferation activity of this active fraction from rat adipose tissue was dose amount-dependent. The near maximal stimulation by the active fraction was comparable to that in the case of PDGF (20 ng/ml), brain FGF (50 ng/ml), or 10% fetal calf serum. This activity was heat- and protease-unstable, but reductant-stable. Furthermore, this activity of the protein factor was not detected in various other cell lines, in particular, Swiss 3T3 cells, from which the 3T3-L1 cell originated and which could proliferate in response to a competent factor (PDGF or FGF) in the same serum-free medium (Figure 5). These results strongly suggest that the protein factor is different from PDGF or FGF and contributes to the formation of new adipocytes by specifically triggering the proliferation of preadipocytes, acting like a competent factor (10). We have termed the protein growth factor **preadipocyte growth factor** (PAGF).

Furthermore, our recent study found that PAGF activity in rat adipose tissue is strongly affected by nutritional conditions such as restricted energy intake and fasting, and the activity exists widely in various mammalian adipose tissues (submitted). These results suggest that the PAGF in adipose tissue contributes to the de novo formation of adipocytes and adipose tissue through stimulation of the proliferation

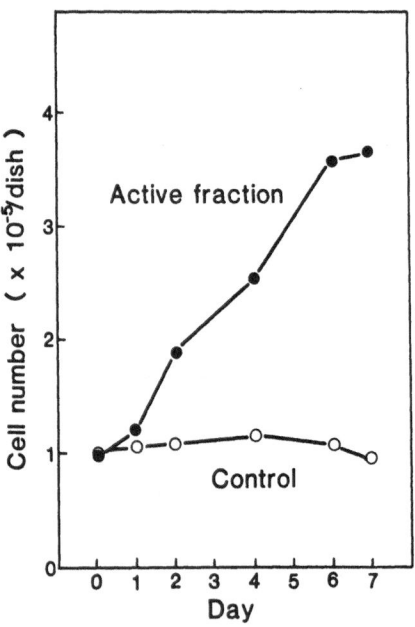

Figure 5. Growth curves for 3T3-L1 cells in ASF301 serum-free medium supplemented with and without the active fraction from rat adipose tissue.

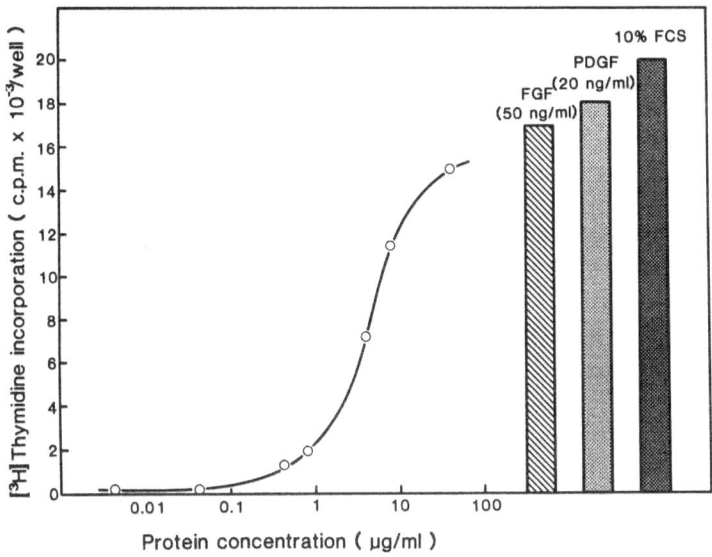

Figure 6.   Dose-dependent effects of the active fraction from rat adipose tissue of [³H]thymidine incorporation into DNA in 3T3-L1 cells.

of preadipocytes.   This factor may provide a useful tool for further elucidation of the development of obesity.

ACKNOWLEDGMENTS

We gratefully thank Professor Gérard AILHAUD (CNRS, France) for providing the Ob1771 cells. We wish to thank Ms. Reiko KAWAI and Ms. Taeko UMEYAMA for their assistance in the experimental work.   This work was supported in part by Grants-in-Aid from the Ministry of Education, Science, and Culture of Japan.

REFERENCES

1. Klyde, B.J. and Hirsh, J. (1979) 'Increased cellular proliferation in adipose tissue of adult rats fed a high-fat diet', J. Lipid Res. 20, 705-715.
2. Björntorp, P. (1981) Adipocyte precursor cells. In: Recent Advance in Obesity Research III, pp58-69, John Libbey, London.
3. Green, H. and Kehinde, O. (1974) 'Sublines of mouse 3T3 cells that accumulate lipids', Cell 1, 113-116.
4. Negrel, R., Grimaldi, P. and Ailhaud, G. (1978) 'Establishment of

preadipocyte clonal cell line from epididymal fat pad of ob/ob mouse that responds to insulin and to lipolytic hormones', Proc. Natl. Acad. Sci. USA. 75, 6054–6058.

5. Smith, P.J., Wise, L.S., Berkowitz, R., Wan, C. and Rubin, C.S. (1988) 'Insulin-like growth factor-I is an essential regulator of the differentiation of 3T3-L1 adipocytes', J. Biol. Chem. 263, 9402–9408.

6. Schmidt, W., Poll-Jordan, G. and Loffler, G. (1990) 'Adipose conversion of 3T3-L1 cells in a serum-free culture system depends on epidermal growth factor, insulin-like growth factor I, cortico-sterone, and cyclic AMP', J. Biol. Chem. 265, 15489–15495.

7. Yamane, I., Kan, M., Hoshi, H. and Minamoto, Y. (1981) 'Primary culture of human deploid cells and its longterm transfer in serum-free medium', 81, 470–474.

8. Hanai, K., Oomura, Y., Kai, Y., Nishikawa, K., Shimizu, N., Morita, H. and Plata-Salaman, C.R. (1989) 'Central action of acidic fibro-blast growth factor in feeding regulation', Am. J. Physiol. 256, R217–R223.

9. Kawada, T., Aoki, N., Kawai, R. and Sugimoto, E. (1990) 'Prolif-eration of 3T3-L1 preadipocytes in a completely defined serum-free medium', Cell Biol. Int. Rep. 14, 567–574.

10. Aoki, N., Kawada, T., Umeyama, T. and Sugimoto, E. (1990) 'Protein factor obtained from rat adipose tissue specifically permits the proliferation of the 3T3-L1 and Ob1771 cell lines', Biochem. Biophys. Res. Commun. 171, 905–912.

NUTRIENT OPTIMIZATION FOR THE PRODUCTION OF BIOLOGICALS FROM ANIMAL CELLS CULTURED AT HIGH DENSITY.

DAVID W. JAYME and STEFAN A. WEISS
GIBCO/Life Technologies, Inc.
Cell Culture Research and Development Laboratory
2086 Grand Island Boulevard
Grand Island, New York 14072 USA

ABSTRACT. The formulation components necessary to maintain animal cells in vitro at high density for biological production may differ quantitatively and qualitatively from the nutrient requirements for routine cell cultivation at low densities. Elimination of serum to facilitate downstream processing and regulatory approval necessitates compensation for the nutritional and biophysical roles of serum in the cell culture environment. Biochemical analysis of spent culture effluents from high density bioreactors offers a helpful insight to the unique nutrient requirements of that specialized system and permits iterative optimization of serum-free, synthetic medium composition to sustain bioreactor productivity. This information may be exploited, either by enhancing the nutrient levels of the basal medium or by providing nutritional supplementation of exhausted metabolites via batch or perfusion feeding of liquid concentrates. This paper focuses upon techniques utilized in this laboratory to assess bioreactor depletion of amino acids, carbohydrates and lipids. These data are applied to the design of serum-free media for biopharmaceutical and biotechnological applications of hybridomas, recombinant CHO cells, and invertebrate cells.

1. Introduction

1.1 ELEMENTS CRITICAL FOR EFFICIENT BIOREACTOR PRODUCTIVITY

The program for this conference has appropriately been subdivided into three elements which require independent and, ultimately, integrated optimization to achieve efficient bioreactor productivity. Multiple papers earlier in the conference described elegant procedures for genetic manipulation of cultured cells to introduce the targeted genetic material and amplify its expression. Subsequent papers will describe the hardware design and monitoring control parameters for a variety of bioreactors suitable for animal cell culture expansion and biological production. The papers presented in this section are targeted towards the third aspect of bioreactor optimization: the design and delivery of nutrient fluids required to sustain bioreactor productivity at high cell density.

R. Sasaki and K. Ikura (eds.), Animal Cell Culture and Production of Biologicals, 205–212.
© 1991 *Kluwer Academic Publishers.*

Too often, we observe the following paradox in biotechnology ventures. Significant resources will have been committed to engineer a technically-elegant genetic construct designed to produce large quantities of the desired biological product. Similarly, great capital investment has been made in the design, purchase and validation of state-of-the-art bioreactors and downstream processing technologies. However, due to time constraints, resource limitations, or lack of understanding or expertise, they may fail to invest the time and effort to optimize the cell culture component of the system. The consequences of this strategy can affect overall yield, manufacturing cost and biochemical integrity of the targeted biological product [Jayme (1991)]. The objective of this paper is to illustrate techniques utilized by the research and new product development group of a media manufacturer to optimize serum-free nutrient formulations for high density biological production applications.

1.2 NUTRIENT MEDIUM DESIGN

The default option is to utilize cell culture techniques and formulations established decades ago. As Richard Ham pointed out at a recent meeting of the U.S. Tissue Culture Association, the most widely selling culture media are based upon nutrient formulations established two or three decades ago [Jayme (1991)]. Practical difficulties of these media for biotechnology applications are illustrated in Table 1.

Table 1. Comparison of historical vs. current cell culture nutrient requirements.

Parameter	Historical	Current
Cell types	Less fastidious	More fastidious
Cell culture use	Proliferation	Biological production
Nutrient optimization	Less analytical	More analytical
Inoculation density	Narrow range	Broad range
Maximal cell density	Relatively low density	Relatively high density
Serum supplementation	Relatively high level	Relatively low level
Bioreactor types	Limited methods	Broader ranging methods
Environmental manipulation	Narrow range	Broader range

Synthetic medium formulations were originally developed to cultivate established cell lines at relatively low density [Jayme and Blackman (1985)]. Nutrient levels were determined by the "limiting toxicity" approach and were augmented by ill-defined supplements, such as serum and organ extracts. Biotechnology applications of cultured animal cells require serum replacement with biochemically-defined components to eliminate concerns over cost, availability and variability and to facilitate downstream processing and regulatory approval of the biological product [Jayme (1991)].

Cell culture functions attributable to serum are manifold [Jayme (1990), Weiss et al. (1980)] and include the provision of hormones and growth factors, cellular adhesion promoting factors, carrier and stabilizing proteins, and factors which protect cells by buffering pH fluctuations and by inactivating toxic or proteolytic reactants. Frequently overlooked, serum addition provides metabolic nutrients which supplement the basal composition of the nutrient medium. From a bioengineering perspective, the bulk protein in serum minimizes absorptive loss of product to bioreactor surfaces, affects nutrient solubility and metabolic clearance, and influences various physical properties of the culture system, such as shear stress, viscosity, oxygen delivery and osmolality. To support the nutritional and biophysical requirements of high density bioreactors under serum-free culture expansion and product harvesting conditions, both qualitative and quantitative changes in basal medium composition have been implemented.

2.   Biochemical Analysis of Spent Medium

2.1   ANALYTICAL TECHNIQUES

This laboratory has developed analytical methods for quantitative assessment of nutrient clearance from spent media. Iterative application of these techniques to high density bioreactors has permitted rational design of novel serum-free and protein-free formulations. These procedures permit customized medium optimization and perfusion concentrates for individual cell types and production applications. Culture medium modifications examined include: optimized hexose, amino acid and lipid concentrations; glutamine stabilization; alternative mechanisms for trace metal delivery; improved pH buffering capacity; improved oxygen delivery and catabolite reduction; and defined additives for selective enrichment of cellular proliferation or biological production.

The power of this analytical technique is illustrated in Figure 1. This graph demonstrates the hybridoma cell density achieved and the monoclonal antibody yield from a small pilot-scale stir tank bioreactor. The nutrient medium monitored in this experiment was PFHM, a protein-free medium designed for hybridoma technology.

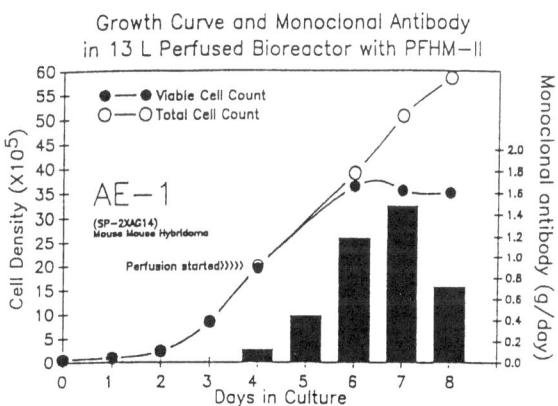

The figure shows that the test hybridoma initially grew very well in PFHM, with monoclonal antibody production exceeding a gram per day from the reactor. However, both cell viability and reactor productivity declined rapidly following day 4, retrospectively attributable to exhaustion of glutamine and other components of the nutrient medium.

Such observations are, of course, not novel; however, the typical response to nutrient depletion is to accelerate the perfusion flow rate or frequency of exchange of a sub-optimal nutrient medium within the culture vessel. Our analysis of spent fluids suggests another approach, dependent upon rational development of optimized nutrient formulations, designed to replenish specifically those nutrients exhausted through bioreactor passage. The net effect of such fortified media and supplemental concentrates may be a reduction in media waste and an increase in bioreactor efficiency.

2.2 SPENT MEDIUM ANALYSIS

2.2.1 *Amino Acid Analysis*. Effluent nutrient fluids from various bioreactors were analyzed by HPLC for residual amino acid composition, based upon a modification of the Waters Pico-Tag protocol [Ogden and Foldi (1987)]. Although there were variations, qualitative and quantitative, in nutrient consumption according to cell type, bioreactor, and even among different clones from the identical parental strain, some generalizations could be derived from this analysis. Typical data from a high density, batch hybridoma culture maintained in PFHM are illustrated in Table 2.

Table 2.   Time Course of Amino Acid Utilization by High Density Hybridoma Culture.

Amino Acid	[Initial]	[Final]	Percent Remaining
Lysine	300	270	90
Threonine	150	105	70
Phenylalanine	130	115	88
Tyrosine	150	105	70
Methionine	70	55	79
Serine	320	310	97
Alanine	415	480	116
Aspartate	20	60	300
Glutamate	150	165	110
Glutamine	300	n.d.*	<1
Proline	315	110	35
Asparagine	325	120	37
Arginine	270	145	54
Valine	170	40	24
Isoleucine	160	5	3
Leucine	310	10	3
Cyst(e)ine	55	n.d.	<1

*n.d. = not detected; Amino acid levels were detected by phenylisothiocyanate (PITC) derivitization and HPLC separation of adducts, with values expressed as mg/L; Initial samples were taken within one hour of medium application, while final samples were removed following three days' medium residence time in the batch reactor.

As illustrated in Table 2, some amino acids were utilized only marginally during the experimental culture residence time. A few amino acids (e.g., alanine, aspartate, glutamate) increased above the initial levels, presumably due to deamidation and transamination reactions. The key element to observe is that a select few nutrients were rapidly utilized by the cells and that the limitation of these essential nutrients may be correlated with diminished culture viability and attenuated biological productivity.

2.2.2. *Hexose Analysis*. In similar fashion, depletion of glucose and other hexoses was monitored from spent culture fluids by HPLC methods. These data, published elsewhere [Jayme (1991)], confirmed the rapid depletion of glucose under typical bioreactor environments. They also demonstrated the utility of substituting alternative hexoses for glucose to minimize lactate formation and microenvironmental acidification without compromising antibody yield. As with any nutrient medium modification, the investigator is cautioned to explore effects of nutrient substitution on product integrity, particularly post-translational glycosylation [Prior et al. (1989)].

3. Lipid Requirements for High Density Bioreactor Productivity

Historically, the biochemical composition of serum-free medium has focused upon buffered salts, appropriately augmented by polypeptide factors, trace elements and supplemental amino acids and sugars [Freshney (1987), Jayme and Blackman (1985)]. Despite the documented role of lipids as membrane components and essential precursors for prostaglandin synthesis, few serum-free formulations adequately account for the cellular lipid requirements of high density bioreactor cultures [Weiss et al. (1989), Spector (1972)]. Many commercially-available lipid delivery mechanisms are ill-defined, suffer from unacceptable lot-to-lot variability, or exhibit limited storage and culture stability.

Our laboratory investigations involving a broad range of cell and bioreactor types has convinced us of the importance of lipid in sustaining cellular integrity and biological productivity at high cell density. Data presented elsewhere have demonstrated that appropriate delivery of lipids during the culture expansion phase can substantially reduce or totally eliminate serum requirement [Jayme (1991)].

3.1 CHROMATOGRAPHICALLY-PURIFIED SERUM ALBUMIN

Bovine serum albumin (BSA) has been widely used as an additive to serum-free medium, since it is a primary constituent of serum and contributes many of serum's useful cell culture effects, including associated lipid materials. The Cohn Fraction V material frequently employed for blood fractionation utilizes a solvent separation procedure [Cohn et al. (1946)] which effectively strips fatty acids and cholesterol from serum albumin. The resultant product is highly suitable for its original applications, but highly-variable in lipid content and generally ill-suited for cell culture use [Jayme (1991)].

We have investigated an alternative method for purifying albumin from bovine blood, eliminating solvent extraction procedures and relying primarily upon ion exchange chromatography. The resultant, chromatographically-purified BSA (tradenamed "Albumax") has been analyzed biochemically and biologically to correlate bound lipid content with cell culture performance. Simple analysis of titratable free fatty acids exhibited no correlation with culture efficacy. However, solvent extraction of BSA samples obtained from various methods, followed by application to reverse phase HPLC, provided excellent correlation of analytical and performance data. Chromatographically-purified samples exhibited less lot-to-lot variation in extractable lipid content, elevated levels of important mono- and poly-unsaturated fatty acids and cholesterol, and superior cell culture performance as supplements for hybridoma and recombinant CHO cell technologies [Airey et al. (1991].

3.2   LIPID CONCENTRATES

In addition to use of serum albumin as a lipid carrier, synthetic lipids and emulsified preparations of marine fish oils have been utilized as cell culture additives [Weiss et al. (1989)]. We have found such natural lipid preparations to perform adequately to enhance cellular proliferation and to sustain bioreactor viability and productivity in several commercially-relevant biotechnologies.

Implementing HPLC techniques analogous to the procedures for analyzing spent culture fluids described previously for other nutrients, we have examined the qualitative and quantitative exhaustion of various lipids from high density bioreactors.   These analytical tools have permitted us to develop chemically-defined lipid concentrates, designed for high density culture applications, and to implement this knowledge into the formulation of novel serum-free media and supplemental concentrates.

4.   Technique Application to Serum-free Medium Design

4.1   MONOCLONAL ANTIBODY PRODUCTION ENHANCER CONCENTRATE

The objective of these spent medium analyses was to investigate common nutrients frequently exhausted by high density hybridoma cultures to improve the nutrient composition of our serum-free and protein-free hybridoma media.

Table 3.   Enhancement of Monoclonal Antibody Production in Serum-free Medium (Hybridoma-SFM) by Addition of a Protein-free Nutrient Supplement.

Condition	Medium	FBS?	Additives[b]	AntibodyProduction[a] Addition	Overall
Positive control	RPMI 1640	10%	None	- 0.4	0.6
Medium control	PFHM	0%	None	(0.0)	(1.0)
Carbohydrate substitution	PFHM	0%	Mannose	0.6	1.6
Amino acid addition	PFHM	0%	Amino acid	1.1	2.7
Inducer addition	PFHM	0%	Butyrate	1.1	3.8
Glutamine stabilization	PFHM	0%	Dipeptide	0.8	4.6

[a]Monoclonal antibody production expressed in micrograms of IgG obtained per milliliter of culture supernatant; Fold antibody production enhancement expressed relative to PFHM medium reference; Individual contributions obtained arithmetically by subtraction of reference performance.
[b]None = no further nutrient changes; Mannose = equimolar replacement of glucose with mannose; Amino acids = Augmented levels of six amino acids identified by HPLC to be rapidly depleted in high density culture of this hybridoma; Butyrate = Sodium butyrate (20 micromolar); Dipeptide = Glutaminyl dipeptide, as glycyl-glutamine, kindly provided by the Ajinomoto Company.

Ultimately, these experiments led us to develop a concentrated nutrient solution, containing supplemental hexose, amino acids and other constituents. This protein-free supplement increased monoclonal antibody production by 50% to 400%, using a broad panel of test hybridomas in various bioreactors. Specific antibody production per cell appeared increased, presumably owing to maintenance of required metabolites at non-production-limiting exogenous levels. Typical results are illustrated in Table 3.

4.2   RECOMBINANT PROTEIN PRODUCTION FROM HIGH-DENSITY, SERUM-FREE CHO CULTURE

Our initial development of a serum-free medium for CHO cell applications had resulted in a formulation which supported suspension culture expansion of CHO cells with first order growth kinetics comparable to serum-supplemented controls and minimal lag time. The maximal cell density approached 3×10^6 cells per mL, which was excellent for a serum-free medium. However, the plateau period at high cell density was abbreviated and the post-plateau decrease in cell viability was dramatic, suggesting depletion of critical nutrients in the synthetic medium. Analysis of spent medium supernatants confirmed that the majority of exogenous glucose and of several amino acids was exhausted by the time maximal cell density was attained. Supplementation of the basal prototype formulation with limiting amino acids reduced the lag period and had a moderate effect on prolonging culture viability. Further addition of glucose synergized with the amino acid additives to accelerate attainment of productive bioreactor cell density and to sustain peak reactor productivity in batch culture for several days beyond unsupplemented controls. Sustained expression of recombinant proteins from CHO cell transfectants in nutritionally-augmented, serum-free CHO cell medium has been field demonstrated in both batch and continuous culture systems [Gorfien and Weiss (1990)].

4.3   RECOMBINANT PROTEIN PRODUCTION FROM HIGH-DENSITY, SERUM-FREE INVERTEBRATE CELL CULTURE USING THE BACULOVIRUS EXPRESSION VECTOR SYSTEM (BEVS)

Initial experiments with IPL-41 Medium, supplemented with yeastolate, lactalbumin hydrolysate and emulsified cod liver oil-derived lipids, demonstrated acceptable performance in cellular expansion and biological production [Weiss et al. (1989)]. However, iterative analysis of spent fluids was highly useful to develop a more biochemically-defined, serum-free medium and to amplify the yield of recombinant protein expressed by BEVS. Glucose and several amino acids were rapidly exhausted from high density lepidopteran cultures [Weiss et al. (1990)]. Batch nutrient augmentation with supplemental glucose and nutritionally-limiting amino acids improved proliferative rate and recombinant protein expression. Similarly, substitution of chemically-defined lipid concentrates, based on HPLC analyses described above, enhanced culture expansion and product yield above levels observed with natural marine fish oil emulsions. Sf-9 cultures maintained in serum-free medium (Sf-900) achieved a cell density of nearly 2×10^7/mL using a biospin filter perfusion system. Several recombinant proteins, including ß-galactosidase and erythropoietin (EPO), have already been expressed in Sf-900 Medium [Weiss et al. (1990)].

5.   Conclusion

Serum-free formulations may be optimized through iterative analysis of spent media. Nutritionally-limiting metabolites may be introduced into the bioreactor to maintain cellular integrity at high density and to augment specific biological productivity. Required nutrients at high density differ qualitatively and quantitatively from cellular requirements under lower density proliferative culture. These analytical techniques offer valuable insights to the development and optimization of serum-free biological production conditions for hybridoma, recombinant CHO cell and invertebrate cell biotechnologies.

6.   References

Airey P, Benny G, Swartzwelder F and Conrad D (1991) "A novel lipid rich albumin supplement for serum-free cell culture." Focus 13: 2-7.

Cohn EJ, Strong LE, Hughes WL Jr, Mulford DJ, Ashworth JN, Melin M & Taylor HL (1946) "Preparation and properties of serum and plasma proteins: a system for the separation into fractions of protein and lipoprotein components of biological tissues and fluids." J. Amer. Chem. Soc. 68: 459-475.

Freshney RI (1987) Culture of animal cells: a manual of basic technique (2nd edition). Liss, New York.

Gorfien S and Weiss SA (1990) "A new, serum-free medium for growth of Chinese hamster ovary (CHO) cells in suspension." Focus 12(3): 75-76.

Jayme DW (1991) "Nutrient optimization for high density biological production applications." Cytotechnology (in press).

Jayme DW (1990) "Alternatives to fetal bovine serum for mammalian cell culture." Focus 12: 3-8.

Jayme DW & Blackman KE (1985) "Culture media for propagation of mammalian cells, viruses, and other biologicals." In: Mizrahi A and van Wezel AL (eds.) Advances in Biotechnological Processes. Vol. 5 (pp. 1-30) Liss, New York.

Ogden G & Foldi P (1987) "Amino acid analysis: an overview of current methods." LC-GC 5: 28-40.

Prior CP, Doyle KR, Duffy SA, Hope JA, Moellering BJ, Prior GM, Scott RW & Tolbert WR (1989) "The recovery of highly purified biopharmaceuticals from perfusion cell culture bioreactors." J. Parent. Sci. Tech. 43: 15-23.

Spector AA (1972) "Fatty acid, glyceride, and phospholipid metabolism." In: Rothblat GH and Cristofalo VJ (eds.) Growth, Nutrition, and Metabolism of Cells in Culture. Vol. II (pp. 257-298) Academic, New York.

Weiss SA, Lester TL, Kalter SS and Vaughn JL (1980) "Chemically Defined Serum-Free Media for the Cultivation of Primary Cells and Their Susceptibility to Viruses." In Vitro 16: 616-628.

Weiss SA, Gorfien S, Fike R, DiSorbo D and Jayme D (1990) "Large Scale Production of Proteins Using Serum-Free Insect Cell Culture." Proceedings of the Ninth Australian Biotechnology Conference, In: Biotechnology: The Science and the Business, pp. 220-231.

Weiss SA, Belisle BW, DeGiovanni A, Godwin G, Kohler J and Summers MD (1989) "Insect Cells as Substrates for Biologicals." In: Proceeding of a Conference on Biotechnology, Biological Pesticides and Novel Plant - Pest Resistance for Pest Management. DW Roberts and RR Granados, eds., Boyce Thompson Institute for Plant Research, Ithaca, NY, pp. 271-189.

GROWTH AND FUNCTION OF BOVINE GRANULOSA CELLS CULTURED IN A SERUM-FREE MEDIUM

H. HOSHI, [1] Y. TAKAGI, [2] K. KOBAYASHI, [1] M. ONODERA, [3]
and T. OIKAWA [1]
[1] Research Institute for the Functional Peptides, Yamagata,
990, Japan, [2] Faculty of Agriculture, Shinshu University,
Nagano, 399-45, Japan, [3] Bio-Science Laboratory, Inc.,
Yamagata, 990, Japan

ABSTRACT
    The granulosa cell has an important physiological role in support-
ing the formation of ovarian follicles and the maturation of oocytes and
early embryos. We have developed an improved serum-free medium to optimize the cell growth of bovine granulosa cells (BGC). BGC seeded on
collagen-coated culture plates proliferated progressively in a serum-
free medium supplemented with insulin, heparin binding growth factor-2
(HBGF-2), lipoprotein (LP), and BSA. The cell doubling time at logarithmic phase and final cell density were equal to those in serum-containing medium. Extensive cell proliferation of BGC was primarlily important
to induce the early bovine embryo development in vitro. When embryos
were co-cultured with BGC in a defined medium, insulin and HBGF-2
stimulated embryonic development to the blastocyst stage as effective
as serum did. Denuded embryos (without granulosa cells) could not enter
into the blastocyst stage even in the presence of insulin and HBGF-2.
These results suggest that insulin and HBGF-2 stimulate early bovine
embryonic development by activating BGC function.

INTRODUCTION
    The bovine granulosa cell culture system has been considered to
take advantage of the understanding of hormonal effects on cell proliferation and function. Growth and maintenance of granulosa cells are
normally supported by a serum-containing medium. Although growth factors,
hormones, binding proteins, spreading factors, and nutrients have been
characterized in serum, serum still contains many undefined components
that are necessary to support the maximal growth of these cells. In some
systems, however, serum severly inhibited the differentiated function of
granulosa cells (1-3). So far, in vitro experiments on granulosa cells
in a serum-free medium have been done primarily with short-term incubation and little is known about the survival and long-term growth of
these cells in the total absence of serum (4-6). To study the mechanisms
involved in the growth and function of granulosa cells, it is important
to develop defined media that can support the maximal cell proliferation
and the long-term maintenance of cell function.

213

R. Sasaki and K. Ikura (eds.), Animal Cell Culture and Production of Biologicals, 213–219.
© 1991 *Kluwer Academic Publishers.*

In cattle, in vitro development of early embryos is generally arrested at the 8-to 16-cell stage (7). These have been recent reports that successful culture of early bovine embryos to the blastocyst stage by transversing the 8-to 16-cell block has been obtained by co-culture with oviductal epithelial cells (8, 9), trophoblastic vesicles (10), and granulosa cells (11-13) in a serum-containing medium. Although a variety of serum preparation from animals are usually required for early embryonic development in vitro, their chemical compositions in serum and bovine serum albumin vary among different preparations or batches (14, 15). Therefore, serum made it difficult to elucidate the function of somatic cells clearly. In this study, we describe a new serum-free medium that allows early bovine embryos matured and fertilized in vitro to grow to the blastocyst stage with the co-culture of granulosa cells.

MATERIALS AND METHODS

Materials
Heparin binding growth factor-2 (HBGF-2), referred to as basic fibroblast growth factor, was obtained from R&D Systems, Inc. (Minneapolis, USA). Bovine lipoprotein (LP)(EX-CYTEⅢ, Miles Inc. Kankakee, USA) is an aqueous mixture of lipoproteins (mostly high density lipoprotein), which bind cholesterol, phospholipids, and unsaturated fatty acids.

Cell isolation and cell culture of bovine granulosa cells (BGC)
BGC were isolated and cultured by the method previously described (15). Briefly, BGC were prepared from ovaries of young nonpregnant heifers and subcultured in the growth medium [DME:F12 (1:1) supplemented with 10% fetal bovine serum (FBS; Flow laboratories)].

Cell growth assays of BGC
Cell growth assays were done with granulosa cell stock cultures at passages between 4 and 9. Cultures were harvested with trypsin solution and cells were counted by a Coulter Counter. Media IFP100 and IFP110 were defined as synthetic media DME:F12 (1:1) or TCM199 plus 500 ng/ml aprotinin (Sigma). Aprotinin was added to the medium to neutralize the residue of trypsin during cell dispersion.

Oocyte collection and maturation in vitro
Ovaries were obtained from Japanese Black cows and heifers killed at a local slaughterhouse. Oocytes with an intact, unexpanded cumulus oophorus were collected and incubated to maturation medium containing TCM199 without HEPES and with 10% FBS on 60 mm culture dishes, which were covered with mineral oil. Those were cultured in a CO_2 incubator (5% CO_2:95% air with high humidity at 38.5℃) for 21-23 h.

Sperm preparation and insemination
Sperm preparation and insemination were done as described before (16).

In vitro development of embryos

After 6 h of insemination, all inseminated oocytes were transferred into 0.35 ml of the various development media tested and cultured in a CO_2 incubator (5% CO_2:95% air with high humidity at 38.5℃). The embryos were denuded from surrounding BGC by gentle pipetting at 48 h (Table 1) or 24 h (Table 2) after insemination, and denuded embryos were co-cultured with the remaining monolayer of BGC, or were cultured alone in a development medium.

Statistical analysis

Statistical analysis in the comparison of the frequencies of bovine embryo development were done by the chi-square test.

RESULTS

Among over 20 purified and crude preparations of hormones and growth factors tested in the presence of limiting amounts of serum (0.5 %), HBGF-2, insulin, LP, and BSA had a significant mitogenic activity for BGC (data not shown). The proliferation rate and final cell density of low density (10 cells/mm²) granulosa cell cultures plated in the serum-free medium or serum-containing medium is shown in Fig.1.

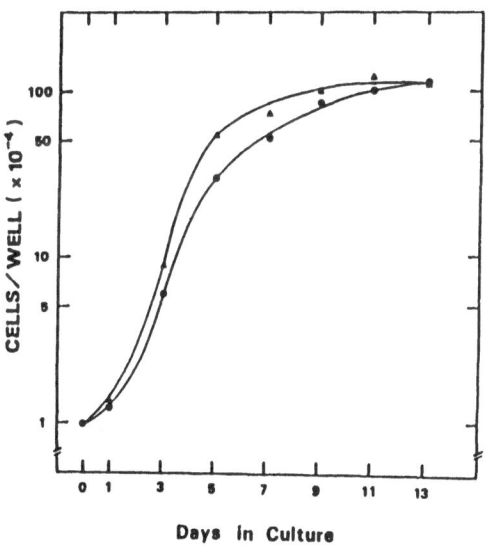

Days in Culture

Fig.1. Cell growth of granulosa cells cultured in the complete serum-free medium or serum-containing medium.

Granulosa cells were seeded at 10¹ cells per 35-mm diameter, collagen-coated dishes. The complete serum-free medium (●—●) contains medium IFP 100, HBGF-2 (10 ng/ml), insulin (5 μg/ml), lipoprotein (25 μg/ml), and BSA (2.5 mg/ml) or serum-containing medium (▲—▲) contains IFP100 and 10% FBS. Each medium was changed every 3 days. The cultures were trypsinized and counted on every other day. Values are expressed as the mean of triplicate dishes.

Population doubling times during their logarithmic growth phase (between 1 to 5 days in culture) exposed to the serum-free medium and serum-containing medium were 20.6 hr and 18.1 hr, respectively. After 13 days in culture, the final cell density (1180 cells/mm²) in serum-free medium was equal to that (1170 cells/mm²) in serum-containing medium. Figure 2 indicates the effects of various individual factors and their combina-

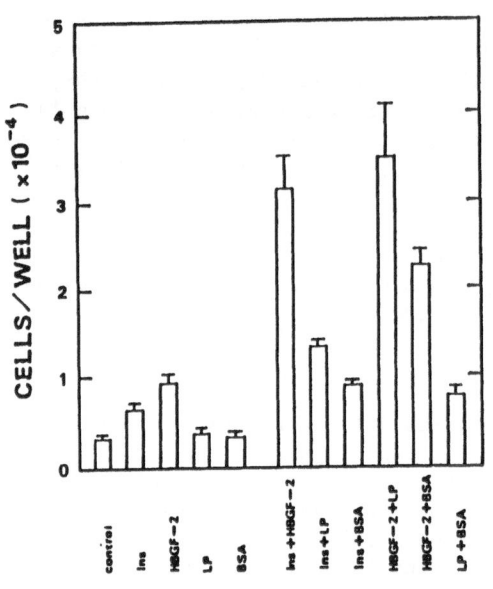

Fig.2. Effects of insulin, HBGF-2, lipoprotein, BSA, and their combinations on the proliferation of bovine granulosa cells cultured in a serum-free medium.

Granulosa cells were seeded at 5×10^3 cells per 24-well, collagen coated culture plates, BGC were exposed to IFP100 medium supplemented with insulin (Ins, 5 μ g/ml), HBGF-2 (10 ng/ml), lipoprotein (LP, 25 μ g/ml), BSA (1 mg/ml), and combinations of these components. After 5 days in culture the cells were trypsinized and counted. Values are expressed as the mean ± S.D of triplicate cultures.

tions on the proliferation of BGC in a serum-free medium. The addition of LP or BSA alone to the nutrient medium IFP100 did not stimulate cell proliferation. Insulin or HBGF-2 alone had a small mitogenic effect of BGC. If LP, BSA, or insulin was added together with HBGF-2, synergistic cell proliferation was observed in all combinations. Insulin, LP, and BSA had additive mitogenic effects on each other.

The effects of insulin, HBGF-2, LP, and BSA on the bovine embryonic development by co-culturing with BGC in a defined medium are shown in Table 1. While the cleavage rates ($\geqq$ 2-cell) did not vary among culture

Table 1. The effects of various growth factors on bovine embryo development to the blastocyst stage with BGC in a defined medium in vitro.

Culture conditions		No. of oocytes examined	No. and % developed to:		
			$\geqq$2-cell	$\geqq$8-cell	$\geqq$Blastocyst
IFP110 alone		131	75 (57%)	44 (34%)	5 (4%) [a]
+Insulin	(5μg/ml)	133	79 (59%)	58 (44%) [a]	25 (19%) [b]
+HBGF-2	(10 ng/ml)	118	69 (58%)	50 (42%)	17 (14%) [b]
+LP	(25μg/ml)	91	55 (60%)	38 (42%)	3 (3%) [a]
+BSA	(1 mg/ml)	86	44 (51%)	27 (31%)	1 (1%) [a]
+4F [*]		87	46 (53%)	30 (34%)	13 (15%) [b]
IFP110+5%FBS		170	76 (45%)	48 (28%) [b]	24 (14%) [b]

4F[*] : Insulin + HBGF-2 + LP + BSA

a vs. b : Means with different superscripts differ (p<0.01)

conditions, the proportion of embryo development to blastocyst stage
was significantly (p<0.01) increased by the addition of insulin or HBGF-
2 and the combination of four factors (4F) in comparison with that in
IFP110 alone. On the other hand, the addition of LP or BSA showed no
significant stimulation of embryonic development. Direct effects of
insulin and HBGF-2 on embryonic development of denuded embryos without
co-culture with BGC were investigated (Table 2). Although the rate of

Table 2. In vitro development of denuded embryos with or without co-cultivation
of BGC in a defined medium.

Culture medium	Co-cultivation of BGC	No. of oocytes examined	No. and % developed to:		
			$\geq$2-cell	$\geq$8-cell	$\geq$Blastocyst
IFP110 alone	−	59	37 (63%)	25 (42%)	0 (0%) [a]
+insulin (5 μg/ml)	−	269	153 (57%)	108 (40%)	0 (0%) [a]
+HBGF-2 (10 ng/ml)	−	56	34 (61%)	21 (38%)	0 (0%) [a]
IFP110+insulin (5 μg/ml)	+	282	177 (63%)	132 (47%)	56 (20%) [b]

a vs. b : Means with different superscripts differ (p<0.001)

cleavage ($\geq$ 2-cell) and 8-cell stage did not differ (>0.05) among dif-
ferant culture conditions, but no further developments of denuded
embryos even in the presence of insulin or HBGF-2 were observed. In
contrast, a high frequency (P<0.01;20%) of denuded embryos developed to
the blastocyst stage was obtained in a co-cultured system with BGC.

DISCUSSION
     We have developed an improved serum-free medium to optimize the
cell growth of BGC and maintain the long-term growth of these cells.
When BGC were cultured in a nutrient medium (IFP100) supplemented with
HBGF-2, insulin, lipoprotein, and BSA on collagen-coated culture plates,
the cell doubling time and final cell density were almost equal to those
in serum-containing medium. These are several reports that granulosa
cells from different species could be maintained and cultured in a serum
-free medium (2, 4, 5, 6). Orly et al. (2) has reported that rat granulosa cells produced steroid hormones in serum-free medium. Furthermore,
serum strongly inhibited follicle stimulating hormone-induced ster-
roidogenesis in these cells, so serum-free culture must be very important to analyze the role of hormones and growth factors on granulosa
cell function. Insulin and insulin-like growth factors are known to
modulate granulosa cell proliferation and function (4, 5, 17, 18). FGF
and high density lipoprotein were essentially required to induce BGC
proliferation in a serum-free medium on undefined extracellular matrix-
coated dishes (4). BSA may be important for the full proliferation of
BGC in a serum-free medium to function as a carrier and storage of
lipids (19).

218

When bovine embryos were co-cultured with BGC in a chemically defined medium, insulin and HBGF-2 independently promoted in vitro development of 1-cell embryos to blastocyst stages with high frequency as serum did. It has been reported that the presence of granulosa cells was important for normal fertilization and further development in serum-containing medium (20-21). Our recent findings also indicate that cell proliferation of BGC surrounding with early bovine embryos might be involved in the high incidence of blastocyst formation in a serum-free medium (16). Moreover, insulin and HBGF-2 are potent mitogens for granulosa cells (4-5). Insulin is known to modulate the cell function of granulosa cells such as steroids and oxytocin production which are important for follicular and oocyte maturation (22-23). There is evidence that insulin acts directly on the embryo development in vitro (24-25) while HBGF-2 and insulin are principal mitogens for granulosa cells, our observation demonstrated that denuded 1-cell embryos failed to transverse until the blastocyst stage without co-culture with BGC even in the presence of both factors. This result suggests that insulin and HBGF-2 may have no direct effect on the stimulation of bovine embryo development in vitro. Our improved serum-free medium is a great tool to characterize the embryotrophic factor produced from granulosa cells as well as to elucidate the granulosa cell proliferation and function.

REFERENCES
1. Erickson, G.F., Wang, C., and Hsueh, A.J.W. (1979), Nature., 279: 336-338.
2. Orly, J., Sato, G., and Erickson, G.F. (1980), Cell., 20: 817-827.
3. Luck, M.R. (1989), J. Exp. Zool., 251: 361-366.
4. Savion, N., Lui, G-M., Laherty, R., and Gospodarowicz, D. (1981) Endorinology., 109: 409-420.
5. Barrano, J.S., and Hammond, J.M. (1985), Endocrinology., 116: 51-58.
6. May, J.V., Frost, J.P., and Schomberg, D.W. (1988), Endocrinology., 123: 168-179.
7. Wright, R.W., and Bondioli, K.R. (1981), J. Anim. Sci., 53: 702-729.
8. Ellington, J.E., Carney, E.W., Farrell, P.B., Smith, M.E., Foote, R. H. (1990), Biol., Reprod., 43: 97-104.
9. Ellington, J.E., Farrell, P.B., Simkin, M.E., Foote, R.H., Goldman, E.E., and McGrath, A.B. (1990), J. Reprod. Fert., 89: 293-299.
10. Heyman, Y., Menezo, Y., Chesne, P., Camous, S., and Garnier, V. (1987), Theriogenology., 27: 59-68.
11. Kajihara, Y., Goto, K., Kosaka, S., Nakanishi, Y., and Ogawa, K. (1987), Jpn. J. Anim. Reprod., 33: 173-180.
12. Fukuda, Y., Ichikawa, M., Naito, K., Toyoda, Y. (1990), Biol. Reprod., 42: 114-119.
13. Fukui, Y. (1990), Mol. Reprod. Develop., 26: 40-46.
14. Kane, M.K., and Headon, D.R. (1980), J. Reprod. Fert., 60: 469-475.
15. Skinner, M.K., and Osteen, K.G. (1988), Endocrinology., 123: 1668-1675.
16. Takagi, Y., Mori, K., Tomizawa, M., Takahashi, T., Sugawara, S., and Masaki, J. (Accepted), Theriogenology.
17. May, J.V., and Schomberg, D.W. (1981) Biol. Reprod., 25: 421-430.

18. Adashi, E.Y., D'Ercole, A.J., Svoboda, M.E., and Van Wyk, J.J. (1985) Endocr. Rev., 6: 400-420.
19. Kan, M., and Yamane, I. (1982), J. Cell. Physiol., 111: 155-162.
20. Staigmiller, R.N., and Moor, R.M. (1984), Gamete Res., 9: 221-229.
21. Fukui, Y., and Ono, Y. (1989), J. Reprod. Fert., 86: 501-506.
22. Modschein, J.S., Canning, S.F., Miller, D.G., and Hammond, J.M. (1989), Biol. Reprod., 40: 79-85.
23. Luck, M.R. (1989), J. Exp. Zool., 251: 361-366.
24. Harvey, M.B., and Kaye, P.L. (1988), Endocrinology., 122: 1182-1184.
25. Heyner, S., Rao, L.V., Jarett, L., and Smith, R.M. (1989), Dev. Biol., 134: 48-58.

Electrically controlled culture of MKN45 cells in serum-free medium

M. Aizawa,[1] J. Kojima,[1] H. Shinohara,[1] S. Morioka,[2] K. Nagaike[2] and
Y. Ikariyama[1]
1. Department of Bioengineering, Tokyo Institute of Technology, 2-12-1 Meguro-ku, Ookayama, Tokyo 152, Japan.
2. Bioscience Laboratory Research Center, Mitsubishi Kasei Corporation, 1000 Kamoshida-cho, Yokohama 227, Japan.

Abstract

Specific protein production of mammalian cells have been promoted by applying a low constant potential to the surface of an electrode on which cells were cultured in serum-free medium. A human carcinoma cell line, MKN45 cells, were cultured on a platinum-coated plate electrode and carbon-mesh electrode with an applied potential below 0.6 V vs. Ag/AgCl in serum-free medium. The carcinoembryonic antigen (CEA) production was dependent on the electrode potential and was rather enhanced at 0.4 V vs. Ag/AgCl on both types of electrode. This optimum potential of 0.4 V vs. Ag/AgCl agreed with the potential that maximally promoted the CEA production in fetal bovine serum (FBS) containing medium. This phenomenon could led us to design a new type of bioreactor which would control the protein production by electrical effects.

Introduction

Mammalian cell culture has become important for producing such proteins as monoclonal antibodies, interleukin, and erythropoietin. One of the strong interests of this field was to enhance and control the protein production of cells.   There are various kinds of approaches including physical methods which were represented by magnetic, optical, thermal, and electrical techniques.[1-3]   Our primary interests were concerned with the electrical stimulation of cellular metabolism, and we have long focused on electrical effects on cellular behaviors on the potential-controlled electrode surface.

When cells were in contact with the electrode, it is reasonable to believe that cells would be extremely affected by electrical effects induced on the electrode surface, namely a change of the plasma membrane potential or a change in ion concentration around the cells. This electric effect could modulate the permeability of membrane or the metabolism of cells. Our group has been researching the electrical effects of electrode surfaces on cellular functions, such as cell morphology, proliferation, differentiation, and protein production,[4-6] and we have found that these cellular functions were significantly dependent on the electrode potential.

221

R. Sasaki and K. Ikura (eds.), Animal Cell Culture and Production of Biologicals, 221–228.
© 1991 *Kluwer Academic Publishers.*

The characteristic feature in mammalian cell culture is the addition of serum to the medium. The role of serum is quite important for growth of cells and preserving the cellular functions, but it includes various unknown substances, which occasionally disturb the research. The study of serum-free culture has been very popular and several special media and cell lines have been developed. In this research, we investigated the electrical effects of electrode surfaces on cells in serum-free medium. This reserch will clarify the electrical effects on cells that are not related to the substances derived from the medium.

MKN45 cells were derived from a human stomach carcinoma and produce carcinoembryonic antigen as a specific protein.[7] It is possible to culture this cell in serum-free RPMI1640 medium. MKN45 cells were cultured on the surface of a platinum-coated plate electrode and a carbon-mesh electrode at a low constant potential and the electrical effects on CEA production were investigated.

Experimental

Cell culture

MKN45 cells were cultured on two types of working electrode. One is a platinum-coated plate electrode and the other is a carbon-mesh electrode. A plastic plate for tissue culture was coated by platinum by ion spattering and used as a working electrode. The electrode system is shown in Fig. 1 and described in previous papers in detail.[5,6] The carbon-mesh was from Mitsubishi Kasei Co.. The carbon fibers, the diameter of which was about 10 microns were woven into cloth and used as a working electrode, too. There were three steps in cell culture.

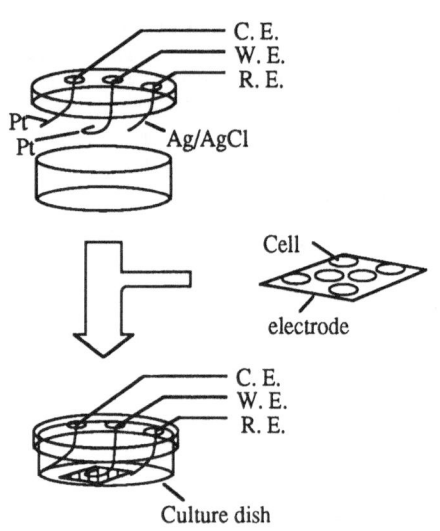

Fig. 1 Experimental set-up for MKN45 cell culture on an electrode.

In the beginning MKN45 cells were cultured in 10% FBS containing E-RDF medium for a hundred hours and after that cells were incubated in serum-free RPMI1640 medium for another hundred hours. Then, the cell-attached plate was moved to the electrode-installed dish and a potential was applied. The electrode potential was controlled with a potentiostat and the electrode-installed dish was placed in a 5% CO_2 incubator.

Measurement of CEA

The secreted CEA into the medium was measured by an immuno-assay system, IMX of Dinabot Co. The membrane-bound CEA was stained by the fluorescence probe Texas Red through immunochemical reactions and the fluorescence intensity was detected and analyzed with an ARGUS of Hamamatsu Hotonics Co.

Observation of cell morphology

The morphology of MKN45 cells cultured on the platinum-coated plate electrode was observed through an inverted microscope once in a day. Cells cultured on the carbon-mesh electrode were identified by electron microscope after finishing a hundred hours of electrically controlled cell culture.

Results

Platinum-coated plate electrode

MKN45 cells were stimulated by a small constant potential by the previously described procedure. We have already investigated the electrical effect on cell proliferation in FBS-containing medium. In that case, cell proliferation was closely related to the applied potential and it was inhibited at the potential range above 0.4 V vs. Ag/AgCl.[6] In this experiment, after 200 hours of total incubation in FBS-containing E-RDF medium for 100 hours and serum-free RPMI1640 medium for 100 hours, the relationship between proliferation and applied potential was studied. The results are shown in Fig. 2, and the total number of cells accumulated every hour was fairly independent of the applied potential. Cell proliferation was inhibited not only at the potential range above 0.4 V vs. Ag/AgCl, but also lower potentials like 0.1 V vs. Ag/AgCl, which did not restrain the cell proliferation in FBS-containing medium. The total cell number was fairly constant.

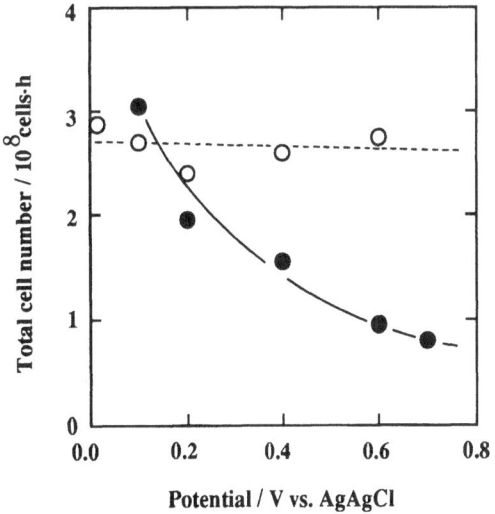

(○) serum-free medium

(●) serum-containing medium

Fig. 2 Total cell number dependence on potential.

The electrical effects on CEA production were investigated in two ways; one is the effect on secreted CEA into the medium and the other is the effect on membrane-bound CEA. After potential application, the medium was sampled every day and the concentration of CEA was measured by enzyme immunoassay. The CEA production per cell and per hour during a hundred-hour incubation is showed in Fig. 3.

At the positive potentials around 0.4 V vs. Ag/AgCl CEA secretion was rather enhanced over that at the lower potentials like a resting potential. The potential of 0.4 V at which CEA secretion was maximally enhanced in serum-free medium agreed with the potential of that in FBS-containing medium. From these results we see that CEA secretion was promoted at 0.4 V not only in serum-containing medium but also in serum-free medium. However, concerning the final CEA concentration in the medium, there was a difference in potential dependency between these cases in medium as shown in Fig. 4. In both cases CEA secretion was

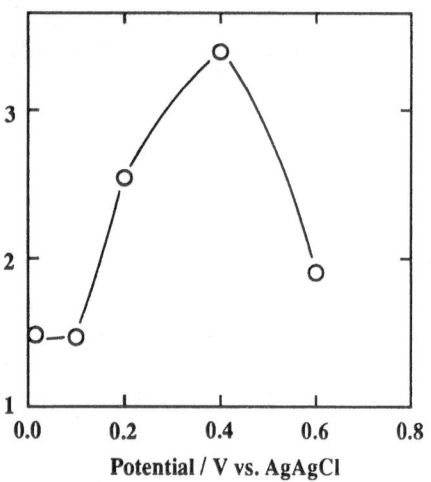

Fig. 3 Dependence of CEA secretion on applied potential in serum-free medium.

promoted at around 0.4 V, but in FBS-containing medium, the cells proliferate more at lower potentials, so there was a difference in total cell number between 0.4 V and resting potential. This phenomenon was reflected and the highest concentration of CEA was detected at 0.2 V in FBS-containing medium. But in the serum-free culture, there were not much difference in the total cell

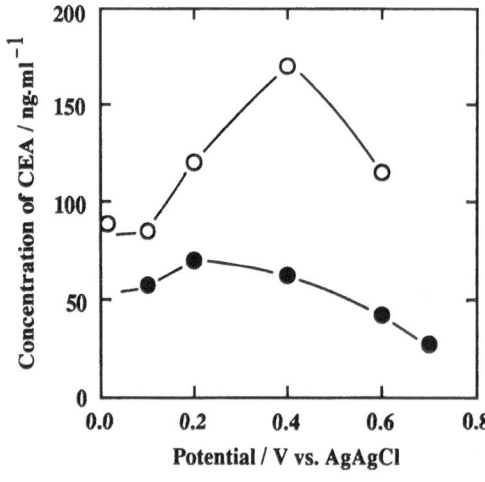

(O) serum-free medium

(●) FBS-containing medium

Fig. 4 Dependence of final CEA concentration on applied potential.

number. So we got the highest CEA concentration medium at 0.4 V, at which the CEA secretion was maximally promoted.

The electrical effect on the amount of membrane-bound CEA was also investigated by the fluorescent probe method. After finishing a hundred-hour electrically controlled culture, the membrane surface of MKN45 cells stained by the fluorescent probe Texas Red through an immunochemical reaction. The amount of membrane-bound CEA was estimated by the fluorescent intensity of Texas Red. The fluorescence intensity of each cell was integrated by an imaging analyzer and displayed in Fig. 5. It was found that the amount of membrane-bound CEA also had potential dependency and it was maximally increased around 0.4 V vs. Ag/AgCl. Judging from these results presented in Fig. 3 and Fig. 5, electrically controlled culture enhanced the protein production at the potential around 0.4 V in serum-free medium.

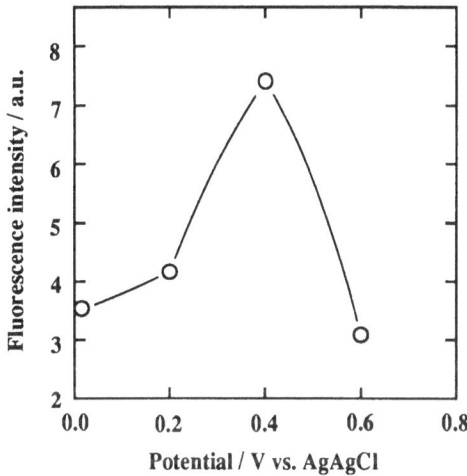

Fig. 5 Dependence of membrane bound CEA on applied potential in serum-free medium.

Carbon-mesh electrode

The platinum-coated electrode was replaced by a carbon-mesh electrode to investigate the material dependency of the electric effect. The cellular adhesion to the carbon was first studied by electronmicroscopy and it was confirmed that MKN45 cells anchored on the surface of carbon. The potential dependency of the final CEA concentration after a hundred-hour electrically controlled culture in serum-free medium was investigated and is displayed in Fig. 6. This figure also showed clear potential dependency and the amount of secreted CEA was greatly increased at the potential around 0.4 V. The tendency of potential dependency was quite similar with the result of a platinum-coated plate electrode. It was demonstrated CEA production of MKN45 cells were promoted on the carbon-mesh electrode by electrically controlled culture in serum-free medium.

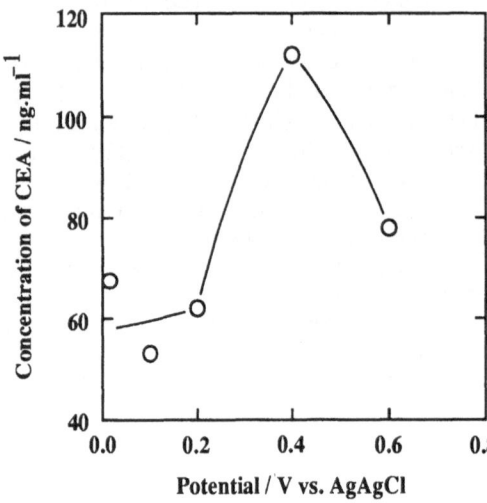

Fig. 6 Dependence of final CEA concentration on appled potential on carbon-mesh electrode.

Discussion

We have studied the electrical effects of electrode surfaces on protein production in FBS-containing medium and found that CEA production of MKN45 cells was greatly enhanced at the positive potentials, but we could not get such higher concentration of CEA solution at 0.4 V that promoted the CEA production maximally (Fig. 4), because cell proliferation was extremely inhibited at the potential and there was a difference in cell population between resting potential and 0.4 V vs. Ag/AgCl. In this research, we kept the cell number constant by using serum-free medium. There were three steps of cell culture in this experimental procedure. At first, cells proliferated in FBS-containing medium for a hundred hours and cells became nearly confluent. Then the medium was changed to serum-free medium. This cell line of MKN45 was able to keep their productivity in serum-free medium and their growth rate became slow although the growth did not stop completely. In this experiment the electrical effect was observed toward the cells that were nearly confluent and cultured in serum-free medium, so the rate of proliferation was quite slow even at the resting potential which cell proliferation was not influenced by potential application. As a result, we got the highest concentration of CEA at the potential which promoted the CEA production maximally.

In the point of discussion about the mechanism of electrical effects on cells, there was a possibility that the substances including in serum were related to the electric effect process. However it was shown in Fig. 3 and 5 that the CEA production was maximally enhanced at the potential range around 0.4 V and the potential dependency was quite similar between FBS-containing medium and serum-free medium. It suggested that this electrical effect of electrode surface on CEA production was independent of chemical substances including in serum.

One of the purposes of this research was to use electrically controlled culture in a large scale bioreactor. It is necessary to use electrodes which have a wide surface area for that purpose. Using a fiber type of electrode was one trial for it. In the mammalian cell culture field, glass and plastic have been used as the base plate material and other materials have not been investigated well. As MKN45 cells strongly adhered to the base plate, it was possible to cultivate these cells on both surfaces of platinum and carbon. This result indicated some types of cell line can be cultivated not only on glass and plastic base plates but also on the metal or other materials. In the experiment with a carbon-mesh electrode, the diameter of the carbon fiber was about 10 microns, so there was a quite large surface area as compared to the plain plate. In this result, as cell did not spread to all the surface of these fibers completely, the final cell number did not increase so much compared to that of tissue culture plate. It is necessary to further investigate and apply this type of electrode but it was confirmed that this carbon was appropriate for CEA production of MKN45 cells.

In this research, it was found that the electrical effects of electrode surfaces promote CEA production of MKN45 cells in serum-free medium both on a platinum-coated plate electrode and a carbon-mesh electrode and it indicated that a carbon-mesh electrode is a possibility for designing a large-scale electrically controlled bioreactor.

References

1. R. Goodman, L. Wei, J. Xu and A. Henderson, (1989) 'Exposure of human cells to low-frequency electromagnetic fields results in quantitative changes in transcripts', Biochim. Biophys. Acta, 1009, 216-220.

2. H. Merivuori, M. Tornkvist and J. A. Sands, (1990) . 'Different temperature profiles of enzyme secretion by two common strains of Tricnoderma Reesei', Biotechnol. Lett., 12, 117-120.

3. H. Matsuoka, S. Matsumoto and Y. Takekawa, (1986) 'Directional control of yeast cell division by electric stimulus', Bioelectrochem. Bioenerg., 16, 235-242.

4. M. Yaoita, Y. Ikariyama, H. Shinohara and M. Aizawa, (1989) 'Electrically regulated cellular morphological and cytoskeletal changes on an optically transparent electrode', Exp. Cell Biol., 57, 43-51.

5. M. Yaoita, Y. Ikariyama, M. Aizawa, (1990) 'Electrical effects on the proliferation of living HeLa cells cultured on optically transparent electrode surface', J. Biotechnol., 14, 321-332.

228

6. J. Kojima, H. Shinohara, Y. Ikariyama, M. Aizawa, K. Nagaike and S. Morioka, 'Electrically controlled proliferation of human carcinoma cells cultured on the surface of an electrode', J. Biotechnol. (in press).

7. T. Motoyama, H. Hojo, T. Suzuki and S. Oboshi, (1979), 'Evaluation of the regrowth assay method as an in vitro drug sensitivity test and its application to cultured human gastric cancer cell lines', Acta Med. Biol., 27, 49-63.

HYBRIDOMA CULTURE IN THE HOLLOW-FIBER SYSTEM
-THE EFFECTS OF GROWTH FACTORS-

Takeshi Omasa,[1] Masaki Kobayashi,[1] Toshio Nishikawa,[1]
Suteaki Shioya,[1] Ken-ichi Suga,[1] Syo-ichi Uemura,[2] and
Yoshio Imamura[2]
[1]Department of Fermentation Technology, Faculty of Engineering,
Osaka University, Yamada oka 2-1, Suita-shi, Osaka 565, Japan
[2]Toyobo Co. Ltd., Katata 2-1-1, Otsu-shi, Shiga 520-02, Japan

ABSTRACT. Based on the data from 60 days' continuous cultivation, the
effects of high molecular weight growth factors on antibody production
were analyzed quantitatively. Transferrin had no effect on antibody
production, while BSA enhanced the antibody production rate
significantly. The antibody production rate increased 4 times and 14
times in the case of the feeding BSA of 2 g/ℓ and 5g/ℓ into the EC space
(space connected to the outer part of the hollow fibers) respectively
compared with the data without addition of BSA. Five g/ℓ of BSA addition
into the IC space (space connected to the inner part of the hollow
fibers) resulted in 2.5 times increase of this production rate compared
with no addition of BSA. The antibody production rate in the hollow-
fiber system was increased 3 times by BSA feeding as much as that in the
perfusion culture system.

1. INTRODUCTION

In hollow-fiber (HF) cell culture systems, It has been reported that
both high cell density ($\simeq 1\times10^8$ (cell/mℓ)) and high productivity
(Altshuler et al. (1986), Hopkinson (1985)) were obtained easily.
However, there were only a few reports that analyzed the experimental
data in HF systems by considering the effects of growth factors. Our
aims of this study are to establish the method for data analysis in HF
system and to investigate the effects of growth factors on cell growth
and antibody production.

2. MATERIALS AND METHODS

2.1. Cell Line and Culture Medium

The mammalian cells used in the experiments were from the mouse-mouse
hybridoma cell line 3A21. The cell line was derived from cell fusion
between a spleen cell from a BALB/c AnNCrj mouse immunized with bovine
pancreatic ribonuclease A (RNase A)(sigma R-5125) and the mouse myeloma
cell (P3-X63-Ag8-U1). This hybridoma produces an anti-RNase A monoclonal
antibody (IgG). The tissue culture medium was serum-free medium, RDF-

R. Sasaki and K. Ikura (eds.), Animal Cell Culture and Production of Biologicals, 229–236.
© 1991 *Kluwer Academic Publishers.*

ITES with BSA (Murakami et al. (1982)), that is, RDF (a 2:1:1 mixture of RPMI 1640, DMEM, F12) without glucose and glutamine (Kyokuto RDF(HO)) supplemented with glucose, glutamine and growth factors, i.e., insulin (Sigma I-5500), transferrin (Sigma T-2252), ethanolamine (Nacalai tesque 234-05), selenium (Nacalai tesque 172-94), (ITES), and bovine serum albumin (BSA) (Sigma A-4503).

2.2. Sample Analysis

Each 1.5ml sample was centrifuged (6800 x g, 10 min) to remove the cells. Glucose was measured by a glucose analyzer (Yellow Spring YSI 27), and lactate by an enzymatic lactate analyzer (Toyobo diagluca HEK-30L). Antibody concentration was measured by an Enzyme Linked Immunosorbent Assay (ELISA) method. Alkaline phosphatase-labeled goat anti-mouse IgG antibody (TAGO 6550) was used as the enzyme-linked second step antibody. The absorbance at 405 nm was measured by an ELISA auto-plate-reader (TOSOH MPR A4). Antibody concentration was calculated from the ratio of both absorbances at A_{405} of the sample and purified antibody.

At the end of hollow-fiber culture, the hollow fibers on which the cells grew were disjointed into pieces and washed by trypsin solution. The cell sample was diluted 1:1 with 0.16% trypan blue – 0.85% NaCl solution, by which only dead cells were dyed. Cell concentration in 0.9µL was estimated on a Burker-Turk hemacytometer (ERMA 4296) then averaged.

2.3. Hollow-fiber (HF) Cell Culture System

Double hollow-fiber units were used in this system. One was for cell culture (50% molecular weight cutoff of 30,000, Amicon Minivitafiber MVF 0004) and another for separation (50% molecular weight cutoff of 5,000, Nipro FB-90T) of some waste small molecules. The fresh medium was continuously fed into the innercapillary (abbreviated as IC) space of hollow fiber unit. In the extracapillary (abbreviated EC) space of the hollow-fiber unit, cells were grown stably by attaching onto the hollow-fiber membrane softly. The medium was withdrawn through the separation hollow-fiber unit. High molecular weight growth factors, such as transferrin and BSA were fed into the IC or EC side properly. The cells were grown in the EC space and the produced antibody was also accumulated in the EC space.

3. RESULTS AND DISCUSSIONS

3.1. Establishment of the Method for Data Analysis

The changes in glucose and lactate concentration in 60 days' continuous cultivation (RUN 1) are shown in Fig. 1.

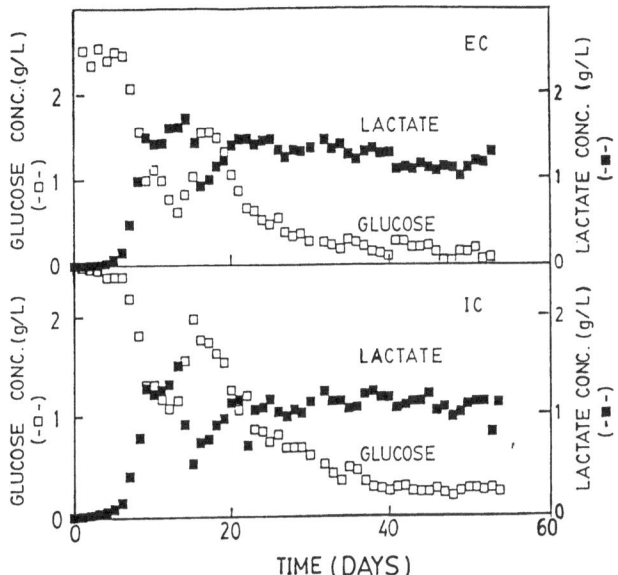

Figure 1 The changes in glucose and lactate concentration in 60 days' continuous cultivation (RUN 1)

After the glucose concentration decreased and lactate concentration increased, both concentrations became constant. The cultivation could be maintained stably during the 60 days without serious fouling in the separation membrane. The fresh medium was fed through the IC space into EC space and the lactate was produced by the hybridoma cells which grew only in the EC space. However, because a significant amount of lactate was observed in the IC space and also glucose concentration in the IC space was not the same as in the fresh medium, mass transfer between EC space and IC space might be considered. We constructed the following model for data analysis.

$$V_I \frac{dG_I}{dt} = F(G_0 - G_I) - \phi_G \tag{1}$$

$$V_I \frac{dL_I}{dt} = -FL_I + \phi_L \tag{2}$$

$$V_E \frac{dG_E}{dt} = F(G_I - G_E) + \phi_G - R_G \tag{3}$$

$$V_E \frac{dL_E}{dt} = F(L_I - L_E) - \phi_L + R_L \tag{4}$$

$$V_E \frac{dAb}{dt} = R_{Ab} \tag{5}$$

$$\phi_G = \alpha(G_I - G_E) \tag{6}$$

$$\phi_L = \alpha(L_I - L_E) \tag{7}$$

If ϕ_G and R_G were assumed to be constant during the time, by integrating both sides of Eqs. 1 and 3, the following equations were obtained,

$$\phi_G\, t = \left| \overline{F(G_0 - G_I)dt + V_I G_I - (V_I G_I)_0} \right. \tag{8}$$

$$(R_G - \phi_G)\, t = \left| \overline{F(G_I - G_E)dt - V_E G_E + (V_E G_E)_0} \right. \tag{9}$$

ϕ_G and $R_G - \phi_G$ values were estimated by the least squares methods from plotting the value of underlined terms in the right hand sides of eq. 8 and 9 against time, respectively. The reaction rate of glucose R_G was calculated from ϕ_G and $R_G - \phi_G$. R_L and R_{Ab} were calculated by the same procedure.

3.2. Estimation of the Cell Concentration

The viable cell concentration in the HF cell culture unit was measured by stopping the cultivation in the middle of log phase and disjointing the hollow fibers. From these results, the kinetic parameters in HF system were calculated (Table 1). Parameters v_G and ρ_L of HF system were agreed with those in the batch culture using a spinner flask. Then, inversely the cell concentration could be evaluated using the overall rate obtained from HF system and specific rates such as v_G, ρ_L evaluated previously from the batch culture. For example, the $(R_G)_{HF\ system}/(v_G)_{batch}$ might give the cell concentration, properly. (Table 2)
But ρ_{Ab} of HF system was higher than the ρ_{Ab} in the spinner flask culture. The reason will be investigated in the following section.

TABLE 1. The kinetic parameters in the log phase of HF cell culture (RUN 2)

	HF culture	spinner batch culture
ρ_{Ab}	2.5×10^{-11}	$0.15 \sim 0.25 \times 10^{-11}$
v_G	1.7×10^{-10}	$1.0 \sim 2.0 \times 10^{-10}$
ρ_L	1.7×10^{-10}	$1.0 \sim 2.0 \times 10^{-10}$
ρ_{Ab}/ρ_L	0.15	$0.015 \sim 0.025$
ρ_{Ab}/v_G	0.15	$0.015 \sim 0.025$

Both estimated cell concentrations in HF system from the ratio R_G/v_G and R_L/ρ_L agreed with the measured cell concentration.

TABLE 2. Estimation of the cell concentration using R_G and R_L values

	from R_G value	from R_L value
cell number (cells)	1.9×10^8 ($= R_G/\nu_G$)	1.7×10^8 ($= R_L/\rho_L$)
estimated cell concentration in HF cell culture unit(4mℓ)(cells/mℓ)	4.75×10^7	4.25×10^7

measured cell concentration in HF cell culture unit (4 mℓ) at the end of cultivation $= 4.25 \times 10^7$ (cells/mℓ)

3.3. Effects of Growth Factors on Cell Growth and Antibody Production

In the HF system, high molecular weight growth factors, i.e., transferrin and BSA had been held in the EC space and the fresh medium containing BSA or transferrin was fed into the IC space continuously. Then, the effects of continuous feeding of BSA or transferrin into the EC space where the cells existed was investigated. Figure 2 shows the course of antibody concentration when the fresh medium containing 5 g/ℓ of BSA was fed into the EC space directly.

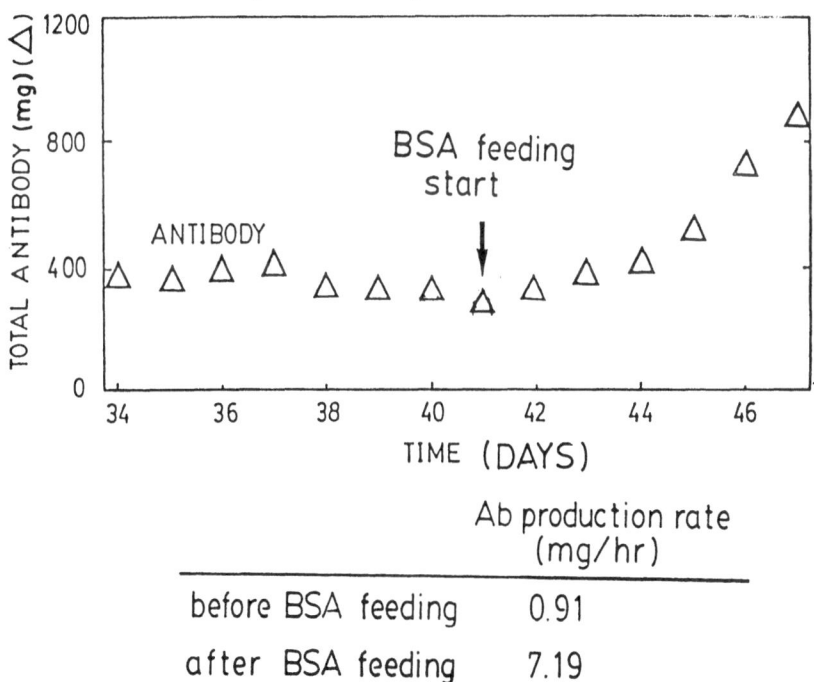

	Ab production rate (mg/hr)
before BSA feeding	0.91
after BSA feeding	7.19

Figure 2 The effects of BSA feeding on antibody production (RUN 1)

234

After the feeding of fresh medium containing 5 g/ℓ of BSA, the antibody concentration in the EC space increased greatly due to the increase of the antibody production rate. BSA might have enhanced the antibody production rate significantly. On the other hand, no significant increase of the antibody production rate by the continuous feeding of transferrin was observed (data not shown). The effects of continuous feeding of BSA into the IC or EC space on cell growth and antibody production were investigated (RUN 3). Moreover, we analyzed the effects of BSA in the perfusion culture system (RUN 4) to compare with the HF system (Fig. 3). The antibody production rate in the perfusion culture system was calculated on the assumption of the same cell number as the HF cell culture system.

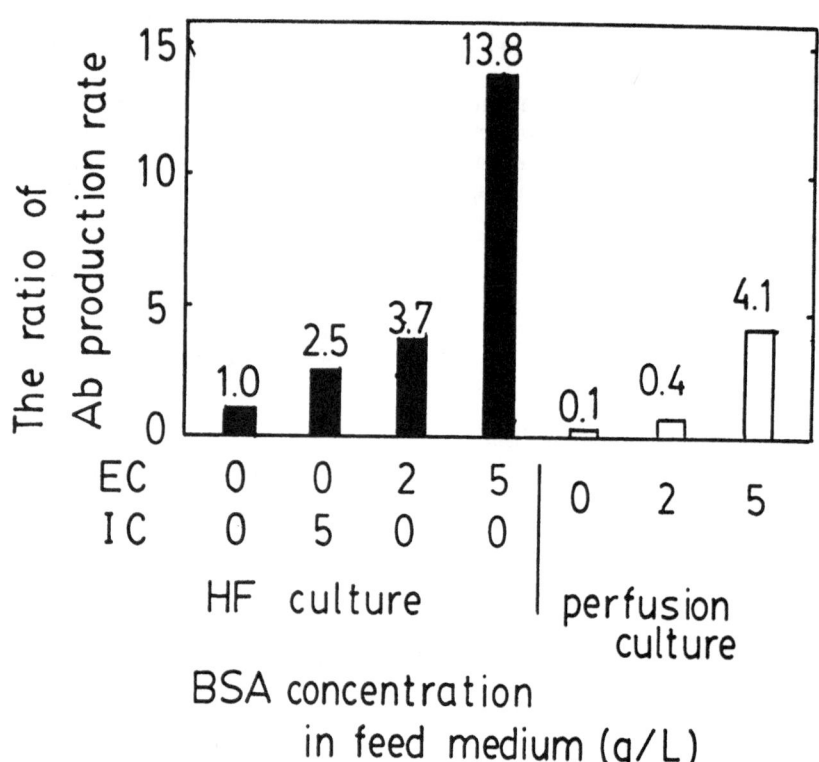

Figure 3 Comparison of BSA effects on antibody production rate in the different cell culture systems (RUN 3 and RUN 4)

The antibody production rate increased 4 times in the case of feeding 2 g/ℓ of BSA into the EC space, and 14 times in the case of 5 g/ℓ into the EC space compared with the controlled data without addition of BSA. Also, addition of 5 g/ℓ of BSA into IC space increased the production rate 2.5 times. However, both R_G and R_L values were not changed significantly during BSA feeding.

The antibody production rate by BSA feeding of same concentration in the HF system was 3 times as much as that by the perfusion culture. The role of BSA has been reported in terms of cell growth (e.g. Nilausen (1978), Rockwell et al. (1980)), but not in terms of antibody production. The role of BSA is now being investigated further.

4. NOMENCLATURE

Ab ; antibody concentration (g/ℓ)
F ; flow rate of fresh medium into IC space and withdrawal rate of waste medium from EC space (ℓ/h)
G ; glucose concentration (g/ℓ)
G_0 ; glucose concentration in feed medium (g/ℓ)
L ; lactate concentration (g/ℓ)
R_{Ab} ; antibody production rate (g/h)
R_G ; glucose consumption rate (g/h)
R_L ; lactate production rate (g/h)
V ; volume (ℓ)
α ; backmixing flow rate from EC space to IC space (ℓ/h)
ν_G ; specific glucose consumption rate (g/cell/h)
ρ_{Ab} ; specific antibody production rate (g/cell/h)
ρ_L ; specific lactate production rate (g/cell/h)
ϕ_G ; mass transfer rate of glucose (g/h)
ϕ_L ; mass transfer rate of lactate (g/h)
subscript E ; extracapillary space (EC space)
I ; innercapillary space (IC space)
0 ; initial value

5. REFERENCES

Altshuler G.L., Dziewulski D.M., Sowek J.A. and Belfort G. (1986) 'Continuous hybridoma growth and monoclonal antibody production in hollow fiber reactors-separators', Biotechnology and Bioengineering 28, 646-658.

Hopkinson J. (1985) 'Hollow fiber cell culture systems for economical cell-product manufacturing', Bio/technology 3, 225-230.

Murakami H., Masui H., Sato G.H., Sueoka N., Chow T.P. and Kano-Sueoka T. (1982) 'Growth of hybridoma cells in serum-free medium: Ethanolamine is an essential component', Proceedings of the National Academy of Sciences of the United States of America 79, 1163-1165.

236

Nilausen K. (1978) 'Role of fatty acids in growth-promoting effect
of serum albumin on hamster cells in vitro', Journal of Cellular
Physiology 96, 1-14.

Rockwell G.A., Sato G.H. and Mcclure D.B. (1980) 'The growth
requirements of SV40 virus transformed Balb/c-3T3 cells in serum-
free monolayer culture' Journal of Cellular Physiology 103, 323-
331.

EFFECTS OF GROWTH FACTORS ON HYBRIDOMA CULTURE IN THE PERFUSION SYSTEM

Takeshi Omasa,[1] Masaki Kobayashi,[1] Toshio Nishikawa,[1]
Suteaki Shioya,[1] Ken-ichi Suga,[1] Syo-ichi Uemura,[2] and
Yoshio Imamura[2]
[1]Department of Fermentation Technology, Faculty of Engineering,
 Osaka University, Yamada oka 2-1, Suita-shi, Osaka 565, Japan
[2]Toyobo Co. Ltd., Katata 2-1-1, Otsu-shi, Shiga 520-02, Japan

ABSTRACT. From the experiments using a perfusion culture system with a cell sedimentation unit and a separation unit composed of a hollow-fiber module, the effects of growth factors such as transferrin and BSA were analyzed. Transferrin enhanced the cell growth but had no significant effects on the specific antibody production rate. BSA promoted cell growth and raised the specific antibody production rate. When the fresh medium containing 2 g/ℓ or 5 g/ℓ of BSA was fed into the flask, both the specific growth rate and specific death rate increased. The specific antibody production rate was increased 2 times and 25 times by feeding 2 g/ℓ and 5 g/ℓ of BSA, respectively, compared with no addition of BSA. In the continuous culture without the separation membrane, BSA addition had no effects on the antibody production.

1. INTRODUCTION

The perfusion culture system with cell sedimentation unit and a separation unit composed of a hollow-fiber module, can keep the high molecular weight growth factors inside the system and remove the low molecular weight waste products from the system. The purposes of this report are to investigate the effects of insulin, transferrin and bovine serum albumin (BSA) on the hybridoma growth and antibody production using the perfusion culture system with separation unit and to compare with these effects in the continuous culture without separation unit.

2. MATERIALS AND METHODS

2.1. Cell Line, Culture Medium, Sample Analysis

The mammalian cells used in the experiments were from the mouse-mouse hybridoma cell line 3A21, which has been described in detail in our previous report (Omasa et al.(1990)). The culture medium contained principally RDF with ITES, which was same as our previous report (Omasa et al.(1990)). The sample analysis also followed our previous report (Omasa et al.(1990)). Insulin was measured by an ELISA method kit (Toyobo immunoball IRI-500).

237

R. Sasaki and K. Ikura (eds.), Animal Cell Culture and Production of Biologicals, 237–243.

238

2.2. Perfusion culture system

A 500 mℓ glass spinner flask (working volume 400∿1100 mℓ) equipped with a sedimentation vessel (200 mℓ) and the hollow-fiber module made of separation membrane (50% molecular weight cutoff of 5,000, Nipro FB-90T) was used for the perfusion culture. The medium was circulated between the sedimentation vessel and the hollow-fiber module at 100 mℓ/h. The fresh medium was fed into the flask at 25 mℓ/h and the medium in the flask was withdrawn through the separation membrane at the same rate. The high molecular weight growth factors i.e., transferrin and BSA could be kept inside this system

3. RESULTS

3.1. The Effects of Transferrin and Insulin on Cell Growth and Antibody Production

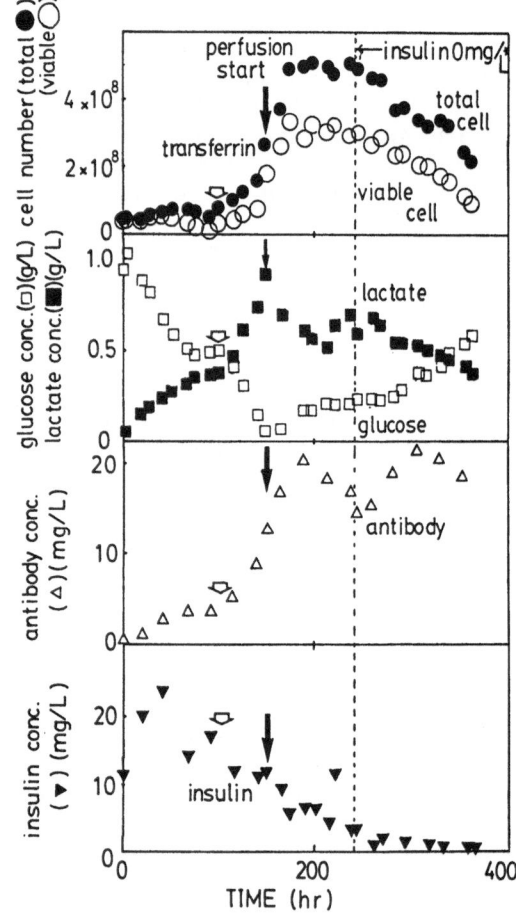

Figure 1 The effects of transferrin on cell growth and antibody production

In perfusion culture system, the cultivation started without transferrin and BSA (Figure 1). After transferrin was added at 101 hr, the cell number increased greatly. When the glucose was completely consumed, the fresh RDF medium which was supplemented with insulin, ethanolamine and selenium was fed into the system and the medium was withdrawn through the separation membrane (perfusion started). The cell number in the reactor was maintained without continuous feeding of transferrin during about 150 hr. When the insulin concentration in the feed medium was changed to zero at 247 hr, the cell number decreased. The insulin concentration in the spinner flask also decreased to zero. From these results, it was seen that continuous feeding of insulin was required to maintain this condition but that of transferrin was not essential. Transferrin might be retained by the membrane in this system. The kinetic parameters before and after the transferrin addition in the batch period were shown in Table 1.

Table 1 The effects of one dose of transferrin on the kinetic parameters

	before transferrin	after transferrin	ratio after/before
specific growth rate (1/h)	0.0192	0.0482	2.5
specific glucose (g/cell/h) consumption rate	1.52×10^{-10}	1.37×10^{-10}	0.9
specific lactate (g/cell/h) production rate	0.81×10^{-10}	1.11×10^{-10}	1.4
specific antibody (g/cell/h) production rate	1.35×10^{-12}	1.35×10^{-12}	1.0

As can been seen from Table 1, the dose of transferrin exerted a noticeable effect on the specific growth rate but not so much on the specific antibody production rate efficiently.

3.2. Effects of BSA on Cell Growth and Antibody Production

3.2.1. Continuous culture without the separation unit

To investigate the effects of BSA on cell growth and antibody production, the BSA concentration in feed medium was changed from 0 to 5 g/ℓ during the continuous operation (Figure 2).

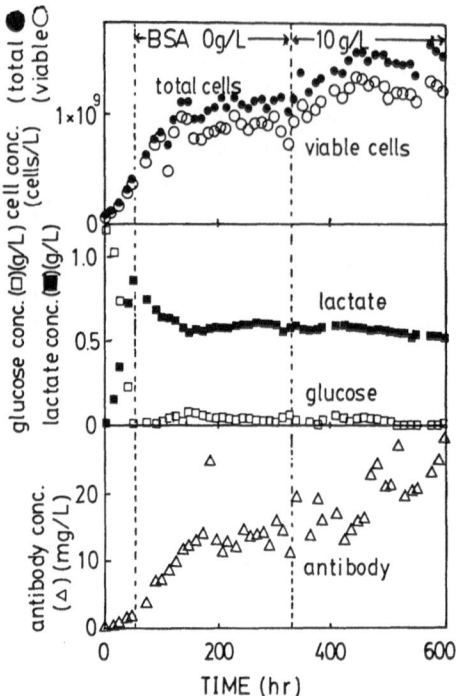

Figure 2 The course of continuous culture without the separation unit

While the feed medium didn't contain BSA, the viable and total cell, lactate, and glucose concentration were constant. When BSA concentration in feed medium was changed from 0 to 5 g/ℓ at 325 hr, the viable and total cell concentration increased but lactate and glucose concentration didn't change. The antibody concentration increased with the increase of cell concentration. The kinetic parameters for this continuous culture were shown in Table 2.

Table 2 The kinetic parameters of continuous culture

BSA concentration in feed medium (g/ℓ)	0	10
specific growth rate (1/h)	0.0312	0.0293
specific death rate (1/h)	0.0071	0.0054
specific glucose (g/cell/h) consumption rate	0.325×10^{-10}	0.232×10^{-10}
specific lactate (g/cell/h) production rate	0.163×10^{-10}	0.110×10^{-10}
specific antibody (g/cell/h) production rate	0.404×10^{-12}	0.425×10^{-12}
cell yield (cells/g-glucose)	9.60×10^{8}	12.6×10^{8}

When BSA was added in continuous culture without a separation unit, the specific growth and death rate didn't change and specific glucose consumption and specific lactate production rate slightly decreased. Therefore, the cell yield based on glucose consumed increased and steady-state cell concentration also increased. While the specific antibody production rate was not altered by addition of BSA. From these experimental results, the BSA had no effects on the antibody and only promoted the cell growth.

3.2.2. Perfusion culture

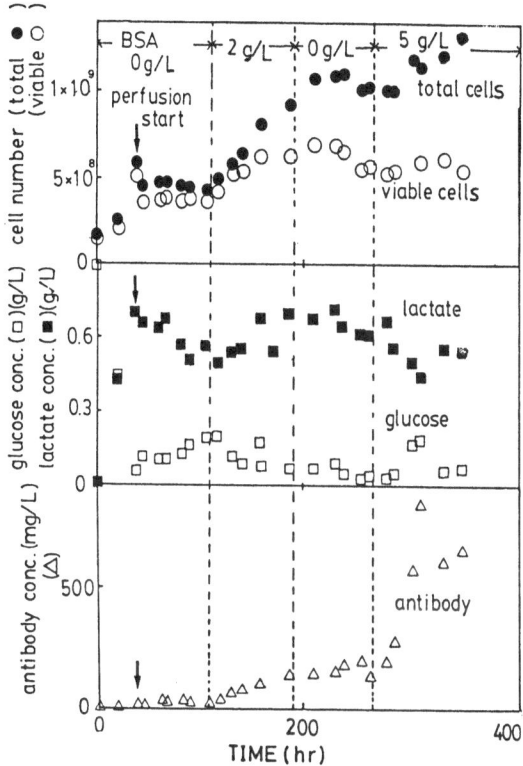

Figure 3 The effects of BSA on the cell growth and antibody production

The effects of changing the BSA concentration in the feed medium in the perfusion period are shown in Fig. 3. The cell concentration increased during the batch periods and became constant during the perfusion period. When fresh medium containing 3 g/ℓ of BSA was fed into the flask, viable and total cell number increased and antibody concentration also increased. However, when the BSA concentration of feed medium changed to zero, the cell growth stopped. During 5 g/ℓ of BSA feeding, the total cell number slightly increased and antibody concentration greatly increased. The kinetic parameters of perfusion culture are shown in Table 3.

Table 3 The kinetic parameters of perfusion culture

BSA concentration in feed medium (g/ℓ)	0	2	0	5
specific growth rate (1/h)	0	0.012	0	0.007
specific death rate (1/h)	0	0.011	0	0.007
specific glucose consumption rate (g/cell/h)	0.359×10^{-10}	0.359×10^{-10}	0.359×10^{-10}	0.221×10^{-10}
specific lactate production rate (g/cell/h)	0.222×10^{-10}	0.222×10^{-10}	0.222×10^{-10}	0.105×10^{-10}
specific antibody production rate (g/cell/h)	0.867×10^{-12}	1.89×10^{-12}	0.737×10^{-12}	21.6×10^{-12}

The specific growth and death rate were both slightly increased during the BSA feeding, and the specific glucose consumption and lactate production rate didn't change. However the specific antibody production rate was increased 2 times and 25 times by feeding of 2 g/ℓ and 5 g/ℓ of BSA respectively, compared with no addition of BSA. In continuous culture without the separation unit, BSA had no effect on antibody production. Therefore, the concentration of BSA and/or materials derived from BSA by separation unit may enhance their effects.

4. DISCUSSION

Transferrin enhanced the cell growth rate but not the specific antibody production rate in the perfusion system. As the molecular weight of transferrin was 80,000, transferrin could be retained in this perfusion system. Continuous feeding of transferrin was not effective for the specific antibody production rate in this system.

Insulin was required for cell growth. Insulin (molecular weight 6000) couldn't be retained by the perfusion membrane used in this system. Moreover, insulin was inactivated during the cultivation (data not shown). Therefore, it was necessary to feed insulin continuously in this system.

Neither transferrin nor insulin promoted the antibody production. While, in the perfusion culture, BSA promoted cell growth and raised the specific antibody production rate. Without the separation membrane in continuous culture, BSA addition had no effects on the antibody production. The role of BSA has been reported in terms of growth promotion (e.g. Nilausen (1978), Rockwell et al. (1980)) but not in terms of enhancement of antibody production. One of the role of BSA has been understood as the storage place of unsaturated fatty acids such as linoleic acid. Under our investigation, however, linoleic acid didn't promote the antibody production. The role of BSA in perfusion culture is now being investigated.

5. REFERENCES

Murakami H., Masui H., Sato G.H., Sueoka N., Chow T.P. and Kano-Sueoka T. (1982) 'Growth of hybridoma cells in serum-free medium: Ethanolamine is an essential component', Proceedings of the National Academy of Sciences of the United States of America 79, 1163-1165.

Nilausen K. (1978) 'Role of fatty acids in growth-promoting effect of serum albumin on hamster cells in vitro', Journal of Cellular Physiology 96, 1-14.

Omasa T., Kobayashi M., Nishikawa T., Shioya S., Suga K. Uemura S. and Imamura Y. (1990) 'Hybridoma culture in the hollow-fiber system -The effects of growth factors-' The 3rd Annual Meeting of Japanese Association for Animal Cell Technology, P-9.

Rockwell G.A., Sato G.H. and Mcclure D.B. (1980) 'The growth requirements of SV40 virus transformed Balb/c-3T3 cells in serum-free monolayer culture' Journal of Cellular Physiology 103, 323-331.

STUDIES ON PHYSIOLOGICAL ASPECT OF IN-VITRO DEVELOPMENT OF CHICK-EMBRYO

A. VENKATARAMAN, P.S.R. BABU and T. PANDA*
Division of Biochemical Engineering
Department of Chemical Engineering
Indian Institute of Technology
Madras 600 036, INDIA

ABSTRACT. Physiological changes of the development of embryo have been studied during entire period of growth. It has been found that the parameters viz. effect of age of incubated eggs, effect of age on transfer and efficiency of transfer operation have more influence on the physiological aspect. Younger embryo did not show any damage to the important blood vessels after the transfer into the artificial medium.

Introduction

For the large scale cultivation of chick-embryo, experimentation has to develop in artificial medium. The conventional development of chick-embryo is a typical batch cultivation mode. It is also questionable for healthy growth of embryos in the conventional process. So a process has been reported in the development of such chick-embryo in artificial medium with the proper maintenance of conditions for growth and development [1]. However, such study is difficult because many of the cell types of interest exist as dense packing of similar cells [2]. Moreover, many fertilized eggs could not be developed to the full embryos stage in artificial medium. In addition to this, development of reactors are practiced as parallel study. This needs supplementation of detailed analysis on physiological aspect of chick-embryo development.

Materials and Methods

FERTILIZED CHICKEN EGG

This was obtained from M/s. Futnani Hatcheries, Madras, India.

CULTIVATION MEDIUM

The medium used for promoting the growth of chick-embryo was developed in

* To whom all correspondence to be made

245

R. Sasaki and K. Ikura (eds.), Animal Cell Culture and Production of Biologicals, 245–249.
© 1991 *Kluwer Academic Publishers.*

our laboratory having the composition for one chick-embryo development: chick ringer solution - 50 ml and fresh egg albumin - 30 ml. All operations were performed under strict aseptic conditions.

CULTURE CONDITIONS

The disection of incubated egg was carried out as described earlier [2]. The embryo was carefully transferred to the growth medium. The top end of the reactor was cleaned throughly with sterile cotton to avoid any contamination of cotton plug upon insertion. The process was carried out at 38°C and 65% humidity was maintained throughout the development. The embryo was observed throughout the incubation stages.

Results

The fertilized eggs were obtained at different stages. These eggs were separately incubated between 1 and 12 days. After that specified time of incubation, the embryo plus attached yolk was transferred to the growth medium in the reator maintained at specified aeration, humidity and temperature conditions. The embryo were followed in the entire stages of development for the ages of the incubated egg at the time of transfer being 1 d and 11 d.

AGE OF INCUBATED EGG AT THE TIME OF TRANSFER INTO REACTOR

For 1 d Old

Following changes were observed when one day old incubated egg was transferred to the growth medium designed for embryo.

At the end of first day of transfer, the yolk was spherical and compact. The medium solution was clear except for tiny structure of the shell membrane. On second day of transfer, there was no change in the condition of yolk. The diameter of the red spot was increased. On the third day of transfer, the shape of the yolk was less spherical and was flattened towards the lateral direction. The red spot developed into a thin film encapsulating the prominent bulge on top of yolk. The size of film was almost the diameter of yolk. The U-shaped body emerged on top of yolk was developed on fourth day of incubation. In case of fifth day, primary optic vesicles with a dark color formed on the U-shaped body (Fig. 1).

Slowly size of the head was increased on sixth day followed by the appearance of fore-limbs and tail on seventh day (Fig.2). Fore-limbs and tail developed from the buds formed at the fourth day stage. The growth of hind limbs was retarded while prominent black eyes were seen on the expanded head (Fig.2). However, the tail size was slowly decreasing and was at right angles to the axis of the body. On the other hand, colour of the embryo changed to pale red on eighth day. After ninth day, there was a total change in the shape of yolk compared to first three days. Also ear slits on developed embryo appeared first time. Again beak buds were seen on the mouth and its development took

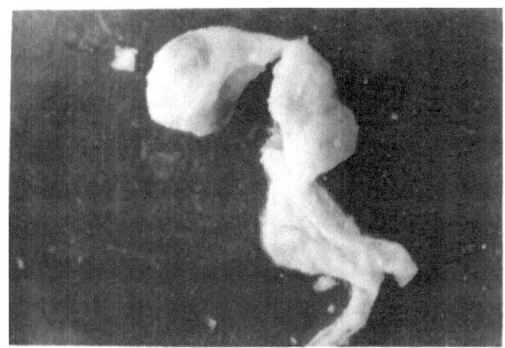

Figure 1. Chick-embryo after five days of cultivation. Age of incubated at the time of transfer was one day.

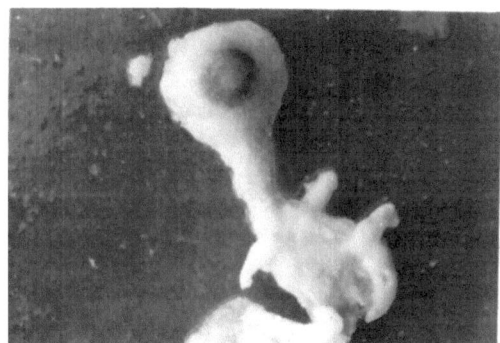

Figure 2. Chick-embryo after 7-days of cultivation for one day age of incubated egg at the time of transfer.

place in later stages which was completed after eleventh day of incubation. Hair was seen on the body of growing embryo. Further neck was prominent during this time. On progressive development, it was found that there was fully developed fore limbs plus body covered completely with hair, proper position of head anterior to yolk, maximization of developing embryo size for thirteenth, fourteenth, fifteenth day of incubation respectively. Afterwards, there was no change in development except for colour of hair between sixteenth and eighteenth days of incubation.

For 11 d Old Fertilized Eggs

After the first day of cultivation of embryo, the body was well developed with all organs. Blood spots were observed at various places. On third day eye vesicles closed body which was highly curved inwards. These vesicles were partially opened and fore-limbs were developed well on fifth day (Fig.3). It was further observed that there was complete hair growth, fully opened eye vesicles and appearance of nail buds on fore-limbs during sixth, seventh and eighth days of cultivation respectively.

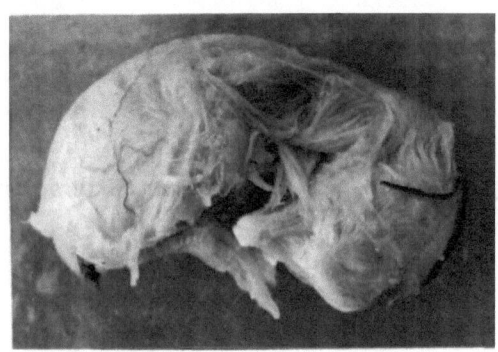

Figure 3.   Chicy-embryo after 5 days of cultivation in in vitro with 11 days old preincubated egg.

Discussion

It has been very easy to transfer one day preincubated egg compared to 11 day preincubated egg. This might be due to more development of blood vessels and other organs on prolonged preincubation (i.e. 11 days old egg). The development has been found to take longer time with 11 days old incubated egg compared to 1-day incubated egg because developed embryo takes longer time to simulate with the artificial medium. It has further been observed that rate of growth appears to be rapid with embryo derived from 1-d old incubated egg in artificial medium compared to the embryo derived from 11-day old egg in in-vitro studies. Also it has been observed that occurence of death of embryo in artificial medium derived from 11 d old fertilized egg was more compared to the embryo derived from 1-d old fertilized egg. As the age of incubation before transfer to the artificial medium increased, there was a gradual development of blood vessel along the shell membrane. These were damaged during transfer.

Acknowledgements

Authors wish to thank M/s. Futnani Hatcheries for uninterrupted supply of fertilized eggs during the study.

References

1. Venkataraman, A. (1989), Studies on the in-vitro development of chick embryo' B.Tech. Project Report, IIT Madras, India.

2. Venkataraman, A., Babu, P.S.R, and Panda, T. (1990) 'Studies on the in-vitro development of chick embryo' in Proc. APBioChEC'90, Kyungju, Korea, pp.149-151.

PILOT SCALE PROTEIN PRODUCTION USING INDUCIBLE GENE AMPLIFICATION

J.M. SEDIVY
Department of Molecular Biophysics and Biochemistry
Yale University School of Medicine
333 Cedar Street
New Haven, CT 06510, USA

ABSTRACT. A cell line, designated BTS (*B*SC-40, *t*emperature sensitive, *S*V40) has been isolated by stably transfecting BSC-40 monkey cells with the *tsA58* allele of the SV40 large T antigen gene. At the nonpermissive temperature (39.5°C) SV40 origin-containing plasmids can be stably transfected to yield cells lines with chromosomal, low copy-number integrants. Upon shiftdown to the permissive temperature (33°C), the large T antigen protein becomes active, the integrated origins begin to replicate, the transfected DNA excises, and continues to replicate episomally to very high copy numbers. This expression cell system is ideally suited for research and development experiments on a pilot scale.

1. Introduction

In many instances, adequate levels of protein production can only be achieved by expressing a gene in heterologous cells. Engineering high gene expression in cultured cell systems is thus a key issue facing the biotechnology industry. On a research scale, transient expression in COS (*C*V-1, *o*rigin-minus, *S*V40) cells (Gluzman, 1981) has been widely used. The chief advantage of this system (see Fig. 1) is that it is very rapid; protein can be harvested within 48 hours of transfection. The major limitation is the transient nature of the assay as well as its small scale. For each assay fresh cells must be transfected, and the transfections are not easily expanded to accommodate $>10^7$ cells.

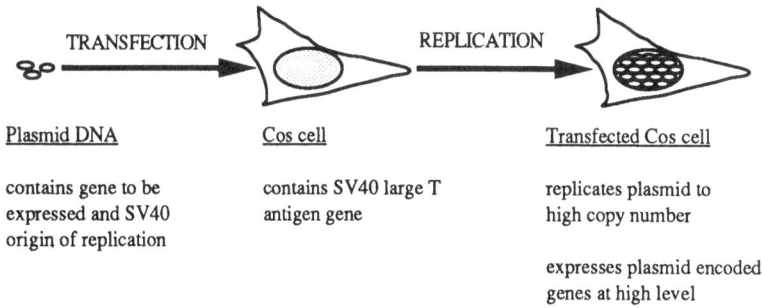

Figure 1. Illustration of the COS cell system of heterologous protein expression. Plasmid DNA is transfected into COS cells, and protein is typically harvested after 48 hours. Only a fraction of the cells take up DNA during transfection; this depends on the method used and can be up to 20 - 30 percent.

R. Sasaki and K. Ikura (eds.), Animal Cell Culture and Production of Biologicals, 251–258.
© 1991 *Kluwer Academic Publishers.*

On a production scale, *in situ* chromosomal amplification in CHO (<u>C</u>hinese <u>H</u>amster <u>O</u>vary) cells (Kaufman and Sharp, 1982) using the dihydrofolate reductase gene is often the method of choice. The chief advantage of this system (see Fig. 2) is the relatively stable expression over prolonged periods; cells can be expanded to large numbers and protein can be harvested up to several months. However, cell lines containing stably amplified genes are time consuming and labor intensive to isolate. Usually, at least several months are needed from the initial transfection, followed by 3 - 4 rounds of amplification, to the final isolate. In addition, highly amplified cell lines can be unstable in the absence of methotrexate, and gene products toxic to the cells cannot be expressed at high levels.

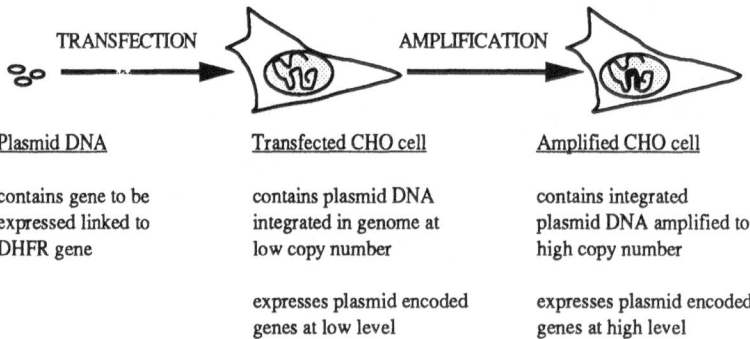

Plasmid DNA	Transfected CHO cell	Amplified CHO cell
contains gene to be expressed linked to DHFR gene	contains plasmid DNA integrated in genome at low copy number	contains integrated plasmid DNA amplified to high copy number
	expresses plasmid encoded genes at low level	expresses plasmid encoded genes at high level

Figure 2. Illustration of the CHO cell system of heterologous protein expression. Plasmid DNA is transfected into CHO cells, and stable integrants are isolated and expanded into cell lines. Such cell lines typically contain a few plasmid copies per cell, and are subsequently subjected to growth in increasing concentrations of methotrexate. After several rounds of such treatments, cell lines containing the transfected DNA amplified to 500 - 1000 copies per cell can be isolated. Amplification often occurs *in situ* within the original integrant chromosome and appears as a homogeneous staining region (HSR).

We have developed (Sedivy and Sharp, 1987; Sedivy, 1988) and further refined a cell expression system based on very rapid, inducible gene amplification (see Fig. 3). Similar systems have been described by others (Rio *et al.*, 1986; Kern and Basilico, 1986; Portela *et al.*, 1986) A cell line, designated BTS (<u>B</u>SC-40, <u>t</u>emperature sensitive, <u>S</u>V40) has been isolated by stably transfecting BSC-40 monkey cells with the *tsA58* allele of the SV40 large T antigen gene. At the nonpermissive temperature (39.5°C) large T antigen activity is not present, and SV40 origins of replication can be maintained stably integrated in the genome. Immediately after shiftdown to the permissive temperature (33°C), however, the large T antigen protein is stabilized and becomes active, and as the cells enter into S phase, the integrated SV40 origins begin to replicate. The SV40 origin is remarkably prolific; thousands of replicative events are initiated during a single S phase period (Botchan *et al.*, 1979). The transfected DNA excises and continues to replicate episomally to very high copy numbers.

2. Materials and Methods

The BTS cell line was derived by transfection of the cell line BSC-40 (Brockman and Nathans, 1974) using plasmids pLTR*tsA58* (Sedivy *et al.*, 1987) and pY3 (hygromycin resistance, Blochlinger and Diggelmann, 1984). The plasmid pLTR*tsA58* was derived from pZipSV40*tsA58* (kind gift of P. Jat), which is analogous to pZipSV40 (Jat *et al.* 1986) but contains the *tsA58* allele of SV40 (Tegtmeyer, 1972), by deleting the *neo* gene and the SV40 origin of replication with *Xho*I. Following cotransfection (10:1 pLTR*tsA58* : pY3), hygromycin resistant colonies (150 µg/ml) were selected at 39.5°C, cloned using cloning cylinders, and expanded in the absence of hygromycin. Individual isolates were tested by *in situ* immunofluorescence for the presence of the SV40 large T antigen after a 24 hour period of growth at 33°C.

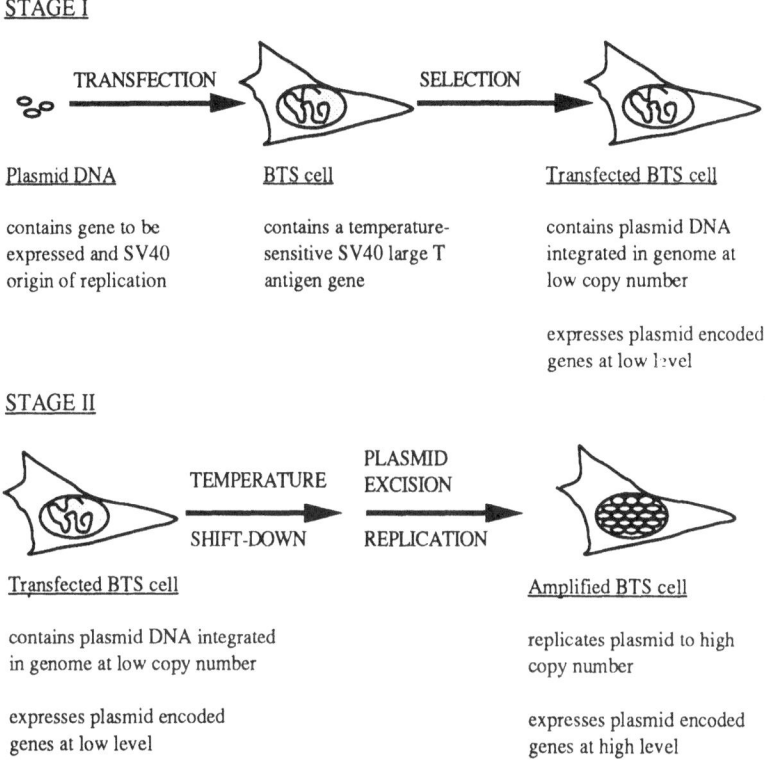

STAGE I

Plasmid DNA	BTS cell	Transfected BTS cell
contains gene to be expressed and SV40 origin of replication	contains a temperature-sensitive SV40 large T antigen gene	contains plasmid DNA integrated in genome at low copy number
		expresses plasmid encoded genes at low level

STAGE II

Transfected BTS cell	Amplified BTS cell
contains plasmid DNA integrated in genome at low copy number	replicates plasmid to high copy number
expresses plasmid encoded genes at low level	expresses plasmid encoded genes at high level

Figure 3. Illustration of the BTS cell system of heterologous protein expression. Stage I. Plasmid DNA is transfected into BTS cells, and stable integrants are isolated and expanded into cell lines at the nonpermissive temperature (39.5°C). Such cell lines typically contain a few plasmid copies per cell. Stage II. Cell line(s) of choice are shifted down to the permissive temperature (33°C). The integrated DNA begins to replicate, excises from the chromosome, and amplifies episomally to thousands of copies per cell. The majority of the cells in the culture undergo amplification.

The BTS cell line is grown in Dulbecco's modified Eagle's medium (DMEM, high glucose with glutamine and pyruvate) with 10 % fetal calf serum in a 5 % CO_2 atmosphere. To avoid excessive alkalinity at 39.5°C the amount of sodium bicarbonate in the medium is decreased to 3.7 g per liter. More recent experiments show that the cells can also be adapted to growth in newborn calf serum. Extra glutamine is added at the time the medium is combined with serum, and complete medium is not stored for more than 3 - 4 weeks.

Trypsinization is accomplished by rinsing cells with a small aliquot of prewarmed trypsin-EDTA solution (0.05 % trypsin, 0.53 mM EDTA in PBS) for 20 - 30 seconds. The solution is sucked off, replaced with fresh solution, and the dishes are replaced in the 39.5°C incubator. The cells are

254

relatively resistant to trypsinization, but should detach by gentle tapping within 5 minutes. The loosened cells are gently pipeted up and down in the trypsin-EDTA solution to break up clumps, aliquoted directly into preequilibrated dishes, and immediately replaced in the 39.5°C incubator.

3. Results

3.1 CONSTRUCTION OF CELL LINES

To construct cell lines using the BTS system, the gene of interest is cloned into an expression vector. The following general criteria should be noted. First, the vector must contain the SV40 origin of replication. Second, the gene to be expressed should be driven by a strong promoter. A variety of promoters have been successfully used, for example: the Rous Sarcoma Virus long terminal repeat, the Adenovirus major late promoter, and the Cytomegalovirus immediate early promoter. The SV40 early promoter should not be used, since it is downregulated in the presence of the large T antigen. Third, to facilitate selection of cell lines, the vector should contain a selectable marker gene, such as *neo*. Any marker gene can be used except *hph* (hygromycin resistance). The presence of a selectable marker in the vector is not an absolute requirement, since cell lines have also been successfully constructed by cotransfection. In that case, however, the proportion of high-producer cell lines among the total number of drug-resistant clones is somewhat lower.

The vector DNA containing the gene of choice is transfected into the BTS cell line and clones are selected at 39.5°C. *Neo* is the marker of choice, and 300 - 350 µg/ml G418 (active drug) should be used for selection. Cell lines are cloned and expanded at the same temperature. Selection for G418 need not be maintained beyond the cloning stage. After an aliquot of each cell line is put into frozen storage, the cells are shifted down and assayed. The most direct method is to assay for the gene product to be overproduced. If a facile assay is not available, cell lines can be screened simply by Southern blot hybridization for plasmid amplification. If the transfected DNA contains a linked selectable marker gene, approximately 10 % of the total drug resistant clones recovered are good producers after shiftdown. If a cotransfection protocol is used, the ratio is 2 - 4 times lower.

3.2 GROWTH CHARACTERISTICS

BTS cells have a doubling time of approximately 22 hours at 39.5°C and 36 hours at 33°C. The cell line and its derivatives do not display complete contact inhibition of growth. In the logarithmic phase of growth, cell cycle occupancy is approximately 40 % G1, 35 % S and 25 % G2. If confluent monolayers are analyzed in the same way, the relative values change to approximately 70 % G1, 15 % S and 15 % G2. Microscopic examination reveals that as cells reach confluency, they first become very tightly packed, and then pile up. The piling up is not constant over the surface of the plate, rather, rope like structures appear, which interlace to give an overall "marbled" appearance. It seems that a fraction of the cells become arrested, or significantly retarded, in the G1/G0 phase of the cell cycle, while some cells continue to cycle, albeit at a reduced rate. In the absence of drug selection overgrowth does not adversely affect the cells unless the medium becomes exhausted.

3.3 HANDLING OF CULTURES

BTS cells metabolize rapidly and should be refed every 4 - 5 days (more frequently if the dishes are crowded, or as soon as the medium turns substantially yellow). For routine maintenance cells can be passaged at a dilution of 1:50 every fifth day at 39.5°C. Gross overgrowth should be avoided in maintenance stocks, since in time it usually results in deterioration of cellular morphology. In particular, in the presence of drugs overgrowth rapidly results in loss of viability. Stocks should not be carried continuously for more than approximately 15 passages.

It must be stressed that the most important precaution is to minimize exposure of the cells to temperatures below 39.5°C. Since the *tsA58* mutant protein is made from a constitutive promoter, a rapid accumulation of active large T antigen at the permissive temperature will destabilize integrated SV40 origins. An easy and convenient procedure is to place a small water bath set at 42°C in the tissue culture hood for holding trypsin (see Materials and Methods). Before trypsinization, the medium is aliquoted into culture vessels which are then placed in a 39.5°C incubator for about 30 minutes to equilibrate. Short periods of exposure to temperatures below 39.5°C, up to 5 - 10 minutes at room temperature, have not been observed to adversely affect the genetic stability of the cells. Routine handling such as trypsinization, aliquoting, and brief periods of microscopic examination, do not therefore present problems or necessitate inconvenient procedures.

If the above simple precautions and procedures are followed, cell lines are genetically quite stable during prolonged culture at 39.5°C. For example, in one case a cell line was carried for 25 passages, at a 1:50 dilution at each passage, and the level of amplification and resultant overproduction remained unchanged. In another instance, the same cell line was carried for 15 passages by another individual under less than optimal conditions, resulting in a decrease in the production level obtained after shiftdown. It is therefore advisable to establish a large freezer stock of early passage cells, and cells should not be subcultured continuously for long periods of time. It is important to note that a very significant level of proliferation can nevertheless be easily achieved: *e.g.*, 10 passages at 1:50 dilution represent about 60 generations, or 10^{18} cells from a single starting cell. In practise, for production purposes on a larger scale, it is advisable to start with freshly thawed cells, expand them continuously to the desired cell mass in a few passages, and shift down the entire culture to induce overproduction.

3.4 INDUCTION OF PRODUCTION BY TEMPERATURE SHIFTDOWN

The following procedure was determined empirically to give the optimal induction by temperature shiftdown. Cells are allowed to become confluent to the point of tight packing, trypsinized well, and dispersed to give approximately 50 % confluent monolayers. The cells are immediately placed in a 33°C incubator, incubated for 24 hours, refed, and incubated further as needed. When cells are treated in this way, over 90 % of the cells attach and produce a loosely packed, uniform monolayer. Within 48 hours after shiftdown, the monolayer will be complete and continuous, and the cells will begin to pack in. Since growth is slow at 33°C, overgrowth of the monolayer ("marbled" appearance) will not begin until about day 6 after shiftdown. Provided that the medium is replenished, considerable overgrowth can be tolerated, until about day 10 - 12 after shiftdown, without significant loss of viability. At about day 12 - 14 the cultures begin to deteriorate. The above discussion assumes, of course, that the overproduced gene product does not affect the growth of the cells. If deleterious or toxic gene products are being overproduced, very little growth occurs after day 3. In that case, it is useful to seed the plates a little more dense.

The fraction of the cells in a given clonal population that are induced and undergo amplification is very high. By *in situ* hybridization to the amplified DNA using a plasmid probe greater than 50 % of the nuclei were observed to be strongly positive. This approaches the maximum resolution limit of the technique applied. In an independent observation using cell lines amplifying gene products that elicited cell cycle arrest, the majority of cells shifted to 33°C stopped growing.

In a number of instances the kinetics of amplification have been examined at the DNA, mRNA, and protein levels (see Fig. 4). The amplification of DNA occurs most rapidly, and reaches a plateau within approximately 48 hours. The copy number of the maximally amplified DNA has been estimated to be approximately 2,000 - 5,000 copies per cell. The vast majority of the amplified DNA is found in an episomal state. The increase in the amount of specific mRNA follows closely the amplification kinetics of the DNA. The levels of mRNA are found usually to peak within 4 days, and hold the plateau for a significant time. The levels of mRNA are profoundly influenced by the particular combination of promoter, vector, and gene organization used. Levels of up to 1 % of total cellular polyA RNA have been achieved. An examination of the parameters influencing gene expression is beyond the scope the current discussion, but it must be emphasized that construction of expression plasmids is still largely an empirical, trial and error process. Fortunately, constructs that express well in COS cells have always been found to express well in BTS cells. The COS cell system is therefore recommended as a rapid way of screening constructs for levels of production. Once the best vector is identified, isolation of BTS cell lines can proceed.

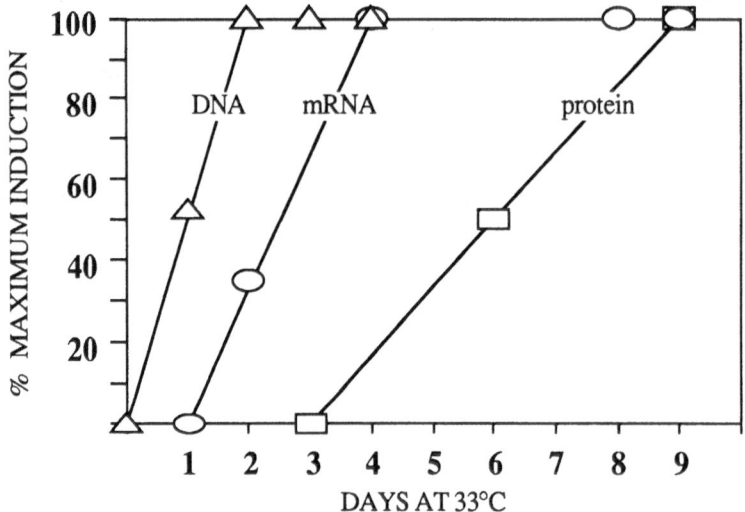

Figure 4. Kinetics of induction after shiftdown. A schematic composite from three individual cell lines showing approximate kinetics of induction of DNA, mRNA, and protein amplification. Triangles: a cell line overproducing an amber suppressor tRNA assayed by Southern blot hybridization. Ovals: a cell line overproducing Factor VIII protein assayed by S1 nuclease digestion. Rectangles: a cell line overproducing Ras protein assayed by immunoprecipitation. All data taken from Sedivy (1988).

Protein production has been observed to increase most slowly. Fig. 4 shows the accumulation of an intracellular protein. Significant accumulation was not seen until the mRNA levels peaked. Production of secreted proteins has also been investigated. Typically, protein production is first detected on about day 4 after shiftdown, close to the the time when mRNA levels peak. Production rate, measured as the amount of protein produced in a 24 hour period, increases over the next 2 - 3 days, and peaks on about days 5 - 6 after shiftdown. Thereafter, the rate of production can be expected to hold constant, provided the medium is replenished, until about day 12, at which time a decrease is observed as a consequence of culture deterioration. An easy and effective culture regimen is to withdraw one half of the medium each 24 - 48 hours and replace it with fresh medium.

In one instance a culture was kept in production until day 10, at which point it became over-grown and the production level was beginning to decline. At that point the cells were trypsinized, dispersed, seeded at 50 % confluence, and incubation was continued at 33°C. After an initial lag period, protein production was seen to resume at the maximal level. The culture was terminated soon thereafter due to contamination. The possibility of ongoing production by extended culture at 33°C is currently under investigation.

BTS cells do not grow or produce well in reduced serum conditions. A reduction to 5 % serum during production has a marginal effect, but 2 % serum resulted, in one case, in a 50 % drop in production. A defined medium, or the use of serum supplements, have not been extensively inves-tigated to date. BTS cells are anchorage dependent. Adaptation to suspension growth has not been attempted. Cells have been successfully grown in dishes, flasks, roller bottles and cell factories. Cell factories have been found most convenient for larger scale production. Roller bottles have also been successfully used, but are more time consuming to handle, and easier to contaminate.

4. Discussion

It is not clear what limits the copy number of the amplified DNA, since SV40 viral genomes can accumulate to much higher levels during the growth cycle of the wild-type virus. It is likely that the level of large T antigen becomes limiting in the BTS cells, since it is being produced at constant levels from a chromosomally located gene. In addition, the *tsA58* mutant protein does not display wild-type levels of activity at 33°C, as evidenced by reduced plaque size of the original mutant virus (Tegtmeyer, 1972). It is also possible that the various vectors used to date do not represent optimal templates relative to the wild-type genome; this possibility is considered less likely since all vectors tested to date accumulate to more or less equivalent copy numbers.

It should be kept in mind that replication only occurs during the S phase of the cell cycle. Since the cell cycle of BTS cells at 33°C is approximately 36 hours, within 48 hours most of the cells in the culture pass through S phase once (only cells in late G1 at the time of shiftdown have the op-portunity to complete the majority of two S phases). The fact that amplification is essentially com-plete in 48 hours suggests that a single S phase is sufficient for the excision of the integrated plas-mid and amplification to maximal levels. This assumption is reinforced by the observation that amplification of plasmids encoding genes eliciting cell cycle arrest proceeds to maximal levels.

The fate of the excised plasmids during mitosis has not been investigated. Cell lines that amplify genes encoding products that do not adversely affect cellular physiology (or control cell lines that amplify vector sequences only) continue to grow at 33°C with unchanged doubling times. This suggests that amplification *per se* is not deleterious to the cell. It has been observed that cells main-tain the fully amplified level of plasmid DNA for at least 3 - 4 generations at 33°C, but the analysis has not been extended beyond that point. As a cell enters mitosis, the nucleus contains 2,000 - 5,000 plasmid copies. It is not known what happens to the plasmids as the nucleus disintegrates at the beginning of mitosis and later reforms, however, even if only a few plasmids are partitioned to each daughter nucleus they should again replicate to the maximum levels as the cells pass through S phase. It is thus possible that a stable, amplified, episomal state could be maintained over numerous generations. It should be kept in mind, however, that since an efficient episome partitioning mechanism is in all likelihood lacking, segregant daughter cells without any plasmids are almost sure to arise at a finite frequency. If such cells posses even a slight growth advantage, they are likely to rapidly overgrow bulk cultures.

The mechanism by which the integrated plasmids excise from the chromosome is also not under-stood in detail. It has been proposed that activation of SV40 origins integrated into cellular chromosomes leads to the production of "onionskin" replicon structures (Botchan *et al.*, 1979). It is evident that such structures are bound to destabilize the chromosome. Structures generated by homologous recombination have been observed in episomes found after excision (Sedivy and Sharp, 1987), but it was impossible to distinguish between intramolecular recombination (occurring *in situ* in the chromosome) and intermolecular recombination (occurring between replicating episomes).

Recombination *in situ* between tandemly repeated copies of the plasmid provides an attractive model that accounts not only for the generation of the episomes but also for the healing of the chromosome. Consistent with this notion is the observation that: 1) integration after transfection often generates such tandem structures, 2) excised episomes often appear identical to monomers of the input plasmid, 3) the excision process does not seem to adversely affect the cell. It should be noted, however, that the structures of the loci containing the integrated plasmid prior to and after amplification, have not been examined in detail. Structures of the excised episomes have been examined, however, and occurrence of nonhomologous recombination is usually also seen. This can range from the observation of rearranged minor species relative to the major species of unit input plasmid, all the way to the major replicating species being obviously rearranged.

It is curious that the profile of the episomal population is not only characteristic of individual cell lines, but is repeatedly and reproducibly generated time after time. This holds not only for the structure of the excised species, but also for the level of amplification. In other words, one particular cell line may always produce a unit size species that amplifies to 5,000 copies per cell, whereas another cell line derived from the same initial transfection may always generate a rearranged species that amplifies to 500 copies per cell. In effect, each cell line has a characteristic Southern blot "signature" after amplification. The reasons for the existence of such "signatures" are not clear, but may well be due to the influence of chromosomal position. effects on the excision process. Certainly, from the point of view of maintaining predictable production levels from culture to culture, these phenomena, although somewhat mysterious, are very fortunate. In addition, the proportion of cells in a culture that undergo amplification is also very reproducible and stable over time.

In summary, this cell expression system is ideally suited to fill the gap between the COS and CHO cell systems. Cell lines can be easily constructed in two weeks by a single transfection. Levels of production are higher than in COS cells. Cell lines are stable, and can be readily expanded into roller bottles or cell factories. In addition, unlike the CHO system, products toxic to cells can be produced. The system is thus ideally suited for research and development experiments on a pilot scale.

5. References

Blochlinger, K. and Diggelmann, H. (1984) Mol. Cell. Biol. **4**: 2929-2931.

Botchan, M., Topp, W. and Sambrook, J. (1979) CHS Symp. Quant. Biol. **43**: 709-719.

Brockman, W.W. and Nathans, D. (1974) Proc. Natl. Acad. Sci. USA **71**: 942-946.

Gluzman, Y. (1981) Cell **23**: 175-182.

Jat, P.S., Cepko, C.L., Mulligan, R.C. and Sharp, P.A. (1986) Mol. Cell. Biol. **6**: 1204-1217.

Kaufman, R.D. and Sharp, P.A. (1982) J. Mol. Biol. **159**: 601-621.

Kern, F.G. and Basilico, C. (1986) Gene **43**: 237-245.

Portela, A., Melero, J.A., de la Luna, S. and Ortin, J. (1986) EMBO J. **5**: 2387-2392.

Rio, D.C., Clark, S.G. and Tjian, R. (1986) Science **227**: 23-28.

Sedivy, J.M. and Sharp, P.A. (1987) Cell **50**: 379-389.

Sedivy, J.M. (1988) Bio/Technology **6**: 1192-1196.

Tegtmeyer, P. (1972) J. Virol. **10**: 591-602.

GENETIC ENHANCEMENT OF PROTEIN PRODUCTIVITY OF ANIMAL CELLS BY ONCOGENES

S. SHIRAHATA,(1) K. TERUYA,(1) T. MORI,(1) K. SEKI,(1) H. OHASHI,(3) H. TACHIBANA,(2) AND H. MURAKAMI (1)

(1)Graduate School of Genetic Resources Technology; (2)Department of Food Science and Technology, Faculty of Agriculture, Kyushu University, 6-10-1 Hakozaki, Higashi-ku, Fukuoka 812, Japan. (3)Pharmaceutical Laboratory, Kirin Brewery Co. Ltd., Souja-machi, Maebashi, Gunma 371, Japan.

ABSTRACT. Effects of oncogenes such as c-*myc*, c-*fos*, v-*myb*, and c-Ha-*ras* were examined on the exogenous protein production by cultured animal cells. Several plasmids expressing oncogenes were constructed as effector plasmids and a plasmid containing the IL-6 gene regulated by the SRα promoter as a reporter plasmid. In transient transfection experiments, only c-*myc* oncogene stimulated the IL-6 production several times in BHK-21 cells. To examine the effects of oncogenes on the permanent expression of the IL-6 gene, the IL-6 gene was introduced into BHK-21 cells. Highly productive clones that secreted 10 times more IL-6 than parental clones were obtained by further transfection of the recombinant BHK-21 cells with oncogenes, suggesting that oncogene products activated the SRα promoter, resulting in the high production of IL-6. These results suggests that it will be possible without gene amplification to get highly productive clones in a short time by using oncogenes for activation of the promoter that regulates a product gene.

Introduction

A problem in large-scale production of bioactive proteins by cultured animal cells is the rather low productivity of animal cells compared to that of bacteria, although animal cells are thought to have potential high productivity of bioactive proteins (Murakami, 1990). Improvement on protein productivity of cultured animal cells will facilitate not only the commercial large-scale production of biologicals, but also the laboratory-scale production for examination of functions of many biologicals produced by animal cells.

There are two main approaches to enhance protein productivity of animal cells in culture. One is physiological enhancement of protein productivity of animal cells. In the physiological approach, optimum conditions must be examined for each cell and product. Another approach is genetic enhancement. This approach have a merit of being able to produce host cells or supercells showing high level of expressions of various exogenous genes. In addition to use of potent promoters and enhancers, gene amplification techniques have been successfully used to enhance exogenous protein production by cultured animal cells. However, it takes a long time (half a year or a year, virtually) to get highly productive clones by gene amplification. Some onocogenes products, especially nuclear oncogene ones, are known to activate transcription (Herrlich and Ponta, 1989). This paper reports an oncogene activated protein production system which can be used to get highly productive clones in a short time.

259

R. Sasaki and K. Ikura (eds.), Animal Cell Culture and Production of Biologicals, 259–266.
© 1991 *Kluwer Academic Publishers.*

Materials and Methods

Materials and chemicals. Human c-*fos*, avian v-*myb*, human c-Ha-*ras*, human c-*myc* genomic oncogenes and the expression vector pCDV1 and pL1 were obtained from Japanese Cancer Research Resources Bank (JCRB). The pSVc-*myc*, pSV2*neo* and pSV2*gpt* plasmids were obtained from the American Type Culture Collection (ATCC). The pRc/CMV vector, containing the CMV promoter and the neomycin-resistant gene, was purchased from Funakoshi Co (Japan). Human interleukin (IL)-6 cDNA and the Bluescript II vector were purchased from Toyobo Co (Japan).The SRα promoter was a generous gift from Dr. Y. Takebe. The pCD-SRα expression vector was constructed using pCDV1 and pL1 as described (Takebe *et al.*, 1988). The pSRα-IL6 was constructed by recombining the human IL-6 cDNA containing the synthetic signal peptide sequence into pCD-SRα vector. The NotI-XbaI fragment of human c-*fos* oncogene, the XbaI fragment of avian v-*myb* oncogene, the BamHI fragment of human c-Ha-*ras* oncogene, the XbaI-EcoRI fragment of human c-*myc* oncogene, and the XbaI-BamHI fragment of mouse c-*myc* oncogene were first cloned into the multicloning site of Bluescript II and then into the multicloning site of the pRc/CMV vector.

Cell cultures. BHK-21 and NIH3T3 cells were grown in Dulbecco's modified Eagle medium (DMEM) supplemented with 10% fetal calf serum (FCS, Hyclone Lab. Co., USA). CHO-DUKXB11 cells were grown in DMEM supplemented with 10%FCS, 10 mM hypoxanthine, and 1.6 mM thymidine. All cells were cultured at 37 °C in a humidified atmosphere of 5% CO_2 in air.

Transfection procedure. Transient and permanent transfections were done by the calcium phosphate coprecipitation method as described before (Shirahata *et al.*, 1990). Briefly, cells (5×10^5 cells per 60-mm plate) were seeded and incubated 24 h before transfection. Medium was changed 1 h before transfection. The DNA-calcium phosphate precipitate was made and added dropwise to the plate. The plate was left undisturbed on the bench top for 1 h, and then returned to the incubator for 6 h, followed by a change to fresh medium. For transient expression, amount of IL-6 was assayed after 3 day culture. For permanent transfection, selections were done after 2 days of culture.

BHK-21 cells cotransfected with pSRα-IL6 and pSVdhfr were selected with 50 nM methotrexate (MTX) after replating in 90-mm plates. A highly IL-6 productive clone (BHK-IL6H) and a low IL-6 productive clone (BHK-IL6L) were selected from among MTX-resistant clones. BHK-IL6H cells were transfected with plasmids containing oncogenes by calcium phosphate precipitation method and selected with 1 mg/ml of G418 in a 48-well microplate, changing medium every 3 days. After 7 days, the medium was changed to one without G418. When cells grew to near confluency, IL-6 in the spent medium was measured by enzyme-liked immunosorbent assay (ELISA) and cell numbers by a cell counter.

For more rapid cloning of highly productive clones, cells after transfection were transferred to a 96-well microplate at cell density of 10^4 cells/well and selected with 1 mg/ml G418. After 7 days, G418 was removed by changing the medium. In these conditions many G418-resistant clones survived and rapidly grew in the wells. Since the cell number per well in these conditions was similar, only the IL-6 in each well was measured to select highly productive clones.

Assays for production of IL-6. IL-6 was assayed by ELISA as described (Shinmoto *et al.*, 1988), using polyclonal sheep anti-IL-6 antibody (Janssen Biochimica, Belgium), biotinylated monoclonal murine anti-IL-6 antibody (Janssen Biochimica, Belgium) and a coloring kit for peroxidase using the 3,3',5,5'-tetramethylbenzidine (TMBZ) substrate solution (Sumitomo Bakelite Co., Japan).

Results

EFFECTS OF ONCOGENES ON THE TRANSIENT EXPRESSION OF THE IL-6 GENE.

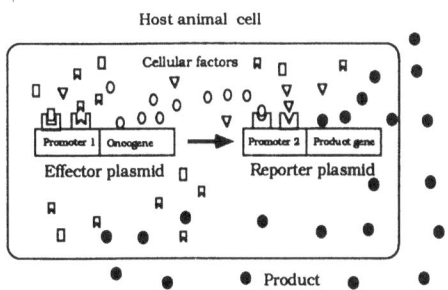

Fig. 1. Concept for oncogene activated production (OAP) system.

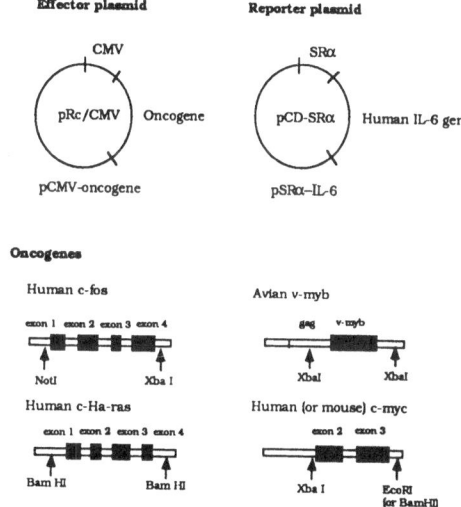

Fig. 2. Construction of the plasmids. Oncogenes were cloned into the pRc/CMV vector containing the CMV promoter and the xanthine guanine ribosyl transferase gene. Human IL-6 cDNA was inserted in the downstream of the SRα promoter.

We intended to enhance protein productivity of cells using oncogenes. This concept is shown in Fig. 1. In this system the oncogene is expressed by promoter 1. Promoter 1 will be regulated by cellular transcriptional activators in the host cells. The product gene is expressed by promoter 2. This promoter 2 will be activated by the oncogene product as well as by cellular transcriptional factors. By continuous high-level expression of the oncogene, the cells will continue to produce a high concentration of protein product. We call this system the oncogene activated production (OAP) system. In this system, it will be important to find an optimum combination of promoter 1, oncogenes, promoter 2, and host cells.

To examine if the OAP system works well or not, we constructed several effector plasmids and a reporter plasmid (Fig. 2). As effector plasmids, various kinds of oncogenes were inserted downstream of the cytomegalo virus (CMV) promoter of pRc/CMV vector. Since this vector contained the xanthine guanine phospho-ribosyl transferase gene regulated by the SV40 early promoter, transformants could be selected by a neomycin-analoge, G418. As oncogenes, human c-*fos*, avian v-*myb*, human normal c-Ha-*ras*, and human or mouse c-*myc* genes were used. Except for these oncogenes, pSV2c-*myc* which expressed mouse c-*myc* under regulation by the SV40 early promoter was also used. As reporter plasmid, the human IL-6 gene was linked to the SRα promoter.

First effects of oncogenes were examined on the transient expression of the IL-6 gene in BHK-21, CHO, and NIH3T3 cells. The cells were cotransfected with oncogenes and

262

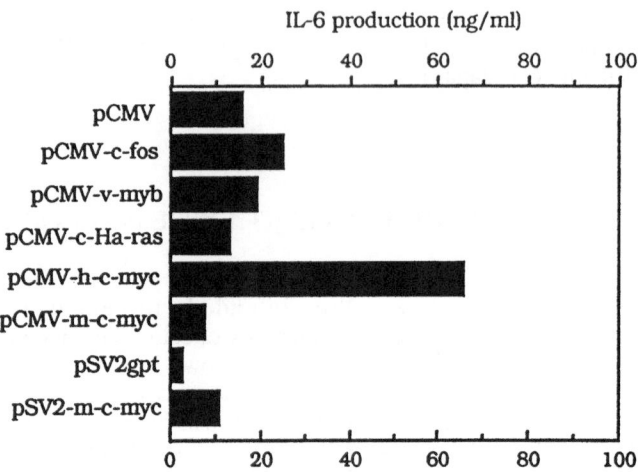

IL-6 production (ng/ml)

Fig. 3. Effects of oncogenes on the IL-6 production by BHK-21 cells. BHK-21 cells (2 x10⁵ cells per 35 mm plate) were cotransfected with 4 μg of the pCMV-oncogene plasmids and 4 μg of pSRα-IL6 plasmid by calcium phosphate precipitation method. On the 3rd day after transfection, transient production of IL-6 was examined by ELISA method. Experiments were triplicated and average values were shown in the figure.

the IL-6 gene by calcium phosphate precipitation method. After 3 days of culture, the amount of IL-6 in culture supernatant was measured by ELISA.

By the cotransfection of BHK-21 cells with the IL-6 gene and human c-*myc* oncogene, the IL-6 productivity was enhanced about four times as high as that of cells transfected with the Il-6 gene only (Fig. 3), but the effects of other oncogenes were not clear. In CHO-DUKXB11, and NIH3T3 cells, no remarkable enhancement of IL-6 productivity of cells was observed upon the cotransfection of the IL-6 gene and oncogenes (data not shown).

ENHANCEMENT OF PERMANENT IL-6 PRODUCTION IN BHK-21 CELLS BY ONCOGENES.

To examine the effects of oncogenes on the permanent expression of the IL-6 gene regulated by the SRα promoter, a highly IL-6 productive clone (BHK-IL6H) and a low IL-6 productive clone (BHK-IL6L) were obtained by transfection of BHK-21 cells with pSRα-IL6 and pSVdhfr. The IL-6 productivity of BHK-IL6H cells was about six hundred times higher than that of BHK-IL6L cells. BHK-IL6H cells were further transfected with various kinds of oncogenes and selected with G418 in 48-well microplates.

When BHK-IL6H cells were transfected with the v-*myb* oncogene, several highly productive clones were obtained (Fig. 4). These clones had about 10 times higher

IL-6 productivity (µg/ml/10^6cells/day)

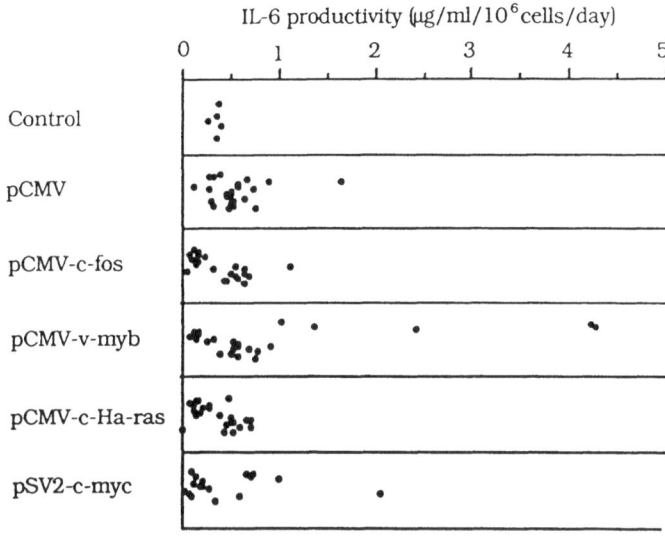

Fig. 4. Appearance of highly IL-6 productive clones by transfection of BHK-IL6H cells with oncogenes. Recombinant BHK-IL6H cells expressing the IL-6 gene highly were transfected with oncogenes and selected with G418 (1 mg/ml) in a 48-well plate. After grown to near confluency in the well, G418-resistant clones were moved to a new 48-well plate and 3 days later, cells were counted by a cell counter and IL-6 secreted in the culture medium were measured by ELISA.

productivity of IL-6 than control cells. The pRc/CMV vector seemed to have a weak effect to produce highly productive clones. Effects of *Fos* and *ras* were not clear.

RAPID SCREENING OF HIGH PRODUCTIVE CLONES.

To screen for highly productive cells more rapidly, 10^4 cells were seeded per well of a 96-well plate after transfection. Under these conditions, many clones survived in each well after G418 selection and the cell number in each well was not very different. Therefore only the IL-6 in the culture supernatant was measured by ELISA.

By introduction of the v-*myb* oncogene into BHK-IL6H, the number of wells containing highly productive cells were increased as shown in Fig. 5. Interestingly, although *fos* and *ras* had little effect by separate transfection, the cotransfection of *fos* and *ras* greatly increased the number of wells containing highly productive cells, suggesting that *fos* and *ras* oncogenes might cooperate to increase the IL-6 productivities of the cells.

When BHK-IL6L cells with low expression of the IL-6 gene were transfected with v-*myb* oncogene, the number of wells containing highly productive cells were increased, suggesting that even in low productive cells, the OAP system worked well (Fig. 6). The effects of combinations of oncogenes were not clear.

Discussion

The purpose of this study was to construct a new system to get highly productive cells in a short time. We noted the transcriptional activation by oncogenes to activate the promoter

IL-6 production (ng/ml)

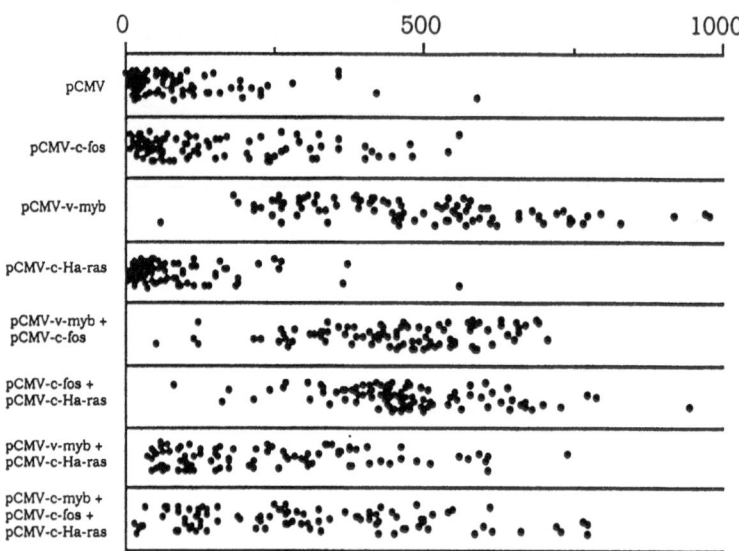

Fig. 5. Rapid screening of highly IL-6 productive clones of BHK-IL6H cells transfected with oncogenes. BHK-IL6H cells were transfected with oncogenes as described in Fig. 4. After cells were transferred to 96-well microplates at the cell density of 10^4/well, G418 (1 mg/ml) selection was continued for 7 days. The amount of IL-6 in the spent medium was assayed 3 days after changing the medium to a fresh one containing no G418.

IL-6 production (ng/ml)

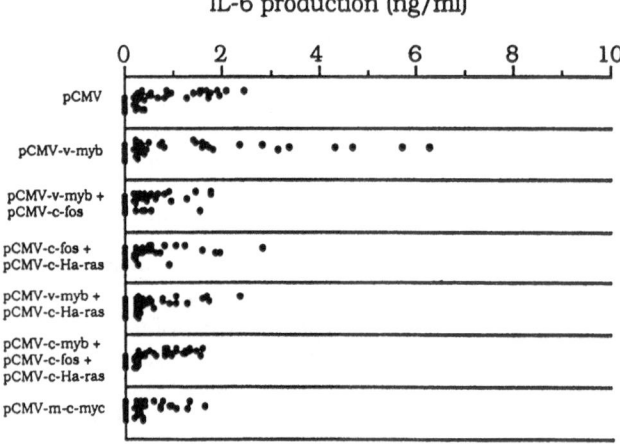

Fig. 6. Rapid screening of high IL-6 productive clones of BHK-IL6L cells transfected with oncogenes. BHK-IL6L cells were transfected and selected as described in Fig. 5.

that regulated the exogenous gene. Nuclear oncogenes such as *myc*, *fos*, *jun*, and *myb* are known to code transcriptional activators (Herrlich and Ponta, 1989). The *Ras* oncogene can
also activate transcription via an indirect mechanism (Wasylyk *et al*.., 1987; Schonthal *et al*.., 1988).

To examine if oncogenes can be used to stimulate exogenous protein productivity of cultured cells, several oncogenes were recombined into an expression vector containing the potent CMV promoter. The IL-6 gene was used as an exogenous protein gene regulated by the SRα promoter, because IL-6 has many biological activities and is secreted into the culture medium (Kishimoto and Hirano, 1988). The SRα promoter is one of the most potent promoters expressed in mammalian cells. It is a fused promoter composed of the SV40 early promoter and the R-U5 sequence of HTLV-LTR and has about 10 to 100 times higher activity than the SV40 early promoter (Takebe *et al*.., 1988).

In the transient expression experiments, only human c-*myc* enhanced the production of IL-6 in BHK-21 cells. No great effect of oncogenes was observed in other cells such as CHO or NIH3T3 cells. Since only average effects of oncogenes could be observed in the transient experiments, permanent transfection was performed to examine effects of oncogenes in each clone.

BHK-21 cells were cotransfected with pSRα-IL6 and pSVdhfr, producing high and low IL-6 productive clones. These BHK-21 cells expressing the IL-6 gene were further transfected with oncogenes. The v-*myb* oncogenes greatly enhanced the IL-6 productivities of both high and low productive clones. The SV40 early promoter is known to contain a specific sequence to be activated by *myb* protein (Nishina *et al*.., 1989; Ibanez *et al*.., 1990). Interestingly, the combination of c-*fos* and c-Ha-*ras* oncogenes was also effective, although their separate use was not. Since c-Ha-*ras* can activate the c-*fos* promoter (Schonthal *et al*.., 1988), *fos* and *ras* may synergistically interact to activate the SRα promoter.

These results suggested that it would be possible to enhance protein productivity of cultured cells by finding an optimum combination of promoter 1, oncogenes, promoter 2, and host cells.

Acknowledgment

We are grateful to Dr. Y. Takebe for his generous gift of the SRα promoter.

References

Friedman, J.S., Cofer, C.L., Anderson, C.L., Kushner J.A., Gray, P.P., Chapman, G.E., Stuart, M.C., Lazarus, L., Shine, J. and Kushner, P.J. (1989) 'High expression in mammalian cells without amplification'. Bio/Technology 7, 359-362.

Herrich, P. and Ponta, H. (1989) 'Nuclear' oncogenes convert extracellular stimuli into changes in the genetic program'. Trends in Genetics 5: 112-115.

Ibanez, C.E. and Lipsick, J.S. (1990) '*Trans* activation of gene expression by v-*myb*'. Mol. Cell. Biol. 10, 2285-2293.

Kishimoto, T. and Hirano, T. (1988) 'Molecular regulation of B lymphocyte response'. Annu. Rev. Immunol. 6, 485-512.

Murakami, H. (1990) 'What should be focused in the study of cell culture technology for production of bioactive proteins'. Cytotechnology 3, 3-7.

Nishina, Y., Nakagoshi, H., Imamoto, F., Gonda, T.J. and Ishii, S. (1989) '*Trans*-activation by the c-*myb* proto-oncogene'. Nucleic Acids Res. 17, 107-117.

Schonthal, A., Herrlich, P., Rahmsdorf, H.J. and Ponta, H. (1988) 'Requirement for *fos* gene expression in the transcriptional activation of collagenase by other oncogenes and phorbol esters'. Cell 54, 325-334.

Shinmoto, H., Murakami, H., Yamada, K., Dosako, S and Omura, H. (1988) 'Immunoglobulin production stimulating and inhibiting factor derived from human lung adenocarcinoma PC-8 cells'. Cytotechnology 1, 295-300.

Shirahata, S., Rawson, C., Loo, D., Chang, Y.-J. and Barnes, D. (1990) '*Ras* and *neu* oncogenes reverse serum inhibition and epidermal growth factor dependence of serum-free mouse embryo cells'. J. Cell. Physiol. 144, 69-76.

Takebe, Y., Seiki, M., Fujisawa, J., Hoy, P., Yokota, K., Arai, K., Yoshida, M. and Arai, N. (1988) 'SRα promoter: an efficient and versatile mammalian cDNA expression system composed of the simian virus 40 early promoter and the R-U5 segment of human T-cell leukemia virus type I long terminal repeat'. Mol. Cell. Biol. 8, 466-472.

Waslyk, C., Imler, J.L., Perez-Mutvl, J. and Wasylyk, B. (1987) 'The c-Ha-*ras* oncogene and a tumor promoter activate the polyoma virus enhancer'. Cell 48, 525-534.

IMMUNOGLOBULIN PRODUCTION STIMULATION BY VARIOUS TYPES OF CASEINS

K. YAMADA,[1] H. NAKAJIMA,[1] I. IKEDA,[1] S. SHIRAHATA,[2] A. ENOMOTO,[3] S. KAMINOGAWA,[3] and H. MURAKAMI[2]
[1] *Department of Food Science and Technology,* [2] *Graduate School of Genetic Resources Technology, Faculty of Agriculture, Kyushu University 46-09, 6-10-1 Hakozaki, Higashi-ku, Fukuoka 812, and* [3] *Department of Agricultural Chemistry, The University of Tokyo, Yayoi, Bunkyo-ku, Tokyo 113, Japan*

ABSTRACT. α-, β-, and κ-caseins are immunoglobulin production stimulating factors (IPSFs), which stimulate proliferation and immunoglobulin production of hybridomas in serum-free media.   To localize the proliferation stimulating and IPSF active sites of these caseins, effects of exopeptidase digestions on these activities of caseins were examined.   Carboxypeptidase Y and aminopeptidase M digestions of caseins showed that both C- and N-terminals are important for their proliferation stimulating and IPSF activities. When synthesized α-casein peptides were added to human-human hybridoma HB4C5 cells, four peptides (1-20, 16-35, 91-110, and 106-125) stimulated their proliferation, but no peptide showed IPSF activity.   When the peptides were added to human lymphocyte cultures in the presence of lipopolysaccharides, four peptides belonging to the 31-95 region stimulated IgG production and four peptides (31-45, 46-65, 76-95, and 151-170) stimulated IgM production.   These results suggest that various parts of caseins are related to the proliferation stimulating and IPSF activities.

INTRODUCTION

Human monoclonal antibodies (MoAbs) are useful for diagnosis and therapy of various human diseases.   Thus, many human-human hybridomas producing MoAbs specific to human antigens have been produced (James and Bell (1987)).   To make highly purified human MoAb massively, serum-free high-density cultures of human-human hybridomas are preferable (Murakami and Yamada (1987)).   However, MoAb productivity of human-human hybridomas in serum-free media is often much lower than that in serum-supplemented medium (Yamada *et al.* (1989a)).   To enhance MoAb productivity of hybridomas in serum-free media, immunoglobulin production stimulating factor (IPSF) had been screened. Various types of IPSFs are found in cell products (Shinmoto *et al.* (1988), Toyoda *et al.* (1990), Yamada *et al.* (1989a)) and in natural resources (Yamada *et al.* (1989b), (1990a)).   Some of the IPSFs

R. Sasaki and K. Ikura (eds.), Animal Cell Culture and Production of Biologicals, 267–274.
© 1991 *Kluwer Academic Publishers.*

screened using human-human hybridomas enhanced Ig production of human lymphocytes cultured in serum-free medium, even in the absence of lipopolysaccharides (Yamada *et al.* (1990b)). Among the IPSFs so far reported, casein is the smallest protein of which the amino acid and nucleic acid sequences were already known. Thus, the relationship between structure and IPSF activity of casein was studied (Yamada *et al.* (1991)).

Bovine casein is mainly composed of α-, β-, and κ-caseins. All these casein components show proliferation stimulating and IPSF activities against human-human hybridoma HB4C5 cells cultured in a serum-free ITES-ERDF medium (Yamada *et al.* (1991)). To localize IPSF-active sites in these casein components, homology analyses are executed on their amino acid sequences. However, their amino acid sequences were completely different. These caseins were digested with endoproteases to isolate IPSF-active fragments of caseins. When these caseins were digested with trypsin, the digests retained weak proliferation stimulating activity, but their IPSF activities were completely lost (Yamada *et al.* (1991)). When they were digested with a milder endoprotease, chymosin, IPSF activities of α- and β-caseins were completely lost, but the activity of κ-casein was recovered in the *p-κ*-casein moiety (Yamada *et al.* (1991)). We report here the effects of exoprotease digestions on proliferation stimulating and IPSF activities of caseins and the effects of synthesized α-casein peptides on proliferation and Ig production of human-human hybridoma and human lymphocytes.

MATERIALS AND METHODS

Materials. α-casein peptides were synthesized with a peptide synthesizer (Applied Biosystems) and purified using a Senshupak H-5251 column (20 x 250 mm, Senshu Sci.), as described previously (Enomoto *et al.* (1990)). Carboxypeptidase Y (CPase) was purchased from Sigma (St. Louis, MO) and aminopeptidase M (APase) from Pierce Chem. Co. (Rockford, IL). For CPase digestion, caseins were dissolved in 10 mM phosphate buffer (PB) (pH 6.0) at 5 mg/ml and CPase was dissolved in the same buffer at 100 IU/ml. Five hundred μl of casein solutions were mixed with 100 μl of CPase solution and 400 μl of 10 mM PB, and then the reaction mixtures were incubated for various periods at 37°C. For APase digestion, 225 μl of the above casein solution was mixed with 25 μl of 20 IU/ml of APase solution dissolved in 10 mM MgCl$_2$ and incubated at 37°C. The reactions were stopped by heating for 5 min in a boiling water bath.

The above casein digests were added to hybridoma cultures (final conc. 50 μg/ml) or electrophoresed on an SDS-polyacrylamide gel (SDS-PAGE). SDS-PAGE of casein digests was done by the method of Laemmli (1970) and the gel was stained with 0.1% Coomassie Brilliant Blue G250. Molecular standards used were ovalbumin (43K), carbonic anhydrase (30K), soybean trypsin inhibitor (20K), and α-lactoglobulin (14.4K).

Cells and cell culture.   Human-human hybridoma HB4C5 cells producing IgM specific to human lung adenocarcinoma cells (Murakami *et al.* (1985)) and tissues (Yano *et al.* (1988)) were cultured in ERDF medium (Kyokuto Pharmacol. Inc., Tokyo, Japan) (Murakami (1989)) supplemented with 10 μg/ml insulin, 35 μg/ml transferrin, 10 μM ethanolamine, and 2.5 nM selenium (ITES) (Murakami *et al.* (1982)).   The cells were cultured at 37°C for 2 days in the presence of casein or casein peptides.   Then, the cells were counted with an electronic cell counter (Toa Medical Electronics, Tokyo, Japan) and the Ig in the culture supernatant was measured by the ELISA method (Engvall and Perlmann (1971)), as described previously (Yamada *et al.* (1989a)). Human lymphocytes were isolated from lymph nodes of mammary cancer patients and cultured for 5 days in ITES-ERDF supplemented with 10 μg/ml of lipopolysaccharides and 50 μg/ml of α-casein peptides, as described previously (Yamada *et al.* (1990b)).

RESULTS

CPase Y digestion of caseins

To localize proliferation stimulating and IPSF active sites of caseins, α-, β-, and κ-caseins were digested with CPase Y for various periods (Fig. 1).   In the case of α-casein, small amounts of intact 30K band still remained after a 30-min digestion, but disappeared after a 60-min digestion.   It gave a major 22K fragment and minor fragments with molecular sizes below 17K by digestions longer than 30 min.   Proliferation stimulating activity of α-casein did not decrease for the first 30 min, decreased rapidly during the next 30 min, and then decreased slowly.   This indicates that the 22K fragment has proliferation stimulating activity weaker than α-casein. This suggests that α-casein has at least two proliferation stimulating sites, one in the C-terminal region and another in the 22K fragment.   On the other hand, IPSF activity of α-casein decreased more rapidly and was lost completely by 2 hr digestion with CPase. This indicates that the C-terminal region was essential for its IPSF activity.
    In the case of β-casein, intact 30K band disappeared within 60 min.   A major 26K fragment and double bands at around 22K were first accumulated, and these bands were disappeared after a 6-hr digestion. Smaller sizes of fragments (20K and 18K) were also produced by 2-6 hr digestions.   Proliferation stimulating activity of β-casein decreased for the first 30 min, retained 70% of the initial activity up to 4 hr, and then decreased to about 40% of the initial activity by the 6 hr digestion.   This suggests that β-casein contains at least three proliferation stimulating sites in the C-terminal region, in the region removed from the 26K fragment, and in the 20K fragment.   IPSF activity of β-casein decreased quickly for the first 30 min and then decreased more slowly than α-casein and weak activity was observed in the 4-hr digest.   This suggests that there are at least two sites related to IPSF activity in the C-terminal region and in the 26K fragments.

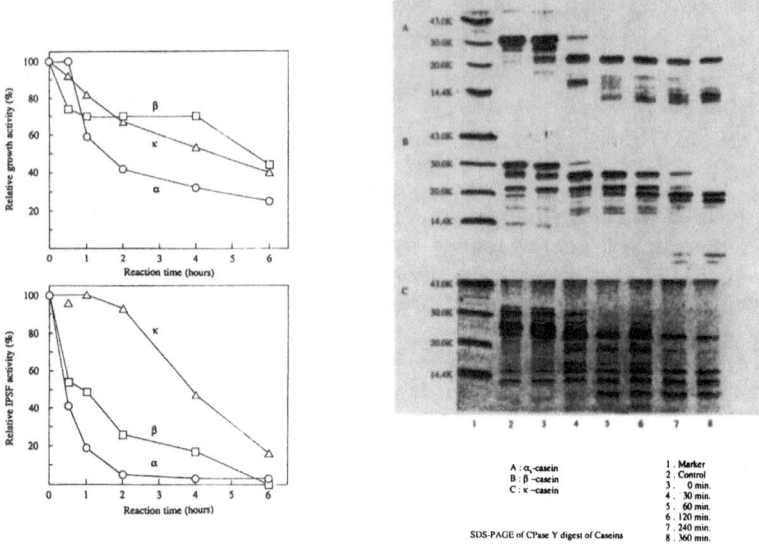

Fig. 1. Effects of CPase digestion on proliferation stimulating and IPSF activities of caseins.

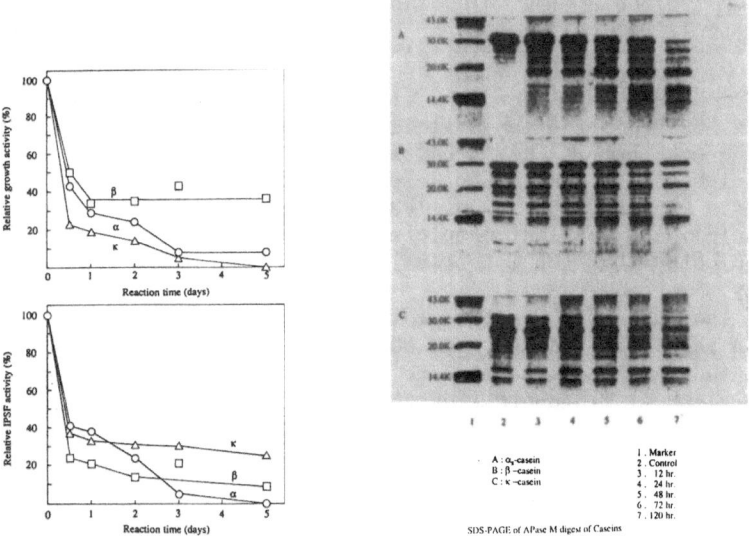

Fig. 2. Effects of APase digestion on proliferation stimulating and IPSF activities of caseins.

Table 1. Proliferation stimulating and IPSF activities of α-casein peptides.

α-casein peptides	HB4C5 cells		Lymph node lymphocytes	
	Relative cell number	Relative IgM productivity	Relative IgG conc.	Relative IgM conc.
α-casein	1.9	3.2	0.8	0.8
1-20	1.2	1.1	0.7	1.0
16-35	1.2	0.9	0.9	1.0
19-26	1.0	0.8	0.9	1.0
31-45	1.1	1.0	1.2	1.4
46-65	0.9	1.2	1.3	1.2
61-80	0.9	1.0	1.3	1.1
76-95	1.1	1.0	1.2	1.5
91-110	1.6	0.9	1.0	0.9
106-125	1.2	1.0	1.0	1.0
121-140	0.9	1.0	1.1	1.0
136-155	1.1	1.1	1.0	1.0
151-170	1.0	1.2	0.9	1.4
166-185	1.0	1.2	1.0	1.1
181-199	1.0	1.1	1.0	1.1

When κ-casein was digested with CPase, the intact 26K band disappeared within 30 min and mainly 23K and 15K fragments accumulated. Proliferation stimulating activity of κ-casein decreased almost linearly during the 6-hr digestion. This suggests that the C-terminal region of κ-casein is related to its proliferation stimulating activity. On the other hand, its IPSF activity was retained for the first 2 hr and then decreased linearly with the decrease of the 26K fragemnt. This indicate that the C-terminal region is not necessary for its IPSF activity.

APase digestion of caseins

When the above caseins were digested with APase, digestion proceeded very slowly under the conditions used here. As shown in Fig. 2, the intensity of intact casein bands decreased very slowly and the intact casein bands were still detectable even after a 5 days digestion. However, both proliferation stimulating and IPSF activities decreased rapidly for the first 12 hr. This suggests that N-terminal regions of these caseins are essential for their proliferation stimulating and IPSF activities.

Proliferation stimulating and IPSF activities of α-casein peptides

To localize proliferation stimulating and IPSF active sites of α-casein more precisely, synthesized α-casein peptides of 20 amino acids were added to HB4C5 cells cultured in a serum-free ITES-ERDF medium.   As summarized in Table 1, a peptide (91-110) showed strong proliferation stimulating activity and three peptides (1-20, 16-35, and 106-125) showed weak activity.   On the other hand, IPSF activity of casein peptides was much weaker than α-casein.

When human lymph node lymphocytes were cultured for 5 days in the same medium supplemented with lipopolysaccharides, α-casein slightly inhibited IgG and IgM syntheses of lymphocytes.   On the other hand, four peptides belonging to the 31-95 region (31-45, 46-65, 61-80, and 76-95) stimulated IgG production of human lymphocytes and four peptides (31-45, 46-65, 76-95, 151-170) stimulated IgM production.

Discussion

Casein is a major milk protein composed of α-, β-, and κ-caseins and all these casein components stimulate proliferation and IgM production of human-human hybridoma HB4C5 cells (Yamada *et al.* (1991)). To localize the proliferation stimulating and IPSF active sites in these caseins, effects of exopeptidase digestions on the activities were examined.   CPase digestion of caseins suggested that these caseins had proliferation stimulating sites in their C-terminal regions and in the inner part of the molecules, and their APase digestion showed that their N-terminals were very important for their proliferation stimulating activities.   When α-casein peptides were added to human-human hybridoma HB4C5 cells, its N-terminal fragments (1-20 and 16-35) showed weak proliferation stimulating activity and a fragment belonging to its central part (91-110) gave a stronger activity.   These results suggest that these caseins have proliferation stimulating sites located at various parts of the molecules.   This indicates that proliferation stimulating activity of caseins can be expressed by casein peptides, as shown previously in endoprotease digests of these caseins (Yamada *et al.* (1991)).

On the other hand, IPSF activity of caseins is lost by endoprotease digestion, except chymosin digestion of κ-casein. Chymosin digests κ-casein at the 105th methionine to produce a hydrophobic p-κ-casein (N-terminal side) and a hydrophilic glycomacroprotein (C-terminal side).   Both fragments stimulate proliferation of the hybridoma, but only p-κ-casein shows IPSF activity (Yamada *et al.* (1991)).   This coincides with the stability of IPSF activity of κ-casein against CPase digestion for the first 2 hr. κ-casein is glycosylated at the 131th threonine and phosphorylated at the 149th serine in glycomacropeptide moiety.   This indicates that the posttranscriptional modifications may be related to the proliferation stimulating activity, but not to IPSF activity. Since IPSF activity of these caseins are highly sensitive to exoprotease and endoprotese digestions, the activity may be dependent on secondary structure of casein melecules.

When α-casein peptides were added to human lymphocytes cultured in ITES-ERDF medium supplemented with lipopolysaccharides, their IgM production was stimulated by four peptides dispersed on the molecule, and IgG production by four peptides belonging to 31-95 region.   These results suggest that α-casein stimulates proliferation of the hybridoma, and IgG and IgM production of human lymphocytes at various sites of the molecule.

REFERENCES

Enomoto, A., Shon, D-H., Aoki, Y., Yamauchi, K. and Kaminogawa, S. (1990) Antibodies raised against peptide fragments of bovine α_{s1}-casein cross-react with the intact protein only when the peptides contain B and T cell determinants. Mol. Immunol. 27: 581-586.

Engvall, E. and Perlmann, P. (1971) Enzyme-linked immunosorbent assay (ELISA), quantitative assay of immunoglobulin G. Immunochemistry 8: 871-874.

James, K. and Bell, G.T. (1987) Human monoclonal antibody production. Current status and future prospect. J. Immunol. Methods 100: 5-40

Laemmli, U.K. (1970) Cleavage of structural proteins during assembly of head of bacteriophage-T4. Nature 227: 680-685.

Murakami, H., Masui, H., Sato, G.H., Sueoka, N., Chow, T.P. and Kano-Sueoka, T. (1982) Growth of hybridoma cells in serum-free medium: ethanolamine is an essential component. Proc. Natl. Acad. Sci. U.S.A. 79: 1158-1162

Murakami, H., Hashizume, S., Ohashi, H., Shinohara, K., Yasumoto, K., Nomoto, K. and Omura, H. (1985) Human-human hybridoma secreting antibodies specific to human lung adenocarcinoma. In Vitro Cell. Develop. Biol. 21: 593-596.

Murakami, H. and Yamada, K. (1987) Production of cancer specific monoclonal antibodies with human-human hybridomas and their serum-free, high-density perfusion culture. In "Modern Approaches to Animal Cell Technology", ed. by RE Spier and JB Griffiths, Butterworths, England, pp. 52-76.

Murakami, H. (1989) Serum-free media used for cultivation of hybridomas. In "Advances in Biotechnological Processes," ed. by A. Mizrahi, vol. 11, Alan R. Liss Inc., New York, pp. 107-141.

Shinomoto, H., Murakami, H., Yamada, K., Dosako, S. and Omura, H. (1988) Immunoglobulin production stimulating and inhibiting factors derived from human lung adenocarcinoma PC-8 cells. Cytotechnology 1: 295-300.

Toyoda, K., Murakami, H., Inoue, K., Yamada, K., Shirahata, S. and Omura, H. (1990) Purification and characterization of immunoglobulin production stimulating factor derived from human B lymphoblastoid HO-323 cells. Cytotechnology 3: 189-197.

Yamada, K., Akiyoshi, K., Murakami, H., Sugahara, T., Ikeda, I., Toyoda, K. and Omura, H. (1989a) Partial purification and characterization of immunoglobulin production stimulating factor derived from Namalwa cells. In Vitro Cell. Develop. Biol. 25: 243-247.

Yamada, K., Ikeda, I., Sugahara, T., Shirahata S. and Murakami, H. (1989b) Screening of immunoglobulin production stimulating factor (IPSF) in foodstuffs using human-human hybridoma HB4C5 cells. Agric. Biol. Chem. 53: 2987-2991.

Yamada, K., Ikeda, I., Sugahara, T., Hashizume, S., Shirahata, S. and Murakami, H. (1990a) Stimulation of proliferation and immunoglobulin M production by lactoferrin in human-human and mouse-mouse hybridomas in serum-free culture. Cytotechnology 3: 123-131.

Yamada, K., Ikeda, I., Maeda, M., Shirahata, S. and Murakami, H. (1990b)Effect of immunoglobulin production stimulating factors in foodstuffs on immunoglobulin production of human lymphocytes. Agric. Biol. Chem. 54: 1087-1089.

Yamada, K., Ikeda, I., Nakajima, H., Shirahata, S. and Murakami, H. (1991) Stimulation of proliferation and immunoglobulin production of human-human hybridoma by various types of caseins and their protease digests. Cytotechnology, in press.

Yano, T., Yasumoto, K., Nagashima, A., Murakami, H., Hashizume, S. and Nomoto, K. (1988) Immunohistochemical characterization of human monoclonal antibody against human lung cancer. J. Sug. Oncol. 39: 108-113.

Effect of dilution rate on the metabolism and product formation of a recombinant mammalian cell line growing in a chemostat with internal recycle of cells

J. Crowley[+], W.L. Marsden and P.P. Gray
Bioengineering Centre
Department of Biotechnology
University of NSW, Kensington 2033
Sydney, Australia

Abstract. A recombinant CHO cell line capable of high level expression of a heterologous protein, human growth hormone, was grown attached to microcarriers. Growth and product formation by the cell line was studied in a continuous culture system with internal recycle of the cells. The culture was fed at dilution rates from 0.03 to 0.19 h^{-1} with a protein free defined medium. The attached cell densities, the percentage of confluent microcarriers and the specific hGH productivities were all functions of the dilution rate, the optimum rate being $0.12h^{-1}$. At dilution rates lower than $0.12h^{-1}$, the metabolic rate of the culture was controlled by the dilution rate. Limitation of glucose or glutamine or end product inhibition by ammonium or lactate were ruled out as possible causes, and it was felt that the supply of amino acids, in particular cystine, serine and aspartic acid was probably limiting metabolism at the low dilution rates.

1. Introduction

There is currently considerable interest in the use of continuous mammalian cell lines as hosts for the production of recombinant biopharmaceuticals. Many potential protein based biopharmaceuticals are complex in structure and heavily modified by glycosylation. Proteins such as these must be produced in mammalian cells, rather than bacteria, to insure a final product that faithfully resembles the natural form.

There are still relatively few papers describing growth and product formation by mammalian cells under closely controlled environmental conditions.

In this paper results obtained with a recombinant Chinese Hamster Ovary cell line (derived from CHO-K1) growing on microcarriers and expressing human growth hormone are described. Cells were grown in an airlift reactor where the microcarriers were retained in the reactor while the reactor was operated in the continuous mode, i.e. a chemostat with internal cell recycle. The affect of dilution rate on the expression of the cloned protein and on cellular metabolism was studied under steady state conditions.

[+] Current address: CSL Ltd., Parkville,
Victoria, Australia.

R. Sasaki and K. Ikura (eds.), Animal Cell Culture and Production of Biologicals, 275–281.

2. Materials and Methods

CHO K1 cells were obtained from the American Type Culture Collection (CCL61). Recombinant cell lines were constructed from the CHO-K1 using an expression system designed to express high levels of a heterologous gene product (human growth hormone) without recourse to gene amplification[1]. The human metallothionein IIA promoter was used to control transcription.

Master stocks of the recombinant cell line CB515-25a were stored in liquid nitrogen, and working stocks were kept at -75°C. Inoculum for the fermenter was prepared as two 300 ml cultures in 500 ml Techne spinner flasks containing 3g/l of Cytodex 2 microcarriers. Cells were grown in the spinner flasks on a custom made DMEM: Coons F12 medium plus 5% heat inactivated foetal calf serum (FCS) to a final cell density of 1-2 x 10^6 cells/ml, reached after 4 to 5 days. The cultures were then transferred to the bioreactor.

The bioreactor was a 2 litre airlift (LH Fermentation). The inoculum, fresh medium and additional microcarriers were added to a final volume of 2.2 litres. An air supply rate of 25-30 ml/min provided gentle circulation of the contents. The pH was maintained at 7.2 by the controlled addition of CO_2 to the air supply. The temperature was maintained at 36.5^{o}C by circulating constant temperature water through a full length jacket on the vessel. The dissolved oxygen level was monitored by a polarographic electrode and was maintained above 75% of air saturation at all times. To prevent foaming, 5 ml of 5% (w/v) of Antifoam C (Sigma) was added at inoculation, followed by controlled addition as required throughout the fermentation.

Microcarriers were retained in the bioreactor by a 100μm stainless steel conical mesh filter extending through the liquid interface at the top of the reactor. During continuous operation, the liquid level was maintained by venting exit gas and medium through a tube placed inside the stainless steel mesh. Another mesh filter placed at the bottom of the reactor was used to totally replace growth medium with production medium.

Cell growth in the bioreactor was monitored daily by determining cell concentration and the percentage of confluent beads. When the cell concentration reached 1-2 x 10^6 cells/ml the growth medium (containing 10% FCS) was replaced with the production medium (DMEM: Coons F12 plus 80mM of zinc sulphate). No antibiotics were used in any of the media.

Continuous culture was maintained by pumping production medium through the reactor at the desired rate. Viability of the cells was at all times > 90%.

Cell concentrations of microcarrier cultures were determined by the nucleii enumeration method developed by Sanford et al (1951)[2] and later modified by van Wezel (1972)[3].

Cell viability was determined by the trypan blue dye exclusion method. Confluence was calculated as the number of beads fully covered with cells as a percentage of the total number of beads.

The hGH concentration in the medium samples was determined by radioimmunoassay carried out by the Garvan Institute of Medical Research, Sydney, Australia. Glucose and lactate concentrations were determined by use of a Technicon autoanalyser[4,5].

Ammonium concentrations were determined by the use of an ammonia probe; glutamine concentrations were obtained by deaminating the glutamine to release ammonium which was measured by the probe.

Amino acid concentrations were analysed by a Waters HPLC.

3. Results and Discussion

Following the growth phase of the cells to a density of approximately 2×10^6 cells/ml, the serum containing medium was replaced with the production medium. Continuous operation was then commenced at an initial dilution rate of $0.042 h^{-1}$. After one day the pump speed was adjusted to give the desired dilution rate. After three volume changes at a particular dilution rate had occured, samples of the culture were taken for assay until three consecutive daily samples indicated a steady state had been reached. Two dilution rates were performed per fermentation run.

The affect of dilution rate on the attached cell density is shown in Figure 1. Theory would predict that the attached cell density would be independent of dilution rate, however the date plotted in Fig. 1 shows that the cell density increased from 1.3×10^6 cells/ml at a dilution rate of $0.033 h^{-1}$ to 3.3×10^6 cells/ml at a dilution rate of $0.124 h^{-1}$. At dilution rates greater than $0.124 h^{-1}$, the attached cell density appeared to decline slightly.

A similar pattern was observed for the percentage of confluent microcarriers. At a dilution rate of $0.033 h^{-1}$ the confluence was a low 18%, increasing to a maximum of 85% at a dilution rate of $0.12 h^{-1}$ then decreasing to 35% at the highest dilution rate studied ($0.19 h^{-1}$) (data not shown).

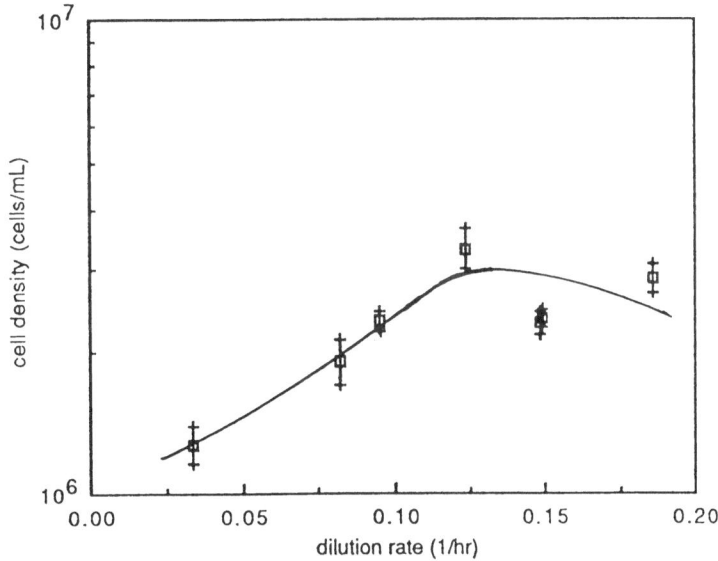

Figure 1. Effect of dilution rate on the numbers of cells attached to the microcarriers in the airlift reactor. Points on the graphs are the mean of three readings and the error bars represent plus or minus one standard deviation.

The steady state specific hGH productivities also varied with dilution rate (Figure 2). At the lower dilution rates the q_{hGH} was relatively constant, increasing by approx. 85% when the dilution rate was increased from $0.095h^{-1}$ to $0.124h^{-1}$. Further increases in dilution rate resulted in small decreases in q_{hGH}.

The results presented in Figs 1 and 2 showed that an optimum existed at a dilution rate of approx. $0.12h^{-1}$. At this dilution rate, maxima in attached cell densities, confluence and specific hGH productivity were obtained; at lower dilution rates there were considerable decreases in the above parameters, at higher dilution rates smaller decreases.

The results indicated that at the lower dilution rates, there was either limitation by the supply of medium component(s), or inhibition by end-product. In the cell recycle system operated, increasing dilution would increase the supply of nutrients per unit volume per unit time and, if cell density was constant, the supply of nutrients per cell per unit time.

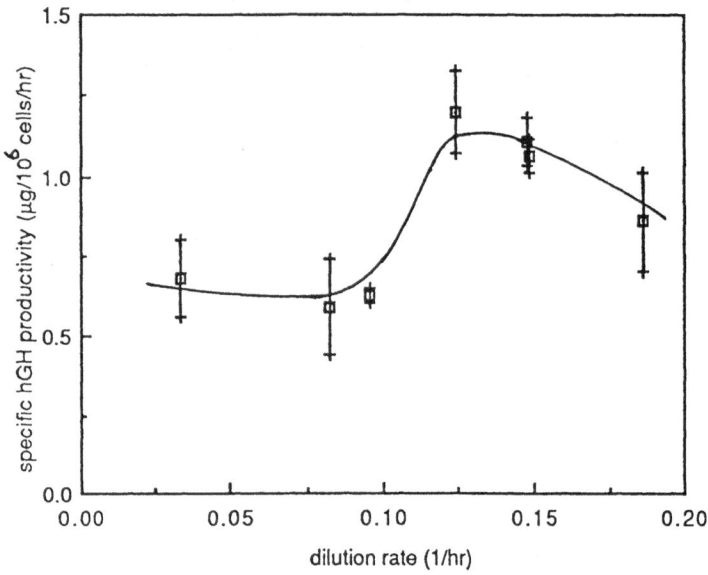

Figure 2. Effect of dilution rate on the specific productivity of hGH (q_{hGH}, $\mu g.10^6$ cells^{-1}.h^{-1}). Error bars as for Figure 1.

Increasing dilution rate would also result in washout of inhibitory substances. Both nutrient supply and waste product accumulation are factors which may limit cell growth and product formation. Glucose and glutamine are the main energy sources present in the media used and lactic acid and ammonium are the main by-products of cellular metabolism. In order to determine if one of these components was responsible for the results shown in Figs 1 and 2, samples from each dilution rate were analysed for these four components. At no dilution rate was the glucose concentration limiting and was usually greater than 2.5 g/l. The exception was at a dilution rate of $0.124h^{-1}$, the glucose level dropped to 1.5 g/l. The lactate concentration mirrored the glucose, with a maximum concentration of 0.83 g/l occurring at a dilution rate of 0.124hr-1. The specific rate of lactate production as a function of the dilution rate is shown in Figure 3. Previous work had shown that the maximum lactate levels observed in these experiments was not sufficient to have inhibitory effects on cell growth or hGH production[6]. Examination of the glutamine levels showed that at all dilution rates the amino acid was present at concentrations greater than 0.23 g/l with a slight increase in concentration as the dilution rate increased. The ammonium ion concentration mirrored the glutamine, decreasing slightly with increasing dilution rate from a maximum value of 0.80mM. Previous work had shown that this maximum value was below the level needed to affect the productivity of the cells[6].

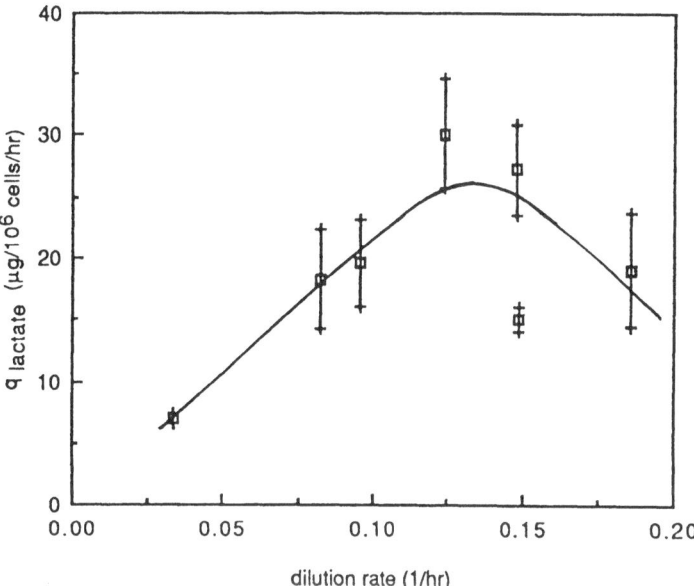

Figure 3: Effect of dilution rate on the specific production rate of lactate (q_{lact},μg.10^6 cells-1.h-1). Error bars as in Figure 1.

These results indicated that across the range of dilution rates tested neither of the major energy sources (glucose and glutamine) nor their metabolic by-products (lactate and ammonium) were inhibiting growth or product formation and that some other component must be limiting metabolism.

Amino acid concentrations in the reactor were determined and the results are shown in Table 1. At the lowest dilution rate of 0.033h-1, aspartic acid, serine and cystine were the only three amino acids to be metabolised to a level of 30% or less of their original concentrations in the medium feed, cystine remaining at low levels of 11-15% at all dilution rates studied. It is possible one of these amino acids, either singly or together, may have limited cell density, and q_{hGH}. Although all the other amino acids were at concentrations greater than 30% of their original concentrations, it has been suggested by Griffiths[7] that amino acid can become limiting before they are exhausted from the medium. This is due to the need for a threshold medium concentration of amino acids in order to maintain the intracellular pool concentration; i.e. other amino acids may also have been limiting.

Table 1: Percentage of Original Medium Amino Acid Concentrations Present During Microcarrier Perfusion Culture

dilution rate (h-1)

Amino acid	Initial Conc. mg/l	0.033	0.082	0.095	0.149	0.186
Asp	15	30	52	45	50	49
Glu	15	100	106	96	79	61
Ser	31.5	31	44	40	66	49
Asn	15	74	54	70	71	81
Gly	23	122	110	94	75	78
His	42	52	51	42	38	39
Thr	59.4	59	53	42	42	48
Ala	9	506	392	340	203	186
Arg	253	59	68	58	55	52
Pro	35	74	80	67	69	66
Tyr	59.83	45	48	39	64	37
Val	58.7	77	74	58	59	53
Met	19.5	80	72	67	62	60
Cys	31.29	11	15	15	11	13
Ile	56.4	58	64	52	53	50
Leu	65.6	56	63	53	53	54
Phe	38	74	75	63	55	53
Lys	109.5	64	63	56	57	49

4. Conclusions

The chemostat with internal recycle of cells was successfully run at dilution rates up to 0.186h-1 (4.46 reactor volumes/day). The method of retaining the microcarriers and cells worked well allowing steady state data to be collected. It was found that the metabolism of the cells was not independent of dilution rate, and that increasing the dilution rate between 0.03 and 0.12h-1 resulted in an increased metabolic activity for the culture. Increases were observed in both the specific uptake rates of substrate and in the specific production rates of products and by-products; the cell density and confluence also increased. These parameters plateaued at approx D = 0.12h-1 and were relatively constant at higher dilution rates.

With the system used for retention of the microcarriers, dead cells, cell debris and any non-attached live cells are flushed from the system through the stainless steel mesh. In the protein free production medium the CHO cells are not actively growing; under ideal conditions there will be a slight increase in the attached cell concentration over several months of operation. Hence the retention method used, by letting free cells pass out of the system whilst retaining microcarriers with cells attached, meant that there was retention in the reactor of the active biomass. It would be expected that, if the supply of all ingredients was in excess and that there was no build-up of inhibitory substances, the metabolism of these cells would be unaffected by the dilution rate. This was not the case, as it was found that at dilution rates less than 0.12h-1, the dilution rate controlled the rate of metabolism of the cells. This could be due to one of two possible reasons: a) as the dilution rate is increased, the supply of nutrients to the cells per unit time also increased, or b) the higher dilution rate coupled with a fixed rate of metabolism of the culture would mean that the concentration of possibly inhibitory end products would be decreased by washout. Assay of the main by-products, ammonium and lactate, showed that at all dilution rates studied the concentrations observed were less than those which had previously been shown to inhibit the cells, ruling out (b) as the explanation. A study of the main substrates, glucose and glutamine, showed that they were always present at non limiting concentrations. Analysis of the amino acid concentrations in the medium showed that several of the amino acids, cystine, serine and aspartic acid, were probably limiting metabolism. At low dilution rates, it is proposed that the rate of supply of the amino acids was not sufficient to satisfy the metabolic needs of the cells and hence the rate of metabolism was controlled by the rate of supply of the amino acids, (proportional to the dilution rate) up to a dilution rate of approx $0.12hr^{-1}$. At higher dilution rates the rate of supply was sufficient to satisfy the demands of the cells.

The data demonstrated the need for careful optimisation of culture conditions in order to maximise the productivity from such perfusion systems.

5. References

[1]   J.S. Friedman, C.L. Cofer, C.L. Anderson, J.A. Kushner, P.P. Gray, G.E. Chapman, M.C. Stuart, L. Lazarus, J. Shine and P.J. Kushner, 'High expression in mammalian cells without amplification'. Bio/Technology, 7, 359 (1989).

[2]   K.K. Sanford, W.R. Earle, V.J. Evans, H.K. Waltz and J.E. Shannon, 'The measurement of proliferation in tissue cultures by enumeration of cell nuclei'. J. Natl. Can. Inst., 11, 773 (1951).

[3]   A.L. van Wezel, 'Microcarrier culture of animal cells'. In "Tissue Culture: Methods and Applications", P.F. Kruse and M.K. Patterson, Eds, (Academic, 1973).

[4]   R. Lachenicht, and E. Bernt, 'Fluorometric determination in blood with automatic analysers', In Methods in Enzymatic Analysis, H.V. Bergmeyer, Ed, (Springer-Verlag, 1974).

[5]   N.J. Hochella and S. Weinhouse, 'Automated lactic acid determination in serum and tissue extracts'. Anal. Biochem., 10, 304 (1965).

[6]   J. Pirhonen. Honours Thesis, UNSW (1986).

[7]   J.B. Griffiths, 'The effects of cell population density on nutrient uptake and cell metabolism: A comparative study of human, diploid and heterodiploid cell lines', J. Cell Sci, 10, 515 (1972).

Immunoglobulin Production Stimulating Factors in Polysaccharides

Mari MAEDA and Makoto TAJIMA
Chugoku National Agricultural Experiment Station, 6-12-1
Nishifukatsu, Fukuyama, Hiroshima 721 JAPAN ,
Koji YAMADA and Hiroki MURAKAMI
Faculty of Agriculture, Kyushu University, 6-10-1, Hakozaki
Higashi-ku, Fukuoka 812 JAPAN

Abstract. We screened IPSFs (Immunoglobulin Production Stimulating Factors) in polysaccharides using human-human hybridoma cells producing monoclonal antibodies cultured in serum-free media, and observed that locustbean gum, pectin, soybean hull hemicellulose (SHH), and chitosan (partially deacetylated chitin) stimulated IgM production. The IPSF activity in SHH was observed in a carbohydrate fraction containing small amounts of protein. Various chitosans showed different IPSF activities according to their deacetylation degree. SHH and chitosan also stimulated IgM production of human lymphocytes.

1. Introduction

Human-human hybridomas producing monoclonal antibodies (MAbs) specific to various antigens have been established for various purposes (James *et al.*,1987) and used for mass production of human MAbs (Murakami *et al.*,1987). Some hybridomas proliferate and produce MAbs in serum-free media, but their MAbs productivities in serum-free media (Murakami *et al.*,1989) are often lower than those in serum-supplemented medium (Aihara *et al.*,1988; Yamada *et al.*,1989b). Thus, we screened for immunoglobulin production stimulating factors (IPSFs), using human-human hybridoma cells cultured in a serum-free medium, and found various proteinous IPSFs in foodstuffs (Yamada *et al.*,1989b; Yamada *et al.*,1990a). This paper describes the IPSF activities in polysaccharides and characterization of the IPSFs.

2. Materials and Methods

2.1 MATERIALS AND CHEMICALS

Various chitosans with different degree of deacetylation were supplied from Katokichi Co.,Ltd.

283

R. Sasaki and K. Ikura (eds.), Animal Cell Culture and Production of Biologicals, 283–290.
© 1991 *Kluwer Academic Publishers.*

Sodium alginate, κ -carageenan, λ -carrageenan, curdlan, pectin, and potassium sulfate were purchased from Wako Jyunyaku. Gum arabic, locustbean gum, and sodium polypectate were purchased from Sigma. Glucomannan was purchased from General Nutrition. Xylan was purchased from Nutritional Biochemicals. Arabinogalactan was purchased from Extrasynthese. Galactan was purchased from Aldrich Chemicals. ERDF medium was purchased from Kyokuto Seiyaku and fetal bovine serum (FBS) from GIBCO. Antibodies were purchased from Tago.

2.2 CELLS AND CELL CULTURE (Fig.1)

Human-human hybridoma HB4C5 cells producing IgM specific to human lung adenocarcinoma cells (Murakami *et al.*,1985), were used to screen IPSF. HB4C5 cells were generated by fusing human lymphocytes of a lung cancer patient with a human fusion partner NAT-30 cells (Murakami *et al.*,1985). The cells were inoculated at a cell density of 0.5×10^5 cells/ml and cultured at 37℃ for 2 days in ERDF medium supplemented with 10μ g/ml insulin, 35μ g/ml transferrin, 10μ M ethanolamine, and 2.5nM selenium (ITES) (Murakami *et al.*,1982). The test samples were added to the medium. Cells were counted using a electric cell counter (Coulter Electric Industry Inc.) and the Ig content in the culture supernatant was measured by the ELISA method as described previously (Yamada *et al.*, 1989a). Ig productivity was calculated by dividing the Ig concentration in the medium by the cell number on each sampling day.

Stimulation of Ig production using human lymphocytes by SHH and chitosan was examined as described previously (Yamada *et al.*, 1990b).

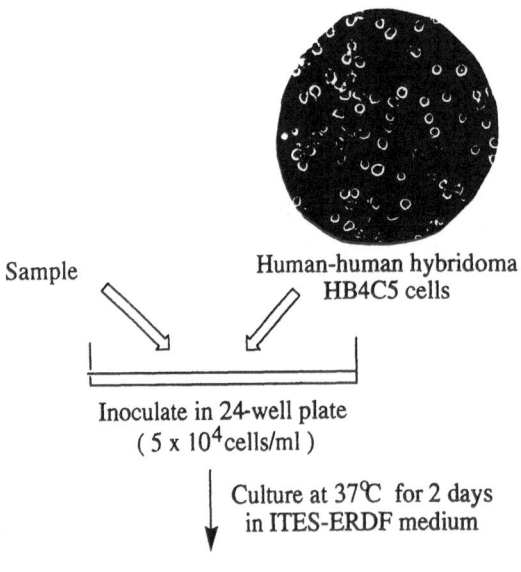

Sample   Human-human hybridoma HB4C5 cells

Inoculate in 24-well plate
(5×10^4 cells/ml)

Culture at 37℃ for 2 days
in ITES-ERDF medium

Count cells and measure IgM by ELISA

Fig.1 Procedure of screening of IPSF in Polysaccharides

2.3 PREPARATION OF SOYBEAN HULL HEMICELLULOSE B (SHH)
(Ayano *et al.*, 1986)

Separated soybean (*Glycine max* var. Tamahomare) hulls were washed with 99% ethanol, and milled to the degree of 30 mesh. SHH was extracted from the milled hull by 1 N NaOH solution under N_2 gas with shaking for 1 day. The extract was neutralized by acetic acid and then tricholoroacetic acid was added to a final concentration of 7%. After standing for 1 hour, it was centrifuged at $400 \times g$ for 20 min. Then the supernatant was dialyzed against distilled water for 3 days. Four times the volume of 99% ethanol was added to the dialyzed supernartant and the resulting precipitate was freeze-dried. SHH was dissolved in PBS and filtered through a 0.22-μm nitrocellulose filter for sterilization.

2.4 COLUMN CHROMATOGRAPHY

SHH was put on a Sepharose CL-4B column (bed volume; 800 ml, column size; 32 i.d. × 1000 mm) equilibrated with 50 mM sodium phosphate buffer (pH 7.7). The elution speed was regulated at 1ml/min and the elution was collected, with 10 ml per tube. The protein content in the fractions was monitored from the absorbance at 280 nm and the carbohydrate content was by phenol-sulfuric acid method (Dubois *et al.*, 1956). IPSF activity of each fraction was measured after dialysis. Pullulan (Showa Denko Co.,Ltd.) was used as the molecular weight marker on the chromatography.

2.5 ENZYMATIC DIGESTION OF SHH

β-Galactosidase was purchased from ICN Biochemical Inc. Five IU of the enzyme was added to SHH solution (500μ g/ml). The same amount of enzyme was also added to the control PBS solution. After incubation at 37℃ for 24 hours, the reactin mixture was added to the culture medium of HB4C5 cells, and the cells were cultured for 2 days to measure IPSF activity of SHH.

3. Results and Discussion

3.1 SCREENING OF IPSFS IN POLYSACCHARIDES

Various polysaccharides were added to the culture medium of HB4C5 cells. As shown in Fig.2, locustbean gum, pectin, soybean hull hemicellulose (SHH), and chitosan stimulated IgM production of the hybridomas.

Fig. 3 shows the dose-dependent stimulation of proliferation and IgM production of HB4C5 cells. At higher SHH concentrations, the IgM concentration increased linearly with the increase in SHH concentration, accompanying weak stimulation of cell proliferation.

286

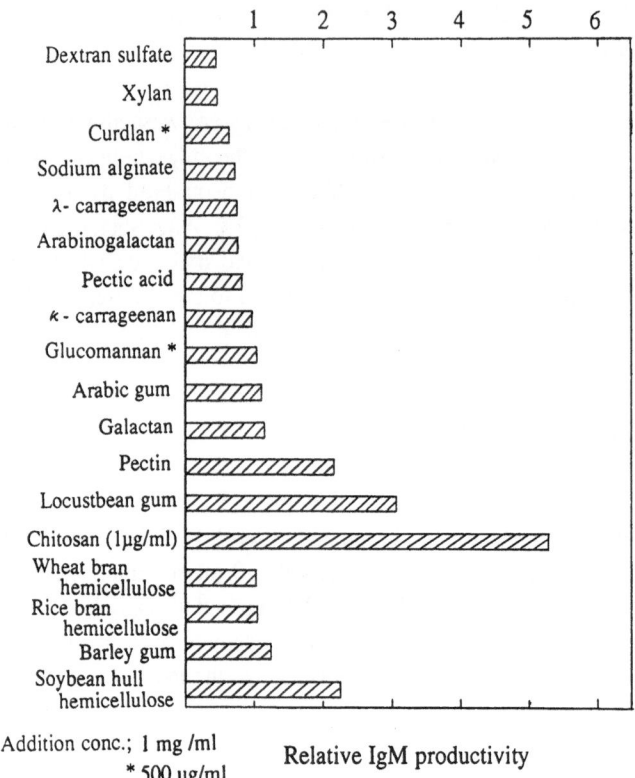

Addition conc.; 1 mg /ml
* 500 μg/ml

Relative IgM productivity

Fig. 2 Screening of IPSFs in polysaccharides using human-human hybridoma HB4C5 cells

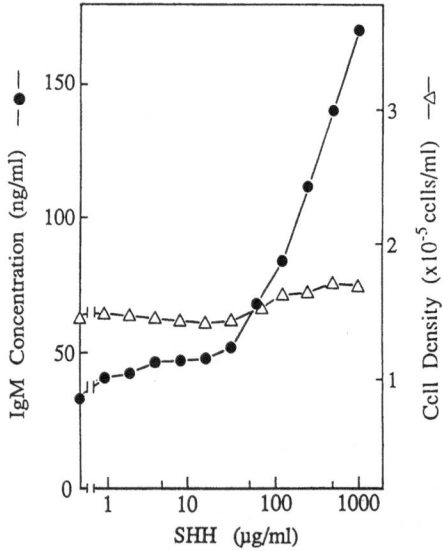

Fig. 3 Dose-dependent stimulation of proliferation and IgM production of HB4C5 cells by SHH

3.2 CHARACTERIZATION OF IPSF IN SHH

To purify IPSF, SHH was gel filtrated using a Sepharose CL-4B column. As shown in Fig. 4, the first peak containing protein and carbohydrate had no IPSF activity. The second carbohydrate peak eluted slightly earlier than the second peak of protein. The highest IPSF activity was observed in a fraction corresponding to the second carbohydrate peak, whose molecular mass was about 32kd.

The IPSF activity of SHH was stable below 60℃, but decreased by the heating at 80℃ and 100℃. However, about 30% of the activity was retained even after the heating at 100℃ for 30 min.

When SHH was treated with various concentration of trypsin for 1 hour, the activity decreased linearly at trypsin concentration over 400×3^{-2} IU/ml and less than 20% of the activity was retained after the treatment with 400 IU/ml of trypsin. It has been reported that soybean hemicellulose is composed of arabinogalactan, xyloglucan, or galactomannan (Kawamura *et al.*, 1967). Thus, the effects of β -galactosidase digestion on IPSF activity of SHH were examined. As shown in Table 1, the activity of SHH was decreased by the treatment. These results suggest that both polygalactopyranoside and protein moieties are involved in the IPSF activity of SHH.

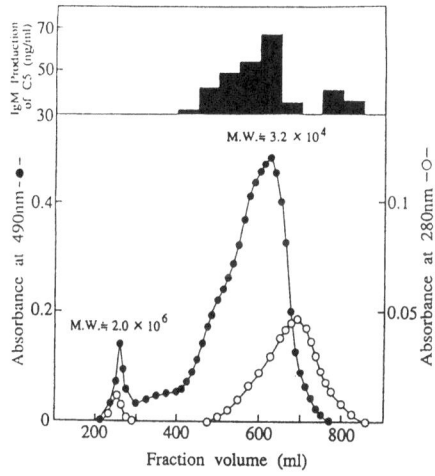

Fig. 4 Gel filtration profile of SHH on Sepharose CL-4B

Table 1. β -Galactosidase digestion of SHH

	Relative activity
PBS+enzyme(50IU)	1.00
SHH	2.18
SHH+enzyme	1.20

* Incubation ; 37℃ , 30 min.

3.3   IPSF ACTIVITY IN CHITOSAN

As shown in Fig. 2, chitosan showed the highest IPSF activity,   stimulating it about 5-fold compared to the control. Chitosan is partially deacetylated chitin ((β 1$\rightarrow$ 4)-linked polysaccharides composed of N-acetyl glucosamine residues). Table 2 also shows the stimulation of IgM production by various chitosans. The highest IPSF activities were observed at addition of DA 70% (70% of amino-group was deacetylaed) or DA 80% chitosan. As shown in Table 2, DA 100% chitosan and CM (carboxyl methyl)-chitin had no IPSF activities. So stimulation of Ig productin by chitosan required proper partial deacetylation.

Table 2. Effect of various chitosans on IgM production of human-human hybridoma HB4C5 cells

Samples* (DA%)	Relative cell number	Relative IgM conc.	Relative IgM productivity
70	1.27	8.20	6.45
80	1.36	8.44	6.20
90	1.19	6.13	5.15
100	0.89	0.89	1.00
CM-chitin	1.01	0.94	0.93

* Addition conc.; 1 μg/ml

**DA; Deacetylation ratio

3.4   EFFECTS OF SHH AND CHITOSAN ON IG PRODUCTION OF HUMAN LYMPHOCYTES

Table 3 shows the effects of SHH and chitosan on Ig production of human lymphocytes. SHH stimulated IgM production of lymphocytes in ERDF medium supplemented with ITES or 5% FBS, but it did not stimulate IgG production. Chitosan stimulatued IgM production of human lymphocytes in ITES-ERDF medium slightly. These indicates that human-human hybridomas cultured in serum-free medium could be very useful for screening of *in vivo* effectors and clarification of mechanism of their stimulation activity of Ig production.

Table 3. Effects of SHH and chitosan on Ig production of human lymphocytes *

Additives	Cell number ($\times 10^5$ cells/ml)	Ig concentration (ng/ml)	
		IgM	IgG
ITES	6.79	10	169
ITES-SHH**	7.12	27	152
5% FBS	9.34	82	173
FBS-SHH**	9.76	139	187
ITES	7.40	50	83
ITES-chitosan***	6.20	80	74
5% FBS	6.54	91	89
FBS-chitosan***	6.70	91	86

* Human lymphocytes (9×10^5 cells/ml) were cultured in ERDF medium
for 5 days.
** SHH was added at 500μg/ml to ERDF medium.
*** Chitosan (DA 90%) was added at 1μg/ml to ERDF medium.

4. Literature Cited

Ayano,Y., Ohta,F., Nozaki,I., Egashira,Y. (1986) ´ Effects of soybean hull and
hemicellulose preparation isolated from soybean hull on the cholesterol metabolism in
rats´ , *Tech. Bull. Fac. Hort. Chiba Univ.* , **38**, 9-18

Aihara,K., Yamada,K., Murakami,H. (1988) ´ Production of human-human
hybridoma secretion monoclonal antibodies reactive to breast cancer cell lines´
in Vitro Cell. Develop. Biol., **24**, 959-962

Dubois,M., Gilles,K.A., Hamilton,J.K., Rebers,P.A., Smith,F. (1956)
´ Colorimetric method for determination of sugars and related substances´
Anal. Chem., **28**, 350-356

James,K., Bell,G.T. (1987) ´ Human monoclonal antibody production. Current status
and future prospect´ , *J. Immuno. Methods* , **100**, 5-40

Kawamura,S. (1967) ´ A review of the chemistry of soybean polysaccharides (1) ´
Nippon Shokuhin Kogyo Gakkaishi (J.Jpn.Soc.Food Sci.Tech.) **14**, 514-523

Murakami,H., Masui,H., Sato,G.H. (1982) ´ Growth of hybridoma cells in serum-
free medium ´ , *Proc. Natl. Acad. Sci. USA* , **79**, 1158-1162

Murakami,H., Hashizume,S., Ohashi,H. (1985) ´ Human-human hybridoma secretion
of antibodies specific to human lung carcinoma ´ , *in Vitro Cell. Develop. Biol.*, **21**,
593-596

Murakami,H., Yamada,K. (1987) ´ Production of cancer specific monoclonal antibodies with human-human hybridomas and their serum-free, high density perfusion culture ´ , *Modern Approaches to Animal Cell Technology*, Ed. by RE Spier and JB Griffiths, Butterworths, England, pp.52-76

Murakami,H. (1989) ´ Serum-free media used for cultivation of hybridomas ´ , *Advances in Biotechnological Processes*, Ed. by A.Mizrahi, vol.11, Alan, R.Liss Inc., New York, pp. 107-141

Yamada,K., Akiyoshi,K., Murakami,H., Sugahara,T., Ikeda,I., Toyoda,K., Omura,H. (1989a) ´ Partial purification and characterization of immunoglobulin production stimulatimg factor derived from namalwa cells´ , *In Vitro Cell Develop. Biol.* , **25**, 243-247

Yamada,K., Ikeda,I., Sugahara,T., Shirahata,S., Murakami,H. (1989b)´ Screening of immunoglobilin production stimulating factor (IPSF) in foodstuffs using human-human hybridoma HB4C5 cells ´ , *Agric. Biol. Chem.*, **53**, 2987-2991

Yamada,K., Ikeda,I., Sugahara,T., Hashizume,S., Shirahata,S.,Murakami,H. (1990a) ´ Stimulation of proliferation and immunoglobulin M production by lactoferrin in human-human and mouse-mouse hybridomas in serum-free culture´ , *Cytotechnology* , **3**, 123-131

Yamada,K., Ikeda,I., Maeda,M., Shirahata,S., Murakami,H. (1990b)´ Effect of immunoglobulin production of human lymphocytes ´ , *Agric. Biol. Chem.*, **54**, 1087-1089

PRODUCTION OF RECOMBINANT PROTEIN C IN A PERFUSION CULTURE

T. SUGIURA and H.B. MARUYAMA
Laboratory for Cell Biology, Pharma Research
Laboratories, Hoechst Japan Ltd.
3-2-1, Kawagoe, Saitama 350
Japan

ABSTRACT. For the development of a perfusion culture producing recombinant human protein C, the effects of fetal calf serum (FCS) and growth factors on cell growth and the recombinant protein production were investigated. Although the growth of recombinant cells was stimulated by serum in a dose-dependent manner, a lower concentration of serum (2%) could support both synthesis and post-translational modification of protein C as efficiently as 10% serum. Among the growth factors tested, transferrin enhanced protein C production to the level comparable with 10% serum, while insulin was effective in maintaining cellular metabolism. Based on these results, a perfusion culture for a scale-up production of recombinant protein C was done using an Opticell culture system. A good productivity of the recombinant protein was obtained in low serum or serum-free medium for more than one month.

Introduction

Recent advances in recombinant DNA technology make the industrial production of useful proteins in mammalian cell culture feasible (Bebbington and Henteschel, 1985). Considering the cost and the complexity of the down-stream process, it is desirable to use serum-free or low serum medium for the cultivation (Barens and Sato, 1980).
We have produced human protein C using recombinant DNA techniques. Protein C is a vitamin K-dependent glycoprotein assumed to be a potential anticoagulant (Kiesel, 1979). This glycoprotein should undergo for its biological activity γ-carboxylation of specific glutamic acid residues, a unique post-translational modification (Sugo et al., 1985). In this study the production-stimulating activity of FCS and growth factors was investigated for the development of a perfusion culture. Finally a sufficient amount of recombinant protein C was obtained in a perfusion culture using an Opticell culture system.

Materials and Methods

R. Sasaki and K. Ikura (eds.), Animal Cell Culture and Production of Biologicals, 291–298.
© 1991 *Kluwer Academic Publishers.*

MATERIALS

Rabbit anti-human protein C polyclonal antibody (α-PC PoAb) was purchased from Dako (Glostrupe, Denmark). Bovine transferrin was obtained from Binding Site limited (Birmingham, UK). Monoclonal antibody specific for Ca^{2+}-dependent conformational change (PC01) was prepared in our laboratory (Kurosawa-Ohsawa et al., 1990). All the other reagents were of analytical grade.

CELLS AND CULTURE

CHO-B3 is a clonal cell line that was engineered to produce the highest level of human protein C (Takeshita et al., 1988). Growth medium was MEMα^- supplemented with 10% FCS, 0.5 μM MTX, and 0.1 μg/ml vitamin K_3. Cells were cultured at 37°C in a humidified 5% CO_2 and 95% air atmosphere.

ELISA

We developed two types of ELISAs that allowed the measurement of the degree of γ-carboxylation of human protein C (Sugiura et al., 1990). One system used commercially available α-PC PoAb and the other, PC01 monoclonal, which specifically reacts with fully γ-carboxylated protein C. The antigen detected by the former was designated T-PC (total protein C) and that by the latter, γ-PC (γ-carboxylated protein C).

MEASUREMENT OF GROWTH

Cells were plated in 24-well plates (Nunc) at 1×10^4 cells per well in 1 ml of MEMα^- including various levels of FCS. Cells were counted every day using a Coulter counter ZM after detachment of cells by treatment with 0.05% trypsin in phosphate buffered saline, pH 7.2 (PBS) containing 0.02% EDTA.

PRODUCTIVITY TESTING

Assays were done on confluent monolayers prepared in 24-well plates as follows. Recombinant cells (1×10^4 per well) in 1 ml of growth medium were plated into each well. After reaching confluency, monolayer cultures were washed twice with PBS and replenished with 1 ml of MEMα^- containing 0.5 μM MTX, 0.1 μg/ml vitamin K_3, and different concentrations of serum or growth factors. The medium was changed daily for 3 days. The harvested media were stored at -20°C for further analysis.

OPTICELL CULTURE

Cultivation was done using an Opticell 5200R (Charles River Biotechnical Service, Wilmington, MA, USA) according to the manufacture's instruction. The culture was started with an inoculation of approximately $2x10^8$ CHO-B3 cells into the Opticore S-51. Cells were grown in ES medium (Nissui Seiyaku) supplemented with 7.5% FCS (Gibco), 2 mM glutamine, 2 g/l glucose, 0.5 μM MTX, and 2 g/l sodium bicarbonate. In the production phase, medium supplements were varied (for details, see legend to Fig. 4). At the medium change, the whole culture system including a ceramic core and connecting tubes was washed twice with the new medium, thoroughly. Temperature was maintained at 36.8°C and the medium volume of the culture was 2 l throughout the cultivation.

Results

EFFECTS OF FETAL CALF SERUM

When we examined the effects of serum on the growth of recombinant cell line, CHO-B3, an apparent dose-dependent effect was observed (Fig. 1).

Next, the effects of serum on the recombinant protein production were evaluated using confluent cultures. The production of protein C was measured daily over 3 days to confirm the consistency and persistency of the activity of serum. Although the absolute value of productivity tended to vary among experiments, the trend of the stimulatory effect of each agent was consistent.

As shown in Fig. 2A, a low level of serum (2%) enhanced T-PC and γ-

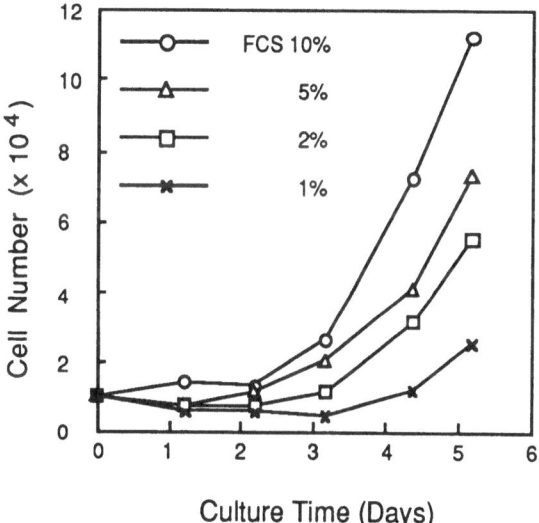

Fig. 1. Effects of serum on the growth of CHO-B3 cells. Cells were plated in 24-well dishes at $1x10^4$/well in 1 ml of MEMα⁻ containing varying levels of serum. Cells were counted daily using a Coulter counter.

294

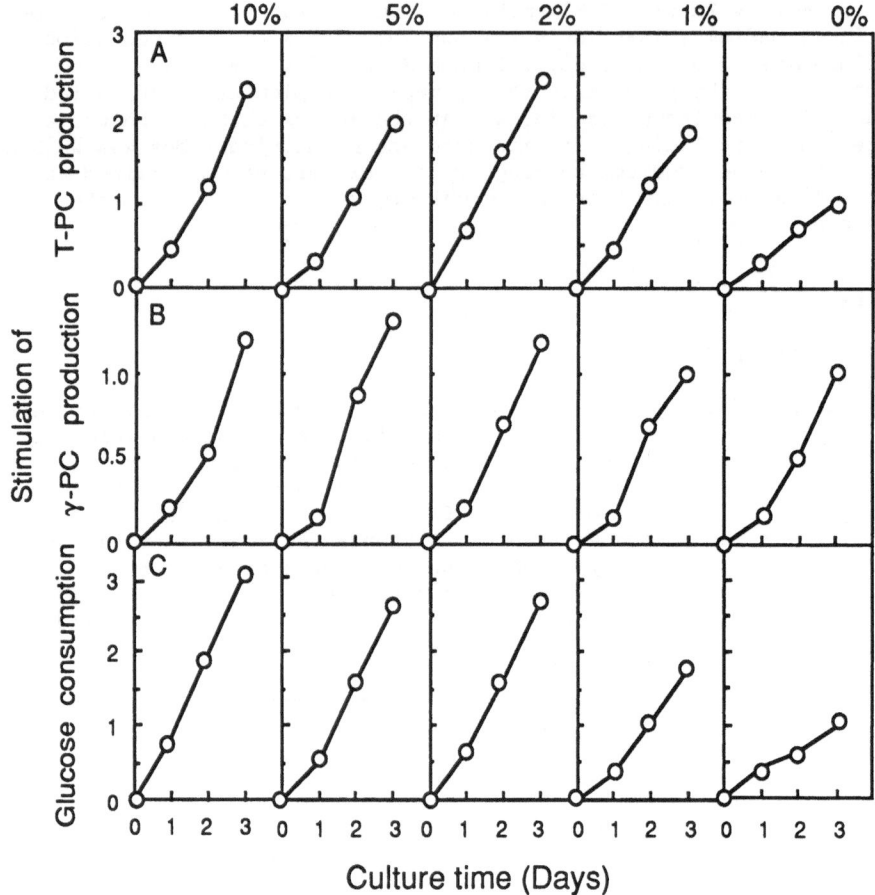

Fig. 2. Effects of serum on the production of T-PC (A), γ-PC (B), and recombinant cells were treated with different levels of serum for 3 days. Experiments were done in duplicate. Each value is expressed relative to the control culture. Value 1 in A, B, and C means 6.3 μg, 4.3 μg, and 0.81 mg, respectively.

PC production as efficiently as 10% serum. The smaller amount of γ-PC relative to T-PC in every culture was explained by the insufficient γ-carboxylating capacity of the recombinant cell line itself (compare Fig. 2, A and B). The metabolism of cells in various cultures was also measured by the glucose consumption rate. As Fig. 2C indicates, serum stimulated the cell metabolism progressively. The amounts of glucose consumed were correlated well with those of T-PC production (r=0.80), suggesting that serum activated overall cellular metabolisms nonspecifically, thereby enhancing the recombinant protein synthesis concomitantly.

EFFECTS OF GROWTH FACTORS

The effects of growth factors on the productivity of protein C were

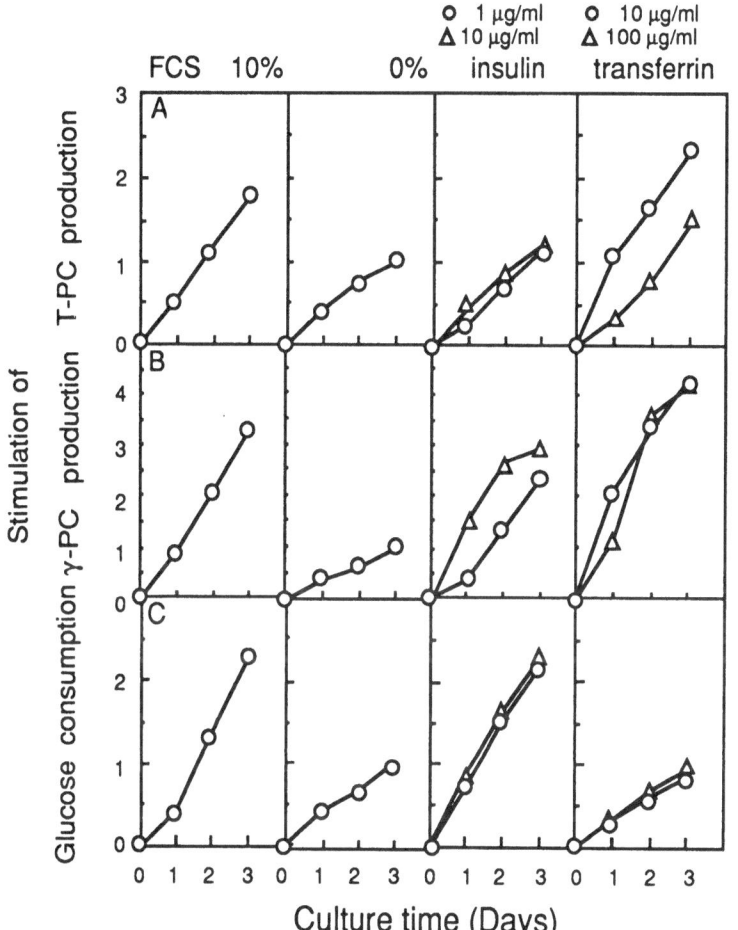

Fig. 3. Effects of growth factors, insulin, and transferrin on the
production of T-PC (A), γ-PC (B), and glucose uptake (C) by CHO-B3
Each value is expressed relative to the control culture. Value 1
in A, B, and C means 2.1 μg, 0.49 μg and 1.1 mg, respectively.

then evaluated by the same methods. Ten percent serum and serum-free cultures were included as a positive and a negative control, respectively. Among various factors tested, transferrin at a concentration of 10 μg/ml increased the amounts of both T-PC and γ-PC to the degree comparable with 10% serum. On the other hand, insulin at 1 or 10 μg/ml had only a marginal effect on T-PC synthesis although effects of γ-PC synthesis were more remarkable (Figs. 3A and B). In terms of cellular metabolism the addition of transferrin gave no stimulatory effect on glucose use (Fig. 3C). The presence of vacuoles in the cytosol indicated that cells became unhealthy in transferrin-treated cultures as was also observed in 0% serum cultures (Johonson and Schwartz, 1976). The effects of insulin on glucose consumption were as great as 10% serum, rendering recombinant cells normal throughout the experiment. Thus, insulin should be included in serum-free medium to sustain the metabolic activity of long-term cell culture.

SERUM-FREE PERFUSION CULTURE USING OPTICELL

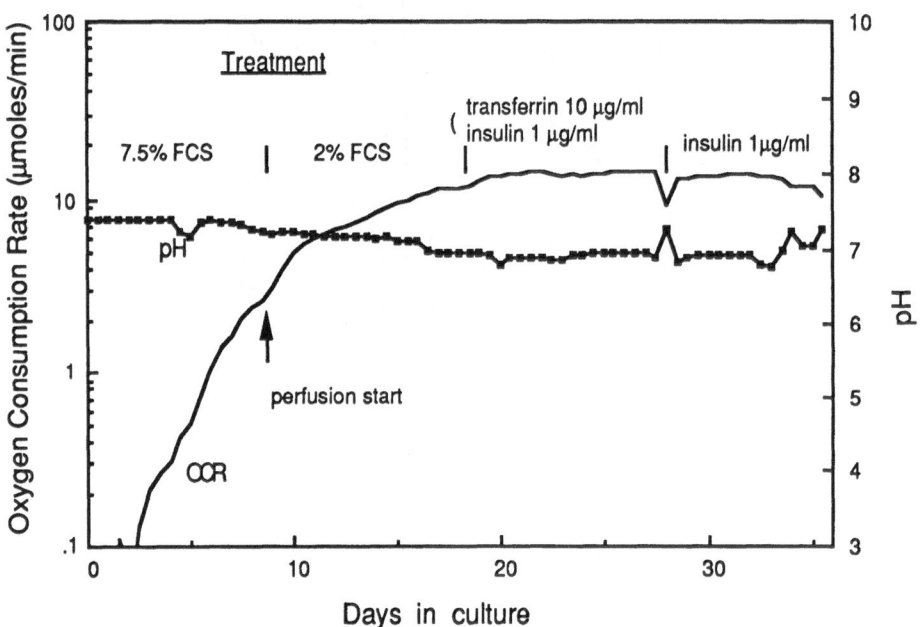

Fig. 4. Cultivation of CHO-B3 cells in an Opticell culture system. The culture was started with an inoculation of about 2×10^8 cells into the Opticore S-51. On day 9 perfusion was started at 2 l/day.

Based on the results obtained above, serum-free perfusion culture was done using the Opticell 5200R system, which can supply exactly the required amount of oxygen (Lyderson et al., 1985). In the growth phase, cells were propagated in ES medium containing 7.5% serum, followed by the reduction of the serum level to 2% in the production phase. Thereafter, serum was completely depleted and replaced with insulin or transferrin. Figure 4 shows the changes of OCR and pH in the Opticell culture. The value of OCR reached about 15 µmoles/min at maximum. Replacement of serum by the growth factors seemed to have no adverse effect on OCR. Furthermore, a favorable productivity was achieved under the serum-free conditions (Table 1). Note that the average ratio of γ-PC to T-PC was higher in Opticell culture (0.83) than that with the conventional dish culture (usually 0.2-0.6). Thus, we could successfully carry out the perfusion culture of the recombinant cell line in both a lower serum and serum-free medium using the Opticell culture system.

TABLE 1. Production of recombinant protein C using Opticell culture system. Culture condition was indicated in Fig. 4. Two types of protein C molecules were measured by a combination of two ELISAs. Glucose was measured with a Beckman glucose analyzer.

Day	Treatment	T-PC	γ-PC	Ratio of γ-PC to T-PC	Glucose Consumption Rate (g/day)
		(mg/day)			
15	2% FCS	3.73	2.25	0.60	2.57
16		3.85	3.33	0.87	2.53
18		8.43	8.53	1.02	2.33
19	transferrin	8.04	7.26	0.90	1.52
22	(10 µg/ml)	6.82	6.40	0.94	2.91
23	plus insulin	8.24	6.21	0.75	3.01
25	(1 µg/ml)	2.46	2.44	0.99	3.38
27	insulin	4.70	4.14	0.88	1.96
28	(1 µg/ml)	5.97	4.65	0.78	2.50
30		3.90	2.08	0.53	2.91

Conclusions

1. A lower level of serum could enhance the productivity of T-PC and γ-PC to almost the same level as 10% serum.

2.   Transferrin at a concentration of 10 μg/ml increased the amounts
       of both T-PC and γ-PC to a degree comparable with 10% serum.
       The effect of insulin on cellular metabolism was as great as 10%
       serum.
3.   A favorable productivity of recombinant protein C was achieved in
       low serum or serum-free culture using the Opticell culture
       system.

References

       Barnes, D. and Sato, G. (1980) 'Methods for growth of cultured
       cells in serum-free medium', Anal. Biochem. 102, 255-270
       Bebbington, C. and Henteschel, C. (1985) 'The expression of
       recombinant DNA products in mammalian cells', Trends in
       Biotecnol. 3, 314-317
       Johonson, G. and Schwartz, J. (1976) 'Effects of sugars on the
       physiology of cultured fibroblasts', Exp. Cell Res. 97, 281-290
       Kisiel, W. (1979) 'Human plasma protein C: isolation,
       characterization and mechanism of activation by α-thrombin', J.
       Clin. Invest. 64, 761-769
       Kurosawa-Ohsawa, K., Kimura, M., Kume-Iwaki, A., Tanaka, T., and
       Tanaka, S. (1990) 'Anti-protein C monoclonal antibody induces
       thrombus in mice', Blood 75, 2156-2163
       Lyderson, B.K., Pugh, G.G., Paris, M.S., Sharma, B.P. and Noll,
       L.A. (1985) 'Ceramic matrix for large scale animal cell culture',
       Bio/Technology 3, 63-67
       Sugiura, T., Kurosawa-Ohsawa, K. Takahashi, M., and Maruyama, H.B.
       (1990) 'Relationship between productivity and γ-carboxylation of
       recombinant protein C', Biotechnol. Lett. 12, 799-804
       Sugo, T., Persson, U., and Stenflo, J. (1985) 'Protein C in
       bovine plasma after warfarin treatment', J. Biol. Chem. 260,
       10453-10457
       Takeshita, S., Tezuka, K.-I., Honkawa, H., Matsuo, A., Matsuishi,
       T. and Hashimoto-Gotoh, T. (1988)' Tandem gene amplification in
       vitro for rapid and efficient expression in animal cells', Gene
       71, 9-18

Optimization of cell culture conditions for G-CSF (granulocyte colony-stimulating factor) production by genetically engineered Namalwa KJM-1 cells

Shinji Hosoi, Kazunari Murozumi, Katsutoshi Sasaki, Mitsuo Satoh,
Tatsuya Tamaoki, and Seiji Sato
Tokyo Research Laboratories, Kyowa Hakko Kogyo Co.,Ltd., 3-6-6
Asahi-machi, Machida-shi, Tokyo 194, Japan.

ABSTRACT

An expression vector for G-CSF, pASLB3-3, was constructed and introduced into Namalwa KJM-1 cells (Hosoi,1988), and cells resistant to methotrexate (MTX) (100 nM) were obtained. Among them, the highest producer, clone SC57, was selected and the productivity of this clone was further characterized. Under the conventional conditions, the maximal production of G-CSF was at the most 1.7 µg/ml/day even though the cell number was above 7×10^5cells/ml. The limiting factors at high density were analyzed : 1. the deficiency of nutrients, such as glucose, cysteine, and serine, 2. pH control. The depression of productivity was overcome by using a modified Biofermenter containing micro-silicone fibers and a dialysis system. ITPSGF medium was modified to elevate concentrations of amino acids and glucose by 2.0 and 2.5 times, respectively. Under the control of pH at 7.4 and dissolved oxygen (DO) at 4 ppm, the specific G-CSF productivity was not depressed even at high cell density (above 1×10^7cells/ml), and the amount of G-CSF reached 41 µg/ml. These results indicated the possibility of finding the optimum culture conditions for the production of recombinant proteins by Namalwa KJM-1 cells.

INTRODUCTION

High density culture is useful for obtaining a large quantity of cellular products and can reduce the cost of the medium. In addition, the differences in physiological conditions between cells at low density and at high density was very significant in devising a new system for high density culture. Thus, we could devise a unique culture technique and culture systems for suspension culture, by analyzing the limiting factors of high density culture (Sato,1983).

We previously reported the way of growth and maintenance of Namalwa KJM-1 cells, a serum-independent subline of Namalwa cells (B lymphoblastoid cells), in suspension culture at high density in chemically defined serum- and albumin-free ITPSGF medium (Hosoi,1988).

Several genes of cytokines were successfully expressed in this cell line and

R. Sasaki and K. Ikura (eds.), Animal Cell Culture and Production of Biologicals, 299–306.
© 1991 *Kluwer Academic Publishers.*

the products were detected in the culture media (Miyaji,1990a,b,c). A method of dihydrofolate reductase (dhfr) gene coamplification was available for efficient expression of foreign genes in this cell line (Miyaji,1990c). We had shown that the expression level of beta-interferon in this cell line was augmented approximately in parallel with the increase of the cell density in perfusion culture (Miyaji,1990a). Thus, Namalwa KJM-1 is a useful host cell for the production of recombinant products.

Molecular cloning and expression of human granulocyte colony-stimulating factor (G-CSF) which stimulates the production of granulocytes almost exclusively from committed precursor cells in soft agar medium, have been reported (Nagata,1986). The production of recombinant G-CSF in *E.coli* (Komatsu,1987) and Chinese hamster ovary (CHO) cells (Oheda,1988) has been reported by several investigators.

In this paper, we show the possibility of finding the optimum culture conditions for the production of G-CSF as an example of recombinant proteins produced by genetically engineered Namalwa KJM-1 cells.

Materials and Methods

Cell and culture medium Namalwa cells, a human lymphoblastoid cell line, were provided from Mr.F.Klein (Frederick Cancer Research Center, Frederick, Maryland, USA). They were adapted to serum- and albumin-free RPMI-1640 medium supplemented with 4-(2-hydroxyethyl)-1-piperazineethanesulfonic acid (HEPES) (10 mM), L-glutamine (4 mM), penicillin (25 U/ml), streptomycin (25 µg/ml), insulin (3 µg/ml), transferrin (5 µg/ml), sodium pyruvate (5 mM), sodium selenite (125 nM), galactose (1 mg/ml), and Pluronic F68 (1 mg/ml) ; we called this ITPSGF medium (Hosoi,1988).

Chemicals and materials MTX was purchased from Sigma Chemicals Co., St.Louis, MO, U.S.A., normal RPMI-1640 medium and modified RPMI-1640 medium were from Kyokuto Seiyaku Kogyo Co.Ltd., Tokyo, Japan. Other chemicals were obtained as described previously (Hosoi,1988; Miyaji,1990c).

DNA manipulations All enzymes were purchased from Takara Shuzo, Kyoto, Japan. Plasmid DNA preparation, DNA fragment purification, *E.coli* DNA polymerase I reaction, ligation, and the introduction of plasmid DNA into *E.coli* were done as described by Nishi (1984).

Construction of an expression vector for G-CSF An expression plasmid, pASLB3-3 (Fig.1), was constructed. It consisted of the following five DNA fragments: i) a 0.45-kilobase pairs (kb) *Sal*I-*Aat*II fragment containing a part of hG-CSF cDNA from pCSF3-3 (Kuga,1990); ii) a 0.2-kb *Aat*II-*Dde*I (blunt) fragment containing a part of hG-CSF cDNA from pCfTA1 (Komatsu,1987); iii) a 8.7-kb *Xho*I-*Sma*I fragment containing an ampicillin-resistance gene, a G418-resistance gene, and a dhfr transcription unit from pSE1βd2-4 (Miyaji,1990c); iv) a 0.3-kb *Xho*I-*Bgl*I fragment containing the SV40 early promoter from pAGE107 (Miyaji,1990a). The *Bgl*I site was converted to a *Ban*II site by insertion of synthetic DNA in the following sequence (A); v) a 0.3-kb *Ban*II-*Sau*3AI fragment containing a segment and part of the U5 sequence of the HTLV-1 LTR from pATK-03 (Seiki,1983). The *Sau*3AI

301

(*Bam*HI) site was converted to *Sal*I site by insertion of synthetic DNA in the following sequence (B)

	*Sau*3AI	*Sal*I
(A) 5' CGGGCT 3'	(B) 5' -G GATCCCCGGTC GACC	
3' GGAGC 5'		

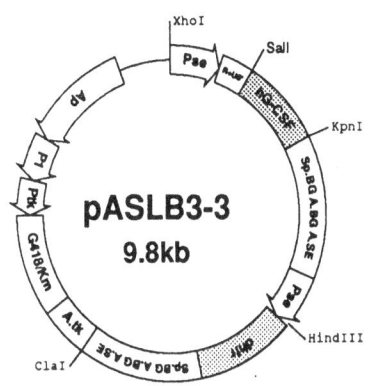

Fig.1 Structure of pASLB3-3

The following abbreviations were used: G418/Km, the Tn5-derived G418 resistance gene; Ap, ampicillin resistance gene; P1, pBR322 P1 promoter: Ptk, the Herpes simplex virus thymidine kinase promoter; P_{SE}, the SV40 early promoter; Atk, the polyadenylation signal from the Herpes simplex virus thymidine kinase gene; A_{SE}, the polyadenylation signal from the SV40 early gene; AβG, the polyadenylation signal from the rabbit β–globin gene; Sp.βG, Splicing signal from the rabbit β–globin gene; R+U5', the R segment and part of the U5 sequence of the HTLV-1 LTR; dhfr, murine dihydrofolate reductase gene.

Electroporation A somatic hybridizer SSH-1 (Shimadzu Seisakusyo, Kyoto, Japan) and an SSH-C13 chamber (distance of electrode; 2 mm) were used ; details of the procedures were previously described by Miyaji (1990a).

Selection of MTX-resistant subclones G418-resistant transformants were plated into 96-well plates at $4x10^4$ cells per well in 0.2 ml of RPMI-1640 medium containing 10% FCS, 0.3 mg/ml G418, and 50 nM MTX. Drug-resistant subclones were obtained two to five weeks after starting selection. 100 nM MTX-resistant subclones were obtained from 50 nM MTX-resistant subclones by the same procedure, except for increasing concentrations of MTX.

Measurement of G-CSF by sandwich-type ELISA G-CSF was measured by sandwich-type ELISA. Purified rabbit polyclonal anti-human G-CSF antibody and purified mouse anti-G-CSF monoclonal antibody KM341(Yoshida,1989), were obtained from Pharmaceutical Research Laboratories, Kyowa Hakko Kogyo, Shizuoka, Japan, and were used for sandwich-type ELISA. The details of the procedures were described previously (Yoshida,1989). As a standard, purified G-CSF obtained from SC57 conditioned medium was used.

RESULTS

Production of G-CSF under conventional conditions

An expression vector for G-CSF, pASLB3-3 (Fig. 1), was constructed and introduced into Namalwa KJM-1 cells by electroporation. The highest G-CSF producer among 100 nM MTX resistant clones, clone SC57, was selected, and the productivity of this clone was further characterized.

SC57 was inoculated at $5x10^5$cells/ml and incubated. Cell count and medium change were done at the indicated time. Under the conventional conditions, the maximal production of G-CSF was at the most 1.7 µg/ml/day, i.e., the specific productivity of SC57 was at the most 2.4 µg/10^6cells/day even though the cell number was above 7×10^5cells/ml (Fig. 2). This limitation of productivity may be solved by use of a perfusion culture system.

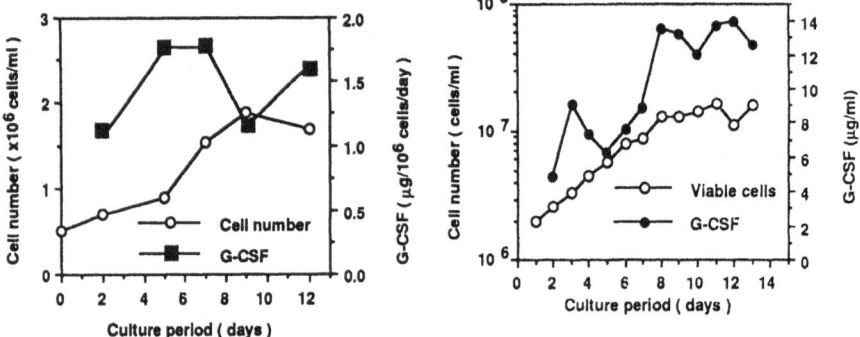

Fig.2 Production of G-CSF under the conventional conditions.

Fig.3 High density culture of SC57 using a Biofermenter with a dialysis membrane.

High density culture of SC57

Using a Biofermenter with a dialysis membrane, the production level was augmented with an increase of cell density (Fig. 3). SC57 was cultured in ITPSGF medium and inoculated at a density of 2×10^6cells/ml (net culture volume was 200 ml). Cell count and sampling were done every day. The perfusion rate was set to 220 ml/day and the dialysis rate was gradually shifted up stepwise from 250 ml/day to 1000 ml/day according to the cell density. DO and pH were not controlled. SC57 proliferated to 1.6×10^7cells/ml and the amount of G-CSF reached 14 µg/ml. The specific productivities were 3.9 µg/10^6cells/day on day 2, 2.8 µg/10^6cells/day on day 7, and 2.3 µg/10^6cells/day on day 12.

Effects of pH control

Compared with the physiological conditions in the flask, those in a modified perfusion culture were significantly different. These differences were assessed from the viewpoint of productivity. Unless otherwise noted, we assessed them by semi-continues culture centrifuged every day (net culture vol. was 500 ml). DO, lactates, and ammonia, the most famous toxic metabolites, have no effect on the productivity of SC57 (data not shown).

Accurately controlled pH by the addition of 7.5% sodium bicarbonate affected on the productivity (Fig. 4). DO was controlled at 4 ppm by flow rate of air through micro-silicone fibers (Nagayanagi Kogyo Co.,Ltd., Tokyo). Higher productivity was observed under the pH controlled at 7.4 rather than at 7.0.

Specific productivity on day 5 was 2.2 μg/10^6cells/day at pH 7.4 and 1.1 μg/10^6cells/day at pH 7.0. In these cases, the cell densities were above 3x10^6cells/ml.

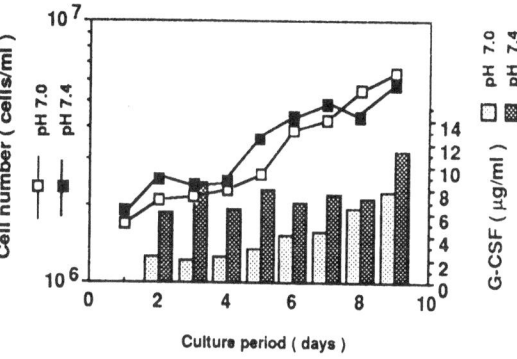

Fig. 4 Effects of pH control

Effects of nutrient

From analyses of glucose and amino acid contents in the conditioned media of SC57, deficiencies of glucose, cysteine, and serine were observed at high density beyond 4x10^6cells/ml (data not shown). Addition of 3 mg/ml of glucose to the ITPSGF medium did not affect the productivity but did affect the growth rate at high densities beyond 4x10^6cells/ml (data not shown).

SC57 was cultured in the ITPSGF medium with 0.1 mM cysteine and serine and 3 mg/ml of glucose added, with the pH controlled at 7.4. SC57 cells were cultured for 6 days (Fig. 5). The specific G-CSF productivity of SC57 was enhanced to 5.6 μg/10^6cells/day at pH 7.4 (day 3), which was 1.8 times higher than that of the control, 3.1 μg/10^6cells/day.

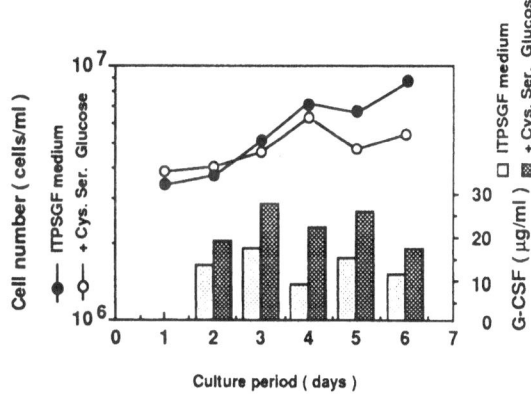

Fig. 5 Effects of nutrients supplementation

High density culture of SC57 using a modified Biofermenter

Using a modified Biofermenter with a dialysis membrane and micro-silicone fibers (net culture vol. was 250 ml), SC57 was cultured for 19 days (Fig. 6). At the beginning of the culture, normal ITPSGF medium was used, and after day 16, a special medium in which contents of amino acids and glucose were elevated twice and 2.5 times, respectively, was used. The perfusion rate was set about to 250 ml/day and the dialysis rate was shifted up stepwise according to cell density from 250 ml/day to 1000 ml/day. DO was adjusted by controlling the flow rate of air through the micro-silicone fibers as described

304

above. The pH was controlled at 7.4 and varied by less than 0.1 pH unit during the culture.

The specific productivity was 2.0 $\mu g/10^6$cells/day on day 3, 1.9 $\mu g/10^6$cells/day on day 7, 2.5 $\mu g/10^6$cells/day on day 11, 2.6 $\mu g/10^6$cells/day on day 15, and 3.3 $\mu g/10^6$cells/day on day 17 (the amount of G-CSF reached 41 $\mu g/ml$). These increases of specific productivity before day15 were due to shifts in the dialysis rate and that of day 17 was due to a change of medium. These results showed that the high density culture of SC57 was optimized without a loss of specific productivity.

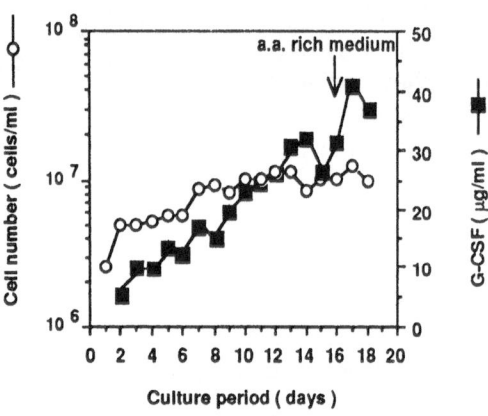

Fig. 6 High density culture of SC57 using modified Biofermenter with a dialysis membrane and micro-silicone fibers

DISCUSSION

High density culture makes it possible to use a smaller vessel to obtain a required amount of target proteins. Gene technology may be applied further to simplify the cellular requirements for nutrients, growth factors, and regulatory factors and to increase productivity. However, the physiological conditions of the cells for production of the target proteins were somewhat different from those of the parent cells at a low density. Thus attention to the physiological and biochemical balance of the growth medium was needed for the technology of maintaining cells at a high productivity for long culture periods.

To achieve high-level expression of recombinant proteins, gene amplification using dhfr is a very effective tool. For this purpose, Chinese hamster ovary (CHO) cells deficient in dhfr have been successfully used as the host cell line (Kaufman,1982; Kaufman,1985). But several cell lines that contain active endogenous dhfr genes can be used for gene amplification with dhfr (Dorai,1987). We previously showed that a foreign gene expression system using recombinant Namalwa KJM-1 cells and a perfusion culture system were preferable to obtain recombinant proteins in large quantities (Miyaji,1990a) and showed that a method of dhfr gene coamplification was available in this cell line (Miyaji,1990c).

The specific G-CSF productivity of SC57 was depressed to less than 2.4 $\mu g/10^6$cells/day when the cell number was above 7×10^5cells/ml (Fig. 1). However, the production of G-CSF might be correlated with the cell density, because G-CSF accumulated in large quantities in a perfusion culture system

(Fig. 2). These results showed that the specific productivity was affected with the physiological conditions, i.e. pH, DO, and the concentrations of nutrients and metabolites.

Analysis of productivity at high density culture showed that the limiting factor was the deficiency of nutrients, such as glucose, cysteine, and serine (Fig. 3), and the specific productivity was depressed at lower than optimum pH (Fig. 4). It is reported that the pH control results in a significant improvement in the IFN production by recombinant CHO cells (Smiley,1989). Our results showed that the range of controlled pH which gave higher productivity was narrower than that of recombinant *E.coli*.

The depression of productivity was solved by using a Biofermenter with micro-silicone fibers and a dialysis membrane in a modified medium in which contents of amino acids and glucose were elevated twice and 2.5 times, respectively (Fig. 5). As a result, the specific productivity was not decreased at a high cell density, and the amount of G-CSF reached 41 μg/ml. This increase of specific productivity before day15 was due to shifts in the dialysis rate and that of day 17 was due to a change of medium. These results showed that the high density culture of SC57 was optimized without a loss of specific productivity.

REFERENCES
1. Dorai H and Moore GP (1987) The effect of dihydrofolate reductase-mediated gene amplification on the expression of transfected immunoglobulin genes, J. Immunol. 139: 4232-4241.
2. Hosoi S, Mioh H, Anzai C, Sato S and Fujiyoshi N (1988) Establishment of Namalva cell lines which grow continuously in glutamine-free medium, Cytotechnology 1: 151-158.
3. Kaufman RJ and Sharp PA (1982) Amplification and expression of sequences cotransfected with a modular dihydrofolate reductase complementary DAN gene, J. Mol. Biol. 159: 601-621.
4. Kaufman RJ, Wasley LC, Spiliotes AJ, Gossels SD, Latt SA, Larsen GR and Kay RM (1985) Coamplification and coexpression of human tissue-type plasminogen activator and murine dihydrofolate reductase sequences in Chinese hamster ovary cells, Mol. Cell. Biol. 5: 1750-1759.
5. Komatsu Y, Matsumoto T, Kuga T, Nishi T, Sekine S, Saito A, Okabe M, Morimoto M, Itoh S, Okabe T and Takaku F (1987) Cloning of granulocyte colony-stimulating factor cDNA from human macrophages and its expression in *Escherichia coli*, Jpn. J. Cancer Res. 78: 1179-1181.
6. Kuga T, Komatsu Y, Mizukami T, Sato M and Itoh S (1990) Effect of N-terminal deletion of signal peptide on the secretion of human granulocyte colony-stimulating factor in mammalian cells, Biotechnol. Lett. 12: 87-92.
7. Miyaji H, Mizukami T, Hosoi S, Sato S, Fujiyoshi N and Itoh S (1990) Expression of human beta-interferon in Namalwa cells which were adapted to serum-free medium, Cytotechnology 3: 133-140.
8. Miyaji H, Harada N, Mizukami T, Sato S, Fujiyoshi N and Itoh S (1990)

Expression of human lymphotoxin in Namalwa cells which were adapted to serum-free medium, Cytotechnology, 4: 39-43

9. Miyaji H, Harada N, Mizukami T, Sato S, Fujiyoshi N and Itoh S (1990) Efficient expression of human beta-interferon in Namalwa cells adapted to serum-free medium by a dhfr gene coamplification method, Cytotechnology, 4: 173-180

10. Nagata S, Tsuchiya M, Asano S, Kaziro Y, Yamazaki T Yamamoto O, Hirata Y, Kubota N, Oheda M, Nomura H and Ono M (1986) Molecular cloning and expression of cDNA for human granulocyte colony-stimulating factor, Nature 319: 415-418

11. Nishi T, Saito A, Oka T, Itoh S, Takaoka C and Taniguchi T (1984) Construction of plasmid expression vectors carrying the Escherichia coli tryptophan promoter, Agric. Biol. Chem. 48: 669-675

12. Oheda M, Hase S, Ono M and Ikenaka T (1988) Structure of the Sugar Chains of Recombinant HUman Granulocyte-Colony-Stimulating Factor Produced by Chinese Hamster Ovary Cells, J. Biochem., 103: 544-546

13. Sato S, Kawamura K and Fujiyoshi N (1983) Animal Cell Cultivation for Production of Biological Substances with a Novel Perfusion Culture Apparatus, J. Tissue Culture Methods, 8: 167-171

14. Seiki M, Hattori S, Hirayama Y and Yoshida M (1983) Human adult T-cell leukemia virus: Complete nucleotide sequence of the provirus genome integrated in leukemia cell DNA, Proc. Natl. Acad. Sci. USA., 80: 3618-3622

15. Smiley AL, Hu W-S and Wang DIC (1989) Production of Human Immune Interferon by recombinant Mammalian cells cultivated on Microcarriers, Biotechnol. Bioeng. 33: 1182-1190

16. Yoshida H and Shitara S (1989) Generation and Characterization of Monoclonal antibodies to Recombinant Human Granulocyte-colony Stimulating Factor (G-CSF) and Its Mutein, Agric. Biol. Chem. 53: 1095-1101

Acknowledgments

Our most sincere appreciation is also due to the Ministry of International Trade and Industry of Japan, and NEDO (New Energy and Industrial Technology Development Organization) for their financial support and excellent advice.

VARIATION IN THE RATIOS AND CONCENTRATIONS OF NUCLEOTIDE TRIPHOSPHATES AND UDP-SUGARS DURING A PERFUSED BATCH CULTIVATION OF HYBRIDOMA CELLS

Thomas Ryll, Volker Jäger, Roland Wagner
Arbeitsgruppe Zellkulturtechnik, Gesellschaft für Biotechnologische Forschung
D-3300 Braunschweig, FRG.

ABSTRACT.   A murine hybridoma cell line, which produces a monoclonal murine IgG$_{2a}$ antibody, was cultivated in a stirred reactor equipped with bubble-free aeration and a continuous perfusion system. During growth of up to $3.1 \bullet 10^7$ viable cells per ml, intracellular amounts of ATP, ADP, AMP, NAD, GTP, UTP, UDP-glucose (UDP-Glc), UDP-N-acetyl-glucosamine (UDP-GlcNac), UDP-N-acetyl-galactosamine (UDP-GalNac) and CTP were measured by the ion-pair HPLC technique after perchloric acid extraction of sedimented cells. Very stable values of the adenylate energy charge were found whereas the ratio of trinucleotides of the purine pool to these of the pyrimidine pool and the ratio between UTP and UDP-N-acetyl-sugars changed rapidly during different growth phases.

1. Introduction

The growth of mammalian cells depends on many physical and chemical parameters. These cells show pleiotypic responses to different culture conditions such that in many cases no direct correlation can be found between nutrients and growth characteristics. Normally, the growth of a culture ceases by limitation of an essential nutrient or by inhibition with a secreted endproduct. In the case of mammalian cell cultures it is often difficult to detect the limiting component. Since the information about nutrients and waste products in culture supernatants is insufficient to explain the behaviour of a cell culture, it is necessary to analyse additional parameters of the intracellular metabolism. Correlations between intracellular conditions and growth characteristics should give better explanations of the cause of growth behaviour in cell cultures.

Nucleotides have been shown to be one of the most important substances for cell metabolism. They are involved in a number of cellular processes and have widespread regulatory potential [1]. They participate as substrates, products, effectors or energy donators in many cellular reactions. Fluctuations of pool size could affect alterations in transport processes, macromolecular synthesis and cell growth. Some evidence has been found, that the pool size of ATP, the adenylate energy charge (AEC) or the ratio of purines to pyrimidines, influences or are dependent upon the cell cycle [2], the stimulation of quiescent cells by serum [3] or colchicine [4] and growth control [5], respectively.

R. Sasaki and K. Ikura (eds.), Animal Cell Culture and Production of Biologicals, 307–317.

Processes for the production of monoclonal antibodies and recombinant proteins with hybridoma and recombinant animal cells, are often well advanced with respect to medium design but there is an absence of information about the cell itself which can be thought of as the real bioreactor.

We have cultured cells in bioreactors with perfusion systems, to achieve high cell densities. In order to predict the behavior of their growth a method for detection of intracellular nucleotides has been established based on perchloric acid extraction followed by ion-pair high-performance liquid chromatography. Data from a perfused batch cultivation of hybridoma cells are presented and it is shown that the growth behavior of these cells correlate with particular patterns of the nucleotide pools.

2. Material and Methods

2.1. CELL LINE AND CELL PROPAGATION

A murine hybridoma cell line (Inst. f. Med. Immunologie, Berlin), which produces a monoclonal murine IgG_{2a} antibody was used. Cells were propagated in a stirred reactor (1.3 l working volume) with bubble-free aeration and a continuous perfusion system [6]. Oxygen content was maintained at 25 % of air saturation (1.7 mg/l). PH was controlled off line and maintained between 6.9 and 7.3. We used serum free medium consisting of a mixture of Iscoves and Ham's F12 supplemented with 3.61 g/l $NaHCO_3$, 0.18 g/l sodium pyruvate, 10 mg/l insulin, 10 mg/l transferrin and 1 g/l HSA (human serum albumin).

2.2. ANALYSIS OF SAMPLES

Viable and dead cells were estimated using the trypan blue exclusion method. Glucose and lactate contents were determined with YSI 27A glucose and lactate analysers (Yellow Springs Instruments, OH). Amino acids were quantified by means of a reversed-phase HPLC system with pre-column derivatisation with o-phthaldialdehyde (OPA, Serva, Heidelberg) [7].

2.3. CELL EXTRACTION PROCEDURE

Intracellular nucleotides were estimated by acid extraction of $2\text{-}3 \cdot 10^6$ cells immediately after sampling from the reactor as described in [8] and [9]. The clear centrifugated extract was filtered through a 0.45 μm filter (Milipore, type SJHVLO4NS) and stored in liquid nitrogen until analysis. The total procedure took about 20 minutes from the fermentor sample to the liquid nitrogen storage stage. Recoveries were tested using standard substances and spiked samples, and were in the range 80 - 90 % for ATP, ADP, AMP, GTP, NAD, UTP, UDP-Glc, UDP-GlcNac, UDP-GalNac and CTP, respectively.

2.4.   ANALYSIS OF CELL EXTRACTS

Cell extracts were analysed by a HPLC system at a detection wavelength of 254 nm based on an ion-pair reversed-phase procedure using tetrabutylammonium hydrogen sulfate (Fluka, Buchs) as ion-pair reagent combined with gradient elution as described in [8] and [9]. Routinely 100 μl of cell extracts, corresponding to $2\text{-}3 \cdot 10^5$ cells, were injected into the column. A typical chromatogram is shown in Fig. 1.

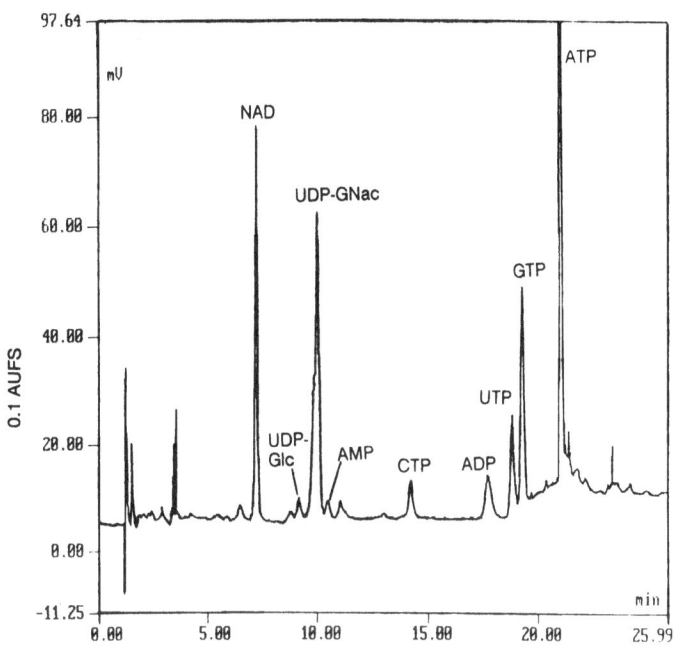

Figure 1. Chromatogram of a cell extract corresponding to $3 \cdot 10^5$ cells taken from the stationary growth phase. UDP-GNac is composed of UDP-GalNac + UDP-GlcNac. A Supelcosil LC-18T column [15 cm $\cdot$ 4.6 mm i.d., 3 μm particle size) combined with a guard column cartridge [5 μm particle size, both from Supelco, Bad Homburg) was used.

3.   Results and Discussion

The reactor was inoculated with $6 \cdot 10^8$ cells from roller bottles (92 % viability). After a short lag-phase of about one day the cells started the exponential growth phase for about four days resulting in a maximum of $1.1 \cdot 10^7$ viable cells/ml. Subsequently, growth slowed down and at day 8 of cultivation the cell concentration reached a stationary phase of approx. $3.1 \cdot 10^7$ viable cells/ml. The perfusion rate was increased up to 3.8 reactor volumes per day. At the end of fermentation the perfusion rate decreased due to membrane clogging. The viability of cells was about 90 % during lag-phase and 90-95 % during log-phase. Subsequently, it decreased to 85 % during the phase of reduced growth and finally resulted in 65 % at the end of the stationary phase (see Fig. 2).

Before the end of the log-phase at day 4, aspartate, glutamate, tryptophane and methionine reached low levels of approx. 10 μmol/l which might cause limited growth. Glucose remained at 4 mmol/l at the end of the log-phase and decreased to below 1 mmol/l at day 8. Lactate concentration increased up to more than 20 mmol/l at the end of the log-phase and thus became potentially inhibitory.

The total amount of nucleotides per viable cell reached highest levels at the beginning of the logarithmic growth phase with 13 fmol/cell (fmol = 10^{-15} mol). During the period of logarithmic growth it was constant at about 11 fmol/cell and decreased after cell growth had ceased. 43-48% of this pool was accounted for the adenyl fraction (ATP + ADP + AMP), whereas 30-35% of the total amount was composed of the uridine fraction (UTP + UDP-Glc + UDP-GalNac + UDP-GlcNac) (see Fig. 3). The ratio between these two different pools remained constant at 1.2-1.5 during cultivation (see Fig. 4). This result was in agreement with investigations of suspension cultured plant cells, where this ratio also remained constant but differed from mammalian cells ranging from 0.3-0.5 [10,11]. The low ratio found in plant cells was caused by hight amounts of UDP-sugars, mainly UDP-Glc, which are used for cell wall synthesis. Additionally, the ratio of the two purine triphosphates (ATP/GTP) and the two pyrimidine triphosphates (UTP/CTP), also remained stable during the different growth phases (see Fig. 4). In plant cells growing in suspension, Meyer and Wagner [10] calculated the intracellular cytoplasmatic concentration of the total nucleotide pool to be 9 mmol/l. Concentrations of uracil nucleotides and ATP were estimated to be 5 mmol/l and 2 mmol/l, respectively. Provided that the average volume of a hybridoma cell can be approximated to 1 pl (volume of a sphere with a diameter of 12.5 μm) the intracellular nucleotide concentrations were estimated to be of the same order of magnitude as shown in table 1.

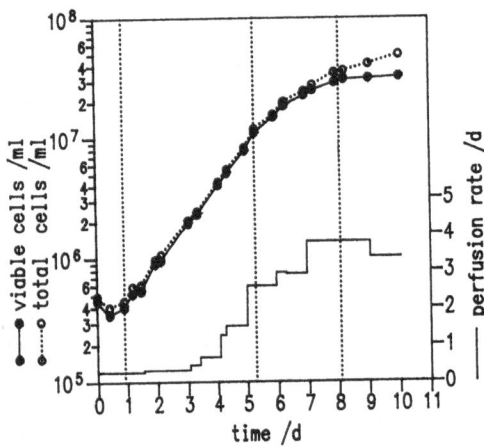

Fig. 2. Growth curve of a murine hybridoma cell line. The perfusion rate is expressed as reactor volumes per day. The growth curve is separated into 4 phases, characterized with dotted lines. Phase 1: lag-phase; 2: log-phase; 3: phase of reduced growth; 4: stationary phase.

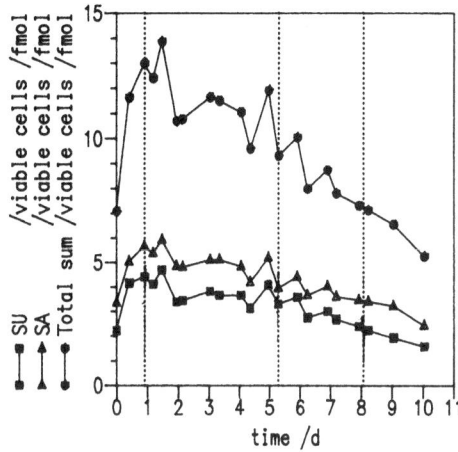

Fig. 3. Total amount and sum of adenosine and uridine nucleotide derivates per viable cell during the growth of a murine hybridoma cell line. fmol = 10^{-15} mol.

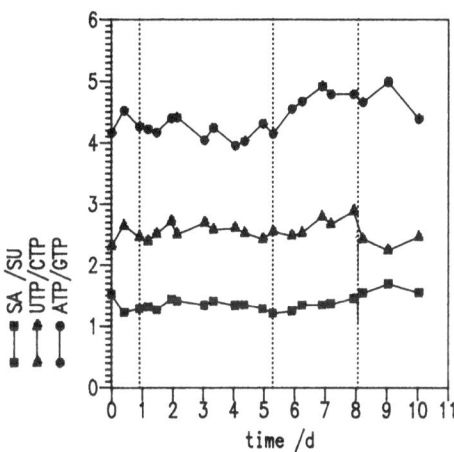

Fig. 4. Ratios between the two purine triphosphates, the two pyrimidine triphosphates and the sum of adenosine (SA) and uridine derivates (SU).

Table 1. Intracellular concentrations of nucleotides in hybridoma cells during the exponential and stationary growth phases. The values were calculated for an estimated cell volume of 1 pl and cell compartimentation was not considered. Values are expressed in mmol/l. SU and SA indicate the sum of uridine and adenosine derivatives.

growth phases	NAD	CTP	GTP	UTP	UDP-GNac	UDP-Glc	ATP	ADP	AMP	SU	SA	total
log phase	0.9	0.7	1.1	1.8	1.7	0.3	4.7	0.3	0.05	**3.8**	**5.1**	**11.6**
stat. phase	0.6	0.2	0.6	0.5	1.3	0.1	2.8	0.2	0.04	**1.9**	**3.0**	**6.3**

The composition of the adenylate pool remained stable during the growth cycle with of 93% ATP, 6% ADP and 1% AMP resulting in a stable and high adenylate energy charge (AEC, ATP + 0.5 ADP / ATP + ADP + AMP; [12]) with values from 0.95 to 0.97. Such a high AEC was also shown to be characteristic for lymphocytes [13] and seems to be specific for mammalian cells. This is in contrast to results gained from some plant cells [10] and *E. coli* [14], for which values of 0.8 or less were detected.

The different changes of distinct nucleotide concentrations during the growth cycle can be better monitored when they were expressed as percentage of the total amount, as seen in Fig. 5-8 and Table 2. Values in Table 2 refer to exponential and stationary growth phases.

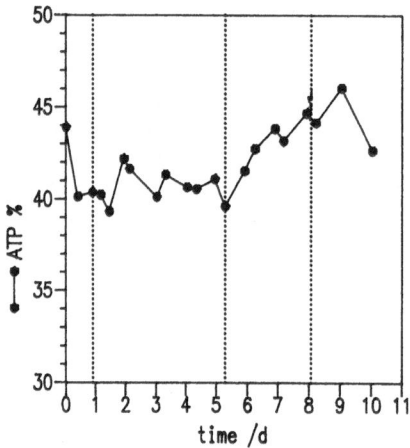

Fig. 5. Relative quantity of ATP referred to the total amount of detected nucleotides during growth of a murine hybridoma cell line.

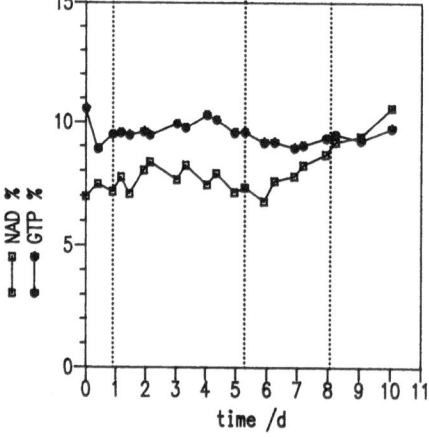

Fig. 6. Relative quantities of GTP and NAD referred to the total amount of detected nucleotides during growth of a murine hybridoma cell line.

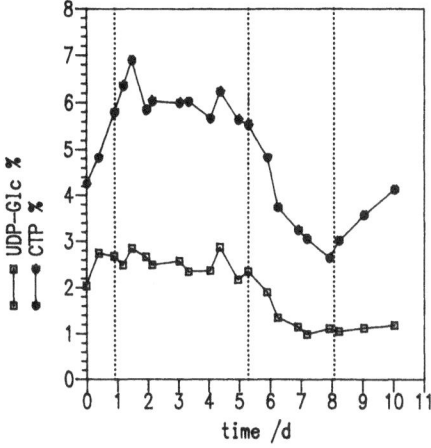

Fig. 7. Relative quantities of CTP and UDP-Glc referred to the total amount of detected nucleotides during growth of a murine hybridoma cell line.

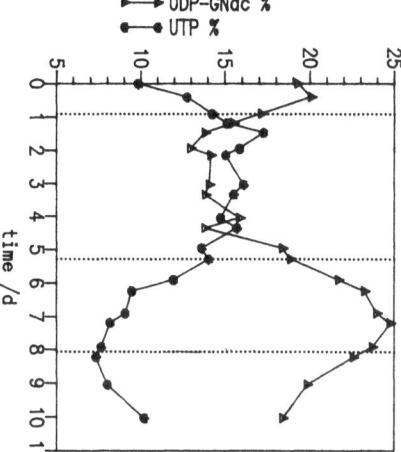

Fig. 8. Relative quantities of UTP and UDP-GNac referred to the total amount of detected nucleotides during growth of a murine hybridoma cell line. UDP-GNac represent the sum of UDP-GalNac and UDP-GlcNac.

Table 2. Percentages of nucleotides from the total detected nucleotide pool. This total pool contained more than 90% of the completely detected UV-active substances. SU and SA indicate the sum of uridine and adenosine derivates.

growth phases	NAD	CTP	GTP	UTP	UDP-GNac	UDP-Glc	ATP	ADP	AMP	SU	SA
log phase	7.8	6.0	9.5	16	15	2.6	41	2.6	0.4	**33.6**	**44**
stat. phase	9.5	3.2	9.5	8.0	21	1.6	44	2.4	0.5	**30.6**	**47**

The relative quantities of purine and pyridine derivatives referred to the total amount of nucleotides remained constant as with GTP or increased after exponential growth ceased as with ATP and NAD. The pools of pyrimidine derivatives showed different changes during the growth cycle. Intracellular concentrations of CTP, UTP and UDP-Glc showed high values during the logarithmic growth phase which decreased to half when cells stopped growing, whereas the UDP-GNac pool show the opposite behaviour. This pool was composed of a stable ratio of 30% UDP-GalNac and 70% UDP-GlcNac. It had its lowest levels during exponential growth and increased to about 60% when the growth phase ceased. This change in the UDP-GNac pool was the reason for the stable ratio between adenosine and uridine derivatives, even when the pyrimidine triphosphate pools decreased to 50% in the stationary phase.

UDP-GlcNac and UDP-GalNac are precursors of the carbohydrate side chains of proteins so that it may be possible that the protein glycosylation process is triggered by fluctuations of pool sizes of the activated sugars. Marx et al. [15] have shown that the relative amount of unspecific antibody secreted by the same cell line as used here increased when cell viability decreased. If this process was induced by altered precursor concentrations has yet to be proven. This assumption was supported by the increase of the relative UDP-GNac pool between day 5 and 7 as shown in Fig. 8 which may be caused by an accumulation based on a decreased protein glycosylation progress during the phase of reduced cell proliferation. When cells reached their stationary growth phase the relative UDP-GNac pool decreased while the relative amount of UTP increased. This behaviour may be a function of reduced production of UDP-GNac after cells stopped growing. Fermentation was terminated at day 10 when cell viability reached 65%. At this point the ratio of UTP/UDP-GNac increased a little after it had passed a minimum at the time when cell growth ceased (see Fig. 9). In conventional batch cultures without perfusion carried out in roller bottles, it could be shown that the pool of UDP-GNac was affected more than the pool of UTP when cell viability decreased resulting in a dramatically increased UTP/UDP-GNac ratio. Even in roller bottles this ratio showed a minimum of 0.4 when cells stopped growing, as found in the fermentation, and increased from 0.4 (85% viability) to 7 (40% viability) within 3 days. This result suggest that under poor culture conditions no protein glycosylation may occur and that inactive antibodys which are secreted in this culture phase lack glycosylation sites.

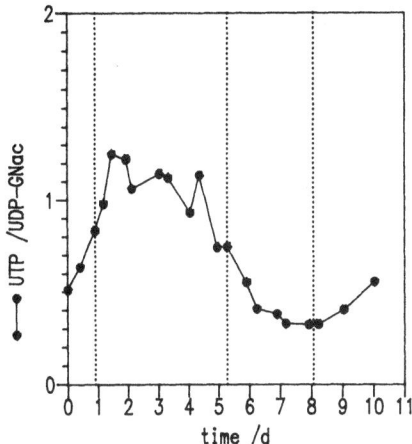

Fig. 9. Ratio between UTP and UDP-GNac during growth of a murine hybridoma cell line.

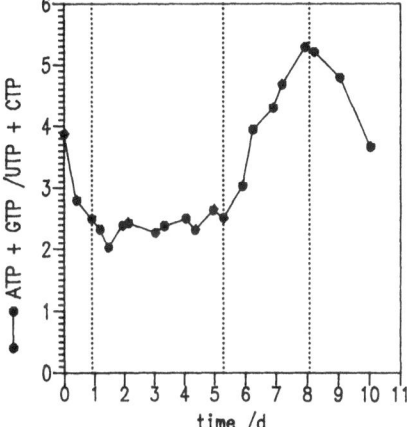

Fig. 10. Ratio of triphosphates from the purine pool and the pyrimidine pool (ATP + GTP / UTP + CTP) during growth of a murine hybridoma cell line.

The ratio of the sum of purine triphosphates referred to the sum of pyrimidine triphosphates (ATP+GTP/UTP+CTP) is a second parameter which was dependent on the different phases of the growth cycle (see Fig. 10). Chou et al. [4] could show that the ATP/UTP ratio decreased in colchicine-induced cells before DNA synthesis was initiated. Results of our experiment suggested that such a decrease of the purine to pyrimidine trinucleotide ratio was necessary to start exponential growth, if starved cells were used as inoculum. When cells harvested from the logarithmic phase of a preculture were used, no such lag-phase was observed.

It can be concluded that two new parameters describing the physiological state of a cell culture have been established. With a knowledge of the ratio between UTP and UDP-GNac and the ratio between the triphosphates from the purine and the pyrimidine pools two parameters are present which can be used a tool for monitoring the physiological state of hybridoma cells during fermentation. These parameters give hints as to further changes in the culture and can be used for efficient prediction of the cell behavior.

4. References

1   Atkinson, D.A. Cellular energy metabolism and its regulation (1977). Academic Press New York

2   Papaport, E.; Garcia-Blanco, M.A.; Zamecnik, P.C. (1979) Regulation of DNA replication in S phase nuclei by ATP and ADP pools. Proc. Natl. Acad. Sci. USA 76, 1643-1647

3   Grummt, F.; Paul, D.; Grummt, I (1977) Regulation of ATP pools, rRNA and DNA synthesis in 3T3 cells in response to serum or hypoxanthine. Eur. J. Biochem. 76, 7 - 12

4   Chou, I-N.; Zeiger, J.; Rapaport, E. (1984) Imbalance of total cellular nucleotide pools and mechanism of the colchicine-induced cell activation. Proc. Natl. Acad. Sci. USA 81, 2401 - 2405

5   Murphree, S.; Moore, E.C.; Peterson D. (1974) Temporal variation of adenine ribonucleotides during the cell cycle of chinese hamster fibroblasts in culture. Experimental Cell Research 83, 189-190

6   Lehmann, J.; Vorlop, J.; Büntemeyer, H. (1988) Bubble-free reactors and their development for continuous culture with cell recycle. In: R.E. Spier and J.B. Griffiths (eds.) Animal Cell Biotechnology Vol. 3, pp. 221-237

7   Ryll, T., Lucki-Lange, M., Jäger, V., Wagner, R. (1990) Production of recombinant human interleukin-2 with BHK cells in a hollow fibre and a stirred tank reactor with protein-free medium. J. Biotechnol. 14, 377-392

8   Ryll, T.; Jäger, V.; Wagner, R. (1990) Intracellular concentration of ATP and other nucleotides during continuous cultivation of hybridoma cells. In: Production of Biologicals from Animal Cells in Culture Research, Development and Achievements (Spier, R.E.; Griffiths, J.B.; eds.) Butterworths, Kent, UK, in press

9   Ryll, T., Wagner, R. (1991) An improved ion-pair HPLC method for the quantification of intracellular bases, nucleosides, nucleotides and sugar-nucleotides in animal cells. J. Chromatogr., submitted

10 Meyer, R.; Wagner, K.G. (1985a) Nucleotide pools in suspension-cultured cells of *datura innoxia*. I. Changes during growth of the batch culture. Planta 166, 439-445

11 Meyer, R.; Wagner, K.G. (1985b) Analysis of the nucleotide pool during growth od suspension cultured cells of *nicotiana tabacum* by high performance liquid chromatography. Physiol. Plant. 65, 439-445

12 Atkinson, D.E.; Walton, G.M. (1967). Adenosine triphosphate conversation in metabolic regulation. J. Biol. Chem. 242, 3239

13 De Korte, D.; Haverkort, W.A.; Van Gennip, A.H.; Roos, D. (1985) Nucleotide profiles of normal human blood cells determined by high-performance liquid chromatography. Analytical Biochemistry 147, 197-209

14 Chapman, A.G.; Fall, L.; Atkinson, D.E. (1971) Adenylate energy charge in *Escherichia coli* during growth and starvation. Journal Bacteriology 108, 1072-1086

15 Marx, U.; Jäger, V.; Kiessig, S.T.; Grunow, R.; von Baehr, R. (1990) Appearance of nonspecific amounts of monoclonal antibodys during fermentation caused by decreased cell viability. In: Production of Biologicals from Animal Cells in Culture Research, Development and Achievements (Spier, R.E.; Griffiths, J.B.; eds.) Butterworths, Kent, UK, in press

Acknowledgements

We are grateful to Prof. Dr. R. von Baehr for providing the murine hybridoma cell line and to Dr. V. Wray for linguistic advice.

PROTEOLYTIC ACTIVITIES IN SERUM-FREE SUPERNATANTS OF MAMMALIAN CELL LINES

WALDEMAR LIND, MONA LIETZ, VOLKER JÄGER, and ROLAND WAGNER
Gesellschaft für Biotechnologische Forschung mbH
Mascheroder Weg 1
D-3300 Braunschweig
Germany

ABSTRACT. [^{3}H]-labelled Casein and chromogenic substrates were used for a quantitative determination of protease activity in cell culture supernatants during long-term cultivation of BHK and hybridoma cells. They produced recombinant human interleukin 2 and an IgG$_{2a}$-antibody respectively in serum- or protein free medium. Characterization of protease activities was performed by inhibitor studies and specific p-NA derivates. Only 20% of the total protease activity in hybridoma and up to 50% in BHK cell supernatants is based on serine type proteases. Supplementation of supernatants with 2-mercaptoethanol or magnesiumchloride caused an increase of protease activities.

1. Introduction

One of the major problems in serum-free cultivtions of animal cells is the proteolytic attack of the synthesized product [10]. Proteolytic enzymes, or proteases, are enzymes which catalyze the cleavage of peptide bonds in other proteins. Whereas terminal peptide bonds are cleaved by exo-peptidases, interior peptide bonds of the protein are broken by endo-peptidases i.e. proteases. Beside their secondary but biotechnological important effect they have many physiological functions, ranging from generalized protein digestion to more specific regulatory functions such as the activation of zymogens, blood coagulation and the lysis of fibrin clots, the release of hormones and pharmacologically active peptides from precursor proteins, and the transport of secretory proteins across membranes [11]. During cultivation of animal cells proteases are released into the medium. This process happens during the proliferation phase of the cells as well as after their death and following cell lysis. Cultures of transformed cells often show a higher protease activity than those of normal cells [2, 4]. Medium supplemented with serum inhibits proteases as the total protein content in serum consists of up to 10% inhibitors. In cultures with serum-free medium proteases can have a negative influence upon cell proliferation as well as causing protein substrate loss [10]. Additionally, a low protein content may cause a higher protease activity [13].

319

R. Sasaki and K. Ikura (eds.), Animal Cell Culture and Production of Biologicals, 319–327.
© 1991 *Kluwer Academic Publishers.*

2. Material and Methods

2.1 MEDIA

The serum-free medium DIF for the cultivation of hybridomas consists of a 1:1 mixture of Iscove's modified Eagle and Ham's F-12 medium (Gibco-BRL) supplemented with 10 mg/l iron saturated human transferrin, (Behringwerke, Marburg), 10 mg/l bovine insulin (I 5500 Sigma) and oleic acid complexed by BSA (Serva 11924). The preparation of these supplements has been described by V. Jäger [5].

The protein-free medium for the cultivation of transformed BHK cells is based on a 1:1 mixture of DMEM and Ham's F-12 medium (Gibco-BRL) and some organic and inorganic supplements as previously described [9].

2.2 CELL LINES

The rat-mouse-hybridoma 412 (Institut für Neurobiologie, Heidelberg) is a hybridoma of the myeloma X63-Ag 8.653 and the spleen of the Spraque-Dawley-rat. It produces an IgG_{2a}-antibody.

The BHK 21 pSVIL2-cell line provided by the genetic engineering department of the GBF was manipulated by genetic methods to produce human interleukin 2 constitutively under the control of the SV 40 promoter [1].

2.3 REACTOR SYSTEMS

For the cultivation of the cells in suspension 1.4-l-double membrane reactors were used [8]. Dense cell cultivation of the hybridoma cells was performed in the hollow fibre bioreactor system according to Jäger [5].

2.4 ANALYSIS OF SAMPLES

Viable and dead cell numbers were estimated by trypan blue exclusion.
Lactate dehydrogenase (LDH) activity was determined as described previously [12].
Immunoglobulin concentrations were determined by using a standard sandwich-ELISA [5].

2.5 PROTEASE-INHIBITORS

SBTI (Soybean Trypsin Inhibitor, Sigma, Deisenhofen) specifically inhibits trypsin. The specific activity of the inhibitor was 6000 U/mg. In the assay the inhibitor was used at a concentration of 1 mg/ml ($\equiv$ 6000 U/ml)

Aprotinin (Bayer, Leverkusen) is a basic protease inhibitor. It inhibits a broad spectrum of serine proteases and shows an unusual stability to proteolytic degradation [7]. The specific activity of aprotinin was 5850 KIU/mg. One KIU (Kallikrein-Inaktivator-Unit) corresponds to an aprotinin amount, which is able to decrease the activity of two biological Kallikrein-Units about 50%. In the assay the inhibitor was used at a concentration of 26 mg/ml ($\equiv$1500 KIU/ml).

ϵ-**aminocaproic acid** specifically inhibits plasmin and plasminogen activators. The inhibitor concentration in the assay was 10 mg/ml.

2.6 SUBSTRATES FOR THE DETERMINATION OF PROTEASE ACTIVITY

The protease assay is based on the measurement of trichloroacetic acid (TCA) soluble peptides which are released from isotopically labelled protein substrates, according to Hatcher *et al.* [3]. The [^{3}H]-labelled casein is prepared by irradiation of casein nach Hammarstan (Serva Feinbiochemica, Heidelberg) with Tritium according to the Wilzbach-Method (Amersham International plc., England), followed by a removal of labile Tritium. This method includes a partly damage of the substrate and it was therefore purified through separation on Sephadex G-25 (PD 10) and additionally on Sephadex G-75 medium columns (Pharmacia, Uppsala). The tritium labelled casein has a specific activity of $1.055 \cdot 10^5$ Bq μg^{-1}. In the assay a casein concentration of 6.66 μg ml^{-1} with a total activity of $3 \cdot 10^5$ cpm was used.

Supernatants of cell cultivation in the hollow fibre system were also tested for protease activity with the chromogenic substrate S-2288 (H-D-Ile-Pro-Arg-p-NA$\cdot$2HCl, Kabi Vitrum, Sweden). This substrate is particularly specific for a broad spectrum of serine proteases such as thrombin, urokinase, plasmin etc.
In the assay S-2288 was used in a 2 mmol$\cdot$l^{-1} concentration. The release of p-NA (p-Nitroaniline) was determined in a kinetik photometer (Ultrospec K, LKB, Freiburg) at a wavelength of 405 nm (t = 1 min; T = 37°C).

3. Results and Discussion

3.1 PROTEOLYTIC ACTIVITIES IN THE PRESENCE OF INHIBITORS

Sample-Nr.	1	2	3	4
$cpm_{[sample]} \cdot 10^{-2}$	383	428	364	530
$cpm_{[sample]} \cdot 10^{-2}$ + SBTI	170 -56%	200 -53%	300 -17%	435 -18%
$cpm_{[sample]} \cdot 10^{-2}$ + Aprotinin	200 -47%	210 -51%	290 -20%	405 -23%
$cpm_{[sample]} \cdot 10^{-2}$ + ϵ-aminocap.	no inhibition			

Table 1: Data are expressed in percent of protease inhibition

Protease activities were characterized with respect to specific inhibitors on supernatants from suspension cultures of the BHK 21 pSVIL2 cell line (sample Nr. 1, 2) and in supernatants of the hetero-hybridoma cell line cultivated in the hollow fibre bioreactor (sample Nr. 3, 4). High inhibitor concentrations were chosen, so that all target proteases would be inhibited.
As shown in table 1 the serine protease inhibitors were able to inhibit only a part of the total protease activity. The transformed BHK cell line released more serine proteases than the hetero-hybridoma cells. Since ϵ-aminocaproic acid showed no inhibitory effect, the serine protease activities are due to other proteases than plasmin or plasminogen activators.

3.2 PROTEOLYTIC ACTIVITIES IN THE PRESENCE OF 2-MERCAPTOETHANOL OR MAGNESIUMCHLORIDE

In some cases a stimulatory [6] or inhibitory [13] effect of $MgCl_2$ and 2-mercaptoethanol on protease activities was detected. The supplements were added within a concentration range of 0.5 mmol·l^{-1} ($MgCl_2$) to 50 mmol·l^{-1} (2-mercaptoethanol).
In order to characterize the sensitivity of the total enzyme activities the influence of salts and SH-groups on proteases was proven on supernatants from suspension cultures of the hetero-hybridoma cell line (sample Nr. 1, 2) and on supernatants of the BHK 21 pSVIL 2 cell line (sample Nr. 4, 5). Additionally, a supernatant of the hetero-hybridoma cell line cultivated in the hollow fibre bioreactor was tested (sample Nr. 3).
The results shown in table 2 indicate clearly that the total protease activity increases in the presence of $MgCl_2$ or 2-mercaptoethanol. The stimulation of protease activities by the reducing SH-groups were much higher than the stimulatory effect of $MgCl_2$. The increase of activities showed different values even when the samples refer to one culture process. This is obviously caused by a qualitative change of the protease composition during cell cultivation.

Sample-Nr.	1	2	3	4	5
$cpm_{[sample]} \cdot 10^{-2}$	130	195	280	74	163
$cpm_{[sample]} \cdot 10^{-2}$ + $MgCl_2$	134 +3%	200 +2,5%	350 +25%	88 +19%	180 +10%
$cpm_{[sample]} \cdot 10^{-2}$ + 2-Mercaptoeth.	183 +40%	368 +88%	733 +162%	105 +42%	218 +33%

Table 2: Data are expressed in percent of protease activity increase

3.3 PROTEOLYTIC ACTIVITIES DURING LONG-TERM CULTIVATION

At the beginning of the cultivation of BHK cells (see fig. 1b) the protease activity was high due to the previous trypsinization step for releasing the cells from the surface of the culture flasks. The cells were transferred under protein-free conditions, which necessitated the termination of trypsin-activity by supplementation with SBTI. Inhibition could not be completely suppressed but when perfusion was started after 2 days the remaining trypsin could be washed out and protease activity which was released from the cells remained at a constant low level. As a result of process control with respect to perfusion a

constant enzyme activity level could be maintained between the fifth and fourteenth day of cultivation (Fig. 1b). The effect of perfusion on medium proteins could be demonstrated after four days when the continuous mode was stopped for a short time and protease activity dramatically increased.

LDH-activity remained constant during cultivation and correlated with the small portion of dead cells. At the end of fermentation when oxygen limitation occured an increase of the toal enzyme activities could be found corresponding to a reduced cell vitality. Whereas the LDH-activity was linearly correlated with the number of dead cells the protease-activity depended on the ratio of viable to dead cells and their cell cycle phase. The higher product concentrations and protease activities as well as the LDH-activity at the end of cultivation were based on membrane clogging of the perfusion system.

It can be concluded that the secretion of proteolytic enzymes is strictly connected to cell growth and vitality. Protease composition secreted into the medium can be regulated by an optimal developed production process in order to influence the special target side of the proteolytic attack.

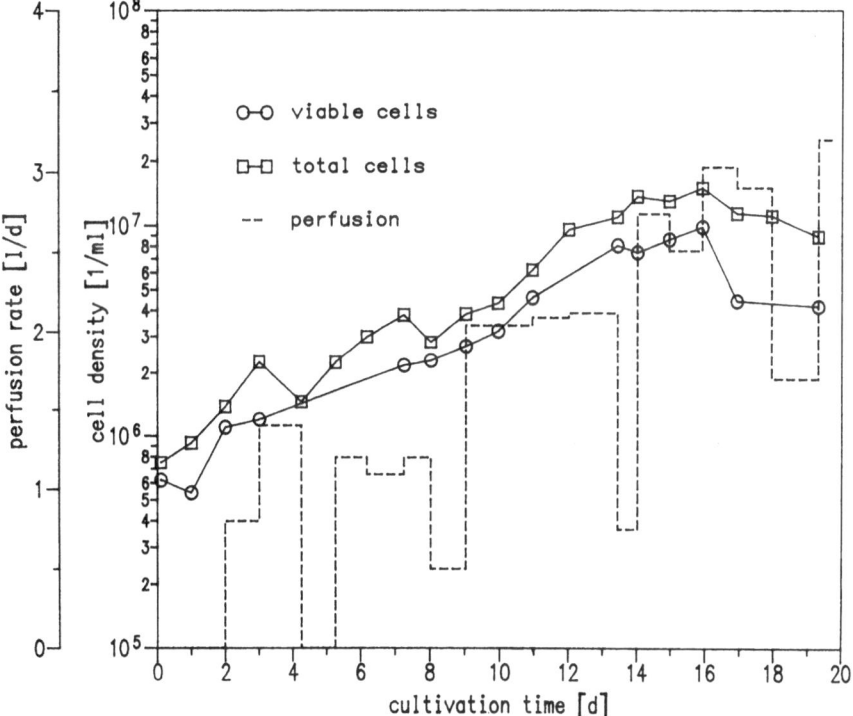

Fig. 1a: Viable and total cells during long-term cultivation of BHK 21 pSVIL2 cells

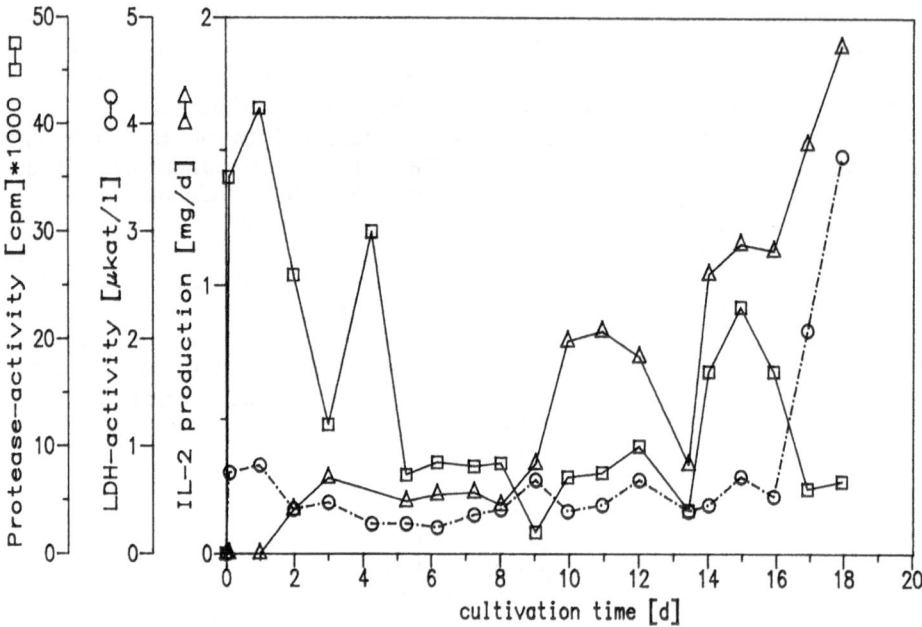

Fig. 1b: Progress of protease-activities, LDH and IL-2 production during cultivation of BHK 21 pSVIL2 cells.

Production process for monoclonal antibodies was characterized by the separation of the cell propagation phase in a 1.4-l-stirred tank and the cell maintenance phase in a hollow fibre reactor. At the beginning of the cell propagation protease-activity was approx. 9000 cpm and with the beginning of perfusion was decreased to approx. 4000 cpm (fig. 2b). With increasing cell density protease activities also increased, whereas the dilution effect of the perfusion caused a maintanence at an intermediate level. After 11 days of culti-vation $1.5 \cdot 10^{10}$ cells were harvested resulting in a reduced mass of $3.66 \cdot 10^6$ cells per ml. At the end of the propagation process protease-activity and LDH-activity increased. In contrast to LDH-activity, which showed a direct cor-relation to the number of dead cells, there is no such correlation between proteolytic activity and the amount of dead cells. At the end of the process protein concentration and enzyme activity increased as a result of membrane clogging as mentioned before. After 18 days of cultivation cells were transferred into the extracapillary space of the hollow fibre cartridge. The high cell density and the ultrafiltration membrane caused higher protein concentrations (see fig. 3). A proteolytic activity of 6.3 μkat/l could be determined within the shell side of the reactor for serine proteases using the specific substrate S-2288. Characterization of protease activities with specific inhibitors demonstrated that only a portion of approx. 20% was caused by serine proteases in hybridoma cells. Due to our results with specific enzyme inhibitors as shown in table Nr. 1 it can be assumed that total protease activity was several times higher and may therefore cause serious problems in high density cell cultivation.

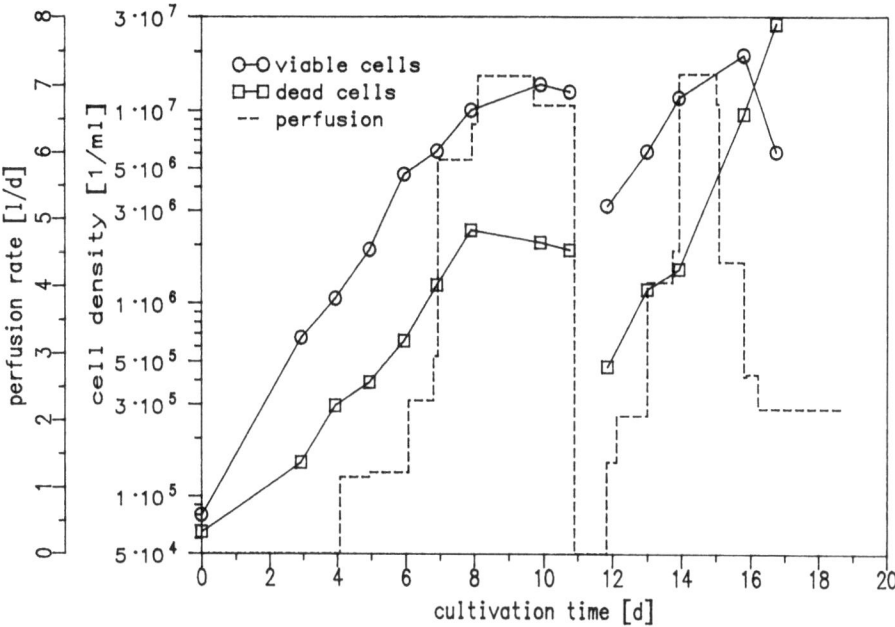

Fig. 2a: Viable and dead cells during the cultivation of the hybridoma cells.

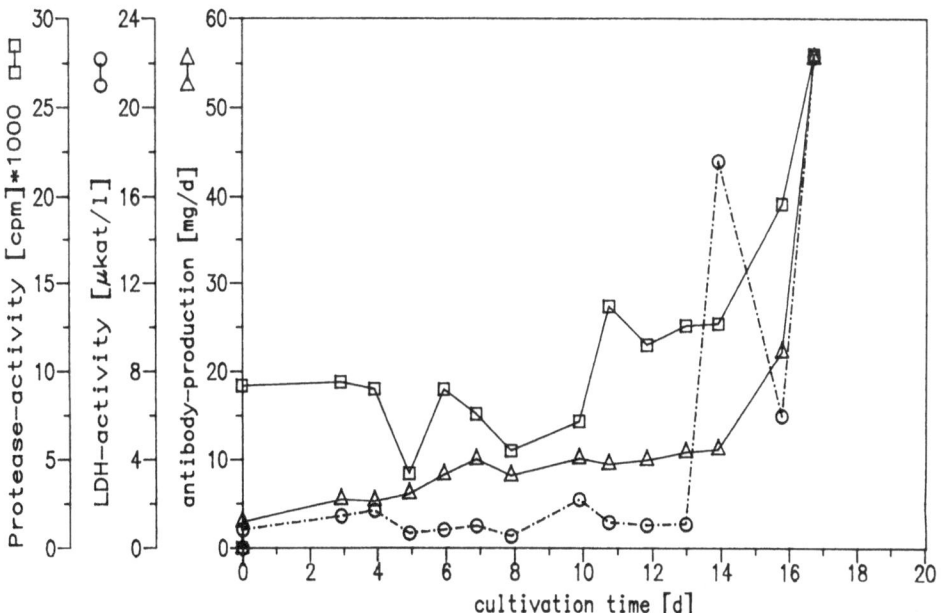

Fig. 2b: Progress of protease-activities, LDH and MAb-production during cultivation of hybridoma cells.

326

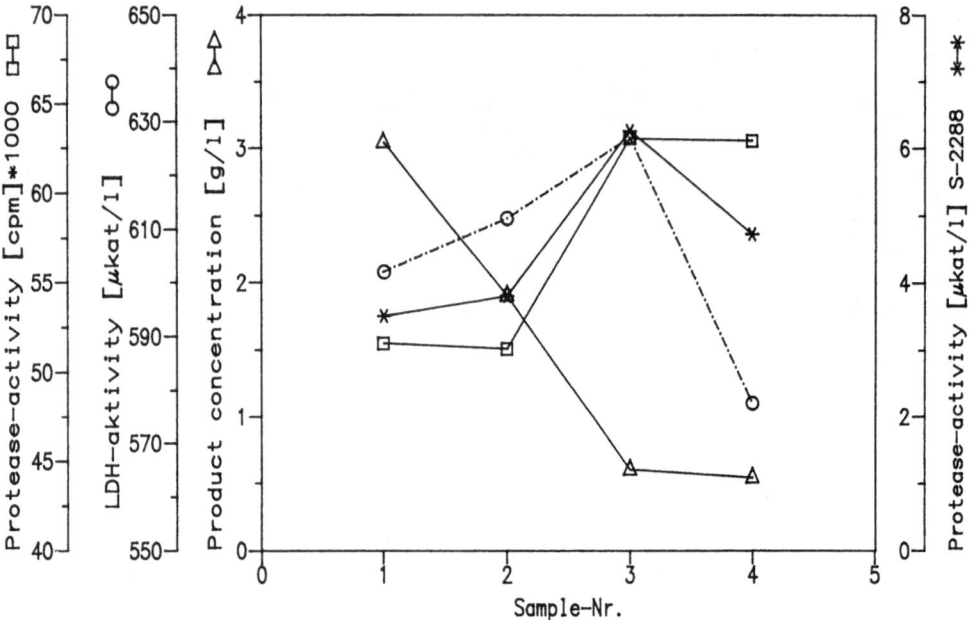

Fig. 3: Cultivation of hybridoma cells in the hollow fibre bioreactor.

4. References

[1] Conradt, H. S.; Ausmeier, M.; Dittmar, K. E.; Hauser, H. J.; Lindenmaier, W. (1986) Secretion of Glycosylated Human Interleukin-2 by Rekombinant Mammalian Cell Lines, Carbohydrate Research **149**, pp 443-450

[2] Goldberg, A. R. (1974) Increased Protease Levels in Transformed Cells: A Casein Overlay Assay for the Detection of Plasminogen Activator Production, Cell **2**, pp 95-102

[3] Hatcher, V. B.; Wertheim, M. S.; Rhee, C. Y.; Tsien, G.; Burk, P. G. (1976) Relationship between Cell Surface Protease Activity and Doubling Time in Various Normal and Transformed Cells, Biochim. et Biophys. Acta **451**, pp 499-510

[4] Hatcher, V. B.; Obermann, M. S.; Wertheim, M. S.; Rhee, C. Y.; Tsien, G.; Burk, P. G. (1977) The Relationship between Surface Protease Activity and the Rate of Cell Proliferation in Normal and Transformed Cells, Biochem. and Biophys. Res. Comm. **76**, pp 602-608

[5] Jäger, V. (1988) Entwicklung eines Hohlfaser-Perfusionsreaktor-Systems zur Produktionsoptimierung monoklonaler Antikörper, Dissertation, Universität Hannover

[6] Kás, J., Rauch, P. (1982) Radioisotopic Method for the Determination of Low and Very Low Proteolytic Activities in Foodstuffs. Z. Lbensm. Unters. Forsch. **174**, 290-293

[7] Kassel, B.; Wang, T.-W. (1970) The Action of Thermolysin on the Basic Trypsin Inhibitor of Bovine Organs; in: Proc. Int. Res. Conf. of Proteinase Inhibitors, München, pp. 89-94, WdG Verlag, Berlin, 1971

[8] Lehmann, J.; Vorlop J.; Büntemeyer H. (1988) Bubble-free Reactors and their Development for Continuous Culture with Cell Recycle. In: Animal Cell Biotechnology (R.E. Spier and J.B. Griffith eds.) Vol. **3**, pp 221-237

[9] Lucki-Lange, M.; Wagner, R. (1990) Conditions for the Production of Recombinant IL-2 in Stirred Suspension Culture Using a Protein-Free Medium. In: Production of Biologicals from Animal Cells in Culture Research, Development and Achievements (Spier, R. E.; Griffiths, J. B.; eds.) Butterworths, in Press

[10]Murakami, H. (1989) Serum-Free Media Used for Cultivition of Hybridomas. In Monoclonal Antibodies: Production and Application, pp 107-141; Alan. Liss Inc.

[11]Neurath, H. (1984) Evolution of Proteolytic Enzymes, Science **224**, pp 350-357

[12]Ryll, T.; Lucki-lange, M.; Jäger, V.; Wagner, R. (1990) Production of Recombinant Interleukin 2 with BHK Cells in a Hollow Fibre and a Stirred Tank Reactor with Protein-Free Medium, J. Biotechnology **14**, pp 377-392

[13]Schlaeger, E. J.; Eggimann, B.; Gast, A. (1987) Proteolytic Activity in the Culture Supernatants of Mouse Hybridoma Cells, Develop. Biol. Standard **66**, pp 403-408

CONTINUOUS PRODUCTION OF ERYTHROPOIETIN USING A RADIAL FLOW BIOREACTOR

H.YOSHIDA, S.MIZUTANI, AND H.IKENAGA
Central Laboratories for Key Technology
Kirin Brewery Co., Ltd.
3, Miyahara-cho, Takasaki-shi, Gunma 370-12
Japan

ABSTRACT. A radial flow packed-bed bioreactor has been developed for continuous mass culture of mammalian cells. Since medium radially flows across the bed, supply of nutrients and oxygen with low shear stress is possible without cell damage. Six-tenths mm diam. porous glass beads as a matrix were packed in a radial flow bioreactor. Using genetically engineered Chinese hamster ovary (CHO) cells, continuous production of erythropoietin (EPO) was studied. The culture conditions were automatically controlled to appropriate values by microcomputer. As the beads had a large attachment surface and the bioreactor decreased concentration gradient, they enabled us to achieve over 1.3×10^8 cells/mL-matrix of high cell density culture in the packed-bed bioreactor. The specific EPO productivity (units/mL-reactor/year) logarithmically increased and reached to 2000 times higher value than that of 175 cm^2 T-flask.

1. INTRODUCTION

Erythropoietin is a hemopoietic factor that regulates the differentiation of erythroid progenitor cells into mature erythrocytes. Human EPO gene was cloned by Lin et al.[1], and mass production of EPO became possible. It is expected that EPO will be used as a therapeutic agent for anemia in patients with renal failure. When EPO is produced using recombinant DNA technology, mammalian cells must be used as a host cell, because EPO is a glycoprotein and glycosylation is important for EPO to be biologically active[2].

Since the growth rate of mammalian cells is slow, it is considered that animal cells should be kept in a reactor and be used to produce useful substances during a long cultivation time. For the successful large scale culture of mammalian cells, a large attachment surface and good chemical environment as well as a low level of shear are required. A packed-bed reactor has high efficiency on a small scale, though it may be more difficult to scale up due to height limitation, and though mixing and mass transfer rates are bad compared to fluidized-bed reactor. So, in the case of applying the

329

R. Sasaki and K. Ikura (eds.), Animal Cell Culture and Production of Biologicals, 329–334.
© 1991 *Kluwer Academic Publishers.*

packed-bed bioreactor to mammalian cell culture, the important problems are the gradients in the concentration of nutrients, inhibitory environment, oxygen supply, and shear stress. In a packed-bed bioreactor, it is also a problem that the cultivation for a long period causes the sedimentation of cells in the bottom of bioreactor by gravitation, and it leads the gradient in the cell concentration. It's a point of success that how necessary nutrients including oxygen are supplied without cell damage. We developed a novel packed-bed bioreactor, the radial flow bioreactor[3], which can overcome the above problems. The purpose of this study is to cultivate mammalian cells at high density using the bioreactor for continuous production of a useful substance, EPO.

2. MATERIALS AND METHODS

2.1. Cells and Medium

In this study, genetically engineered CHO cells were used, which produced human EPO. The medium consisted of DMEM (Gibco) and Ham's F12 (Gibco) mixture supplemented with 5%(v/v) heat-inactivated fetal bovine serum (Flow laboratories) and methotrexate (MTX, Lederle(Japan) Ltd.). For the reactor cultures, some amino acids and 5 g/L glucose were added.

2.3. Matrix

The 0.6-mm diam. porous glass beads: Siran (Schott glaswerke) were packed in the bioreactor. They have a large surface area of 90 m^2/L based on the vendor's data. If the beads are not porous, the surface area of the beads is calculated to be 10 m^2/L from the diameter. The rest of surface area, 80 m^2/L is in the pores.

2.4. Radial-flow bioreactor

The prototype of the radial flow bioreactor is shown in Fig. 1. A 2900-mL reactor contained a central tube that was wrapped with stainless-steel meshes. The 900-mL matrix was packed in the central tube. Around the central tube, there were 24 syringes of variable length. The medium was supplied to the bioreactor through the syringes, radially flowed across the bed, and recovered through the 7 tubes made of metal located in the middle of the bed. The tubes prevented the cells from going out of the reactor.

2.5. Continuous cultivation

Figure 2 shows the experimental apparatus. The bioreactor system consists of a radial flow bioreactor, control vessel (fermentor), and the instruments to monitor and control. Cells used to inoculate continuous cultures were trypsinized from 64 T-flask (175 cm^2) cultures. To inoculate the radial flow bioreactor, about 1×10^9 cells

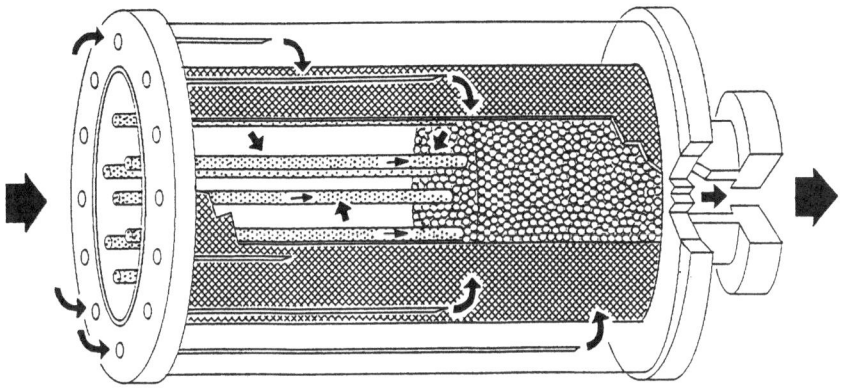

Fig. 1　Radial flow bioreactor
This bioreactor is a prototype (2.9 L), in which
900-mL of the matrix can be packed. If it is
improved, more matrix can be packed.

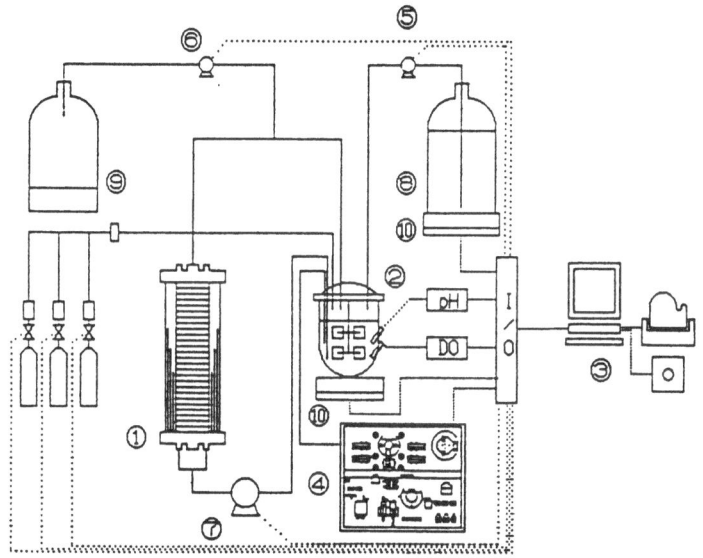

1　Radial flow bioreactor	6　Recovery pump of product
2　Control vessel	7　Circulation pump
3　Microcomputer	8　Fresh medium
4　On-line glucose analyzer	9　Product
5　Feed pump of fresh medium	10　Load cell

Fig. 2　Schematic diagram of experimental apparatus
After cell inoculation, the bioreactor system is full-automatically
controlled by microcomputer. It is not necessary for a man to operate
the bioprocess.

were inoculated the fermentor and the broth was pumped to the bioreactor. Until the cells attached to the matrix, the circulation rate of the medium was slow. When all cells were attached or entrapped in the matrix, the circulation rate was elevated. Dissolved oxygen (DO) concentrations in the fermentor and at the exit of the bioreactor were measured by DO electrodes (Ingold), and kept at appropriate values so as not to be depleted at the exit of the bioreactor. The pH in the fermentor was measured by a pH electrode (Ingold) and kept at 7.4. At the start of the cultures, 95% air and 5% CO_2 was supplied into the head space of the fermentor. As the cells grew, CO_2 and air flow rates were decreased, and the O_2 flow rate was increased. After the partial pressure of CO_2 was zero, 2 N NaOH was added to maintain the pH. Feed rate of fresh medium was gradually increased with cell growth. Circulation rate of the medium was also gradually increased. Bioreactor system was full-automatically controlled by microcomputer. (Control strategy will be reported in another paper.)

2.6. Assays

Glucose and lactic acid were analyzed by enzymatic assays. EPO titer in the samples was measured by radioimmunoassay (RIA).

3. RESULTS AND DISCUSSION

Continuous EPO production using radial flow bioreactor was shown in Fig. 3. The medium feed rate and circulation rate were controlled by the micro-computer, so the glucose and oxygen concentrations were not depleted. The medium feed rate attained 10 L/d after 21 days. In spite of the medium feed rate being increased, the glucose concentration decreased, and the lactic acid concentration increased in proportion to consumed glucose. Lactic acid yield from glucose was not changed through the

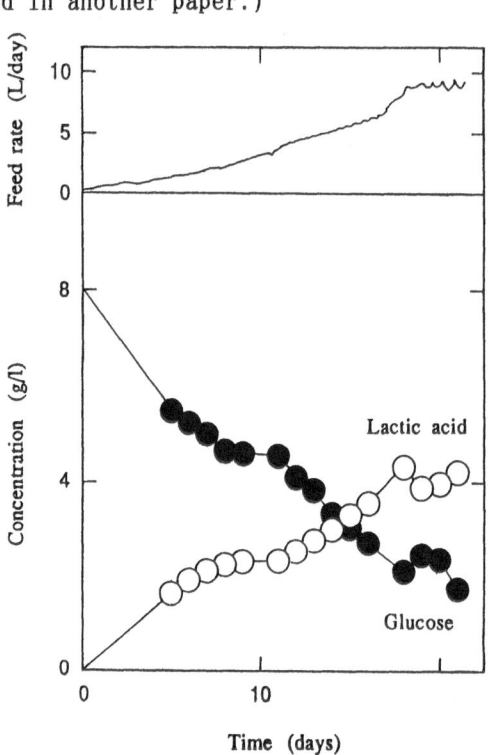

Fig. 3 High cell density culture for continuous EPO production

culture. The yield is affected by the environment such as pH or DO[4]. However, it was not observed in the radial flow bioreactor, because the environment in it was kept at the optimum. Though the glucose concentration in fresh medium was 8 g/L, the final glucose concentration in broth was lower than 2 g/L. Glucose starvation may lead to abnormal glycosylations of EPO[5], and differences in oligosaccharide structure affect the biological activity of EPO. So we added glucose to the medium. Lactic acid concentration was over 4 g/L after 21 days, but cell damage was not noted at this level(data not shown). The final EPO concentration was as much as that of batch culture in T-flask.

The consumption rate of glucose per day is shown in Fig.4. The consumption rate increased to over 55 g/day. The oxygen consumption rate increased at the same rate(data not shown). It was suggested that some metabolic change to consume much glucose did not occur. The cell count calculated from glucose and oxygen consumption rate is over 1.3×10^8 cells/mL-matrix. Using 0.6-mm diam. small porous glass beads had a large attachment surface, high cell density could be easily achieved. The smaller the beads diameter, the higher the shear stress. However, since the linear velocity of medium in radial flow bioreactor is kept at a low level, shear stress for cells is less than that of conventional packed-bed bioreactor.

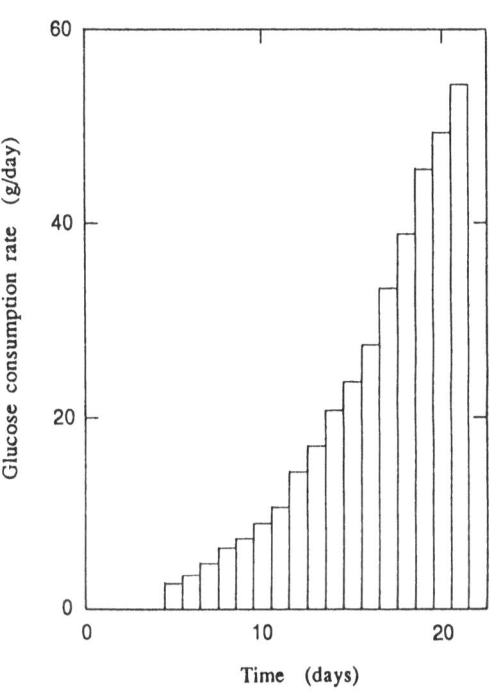

Fig. 4 Daily glucose consumption

The daily EPO production rate is shown in Fig. 5. The EPO production rate logarithmically increased the same as glucose consumption rate. EPO production was sensitive to such cultivation conditions as pH and DO. If DO is lower, specific glucose consumption rate may be higher [4], but the specific EPO production rate was dramatically lower. In the radial flow bioreactor, they can be easily kept in good condition without giving large shear stress to the cells. Finally, 10 L of broth per day was obtained. In this case, specific EPO productivity (units/mL-reactor/year) is 2000 times higher than that of 175 cm^2 flask. In this prototype, 900-mL of the matrix can be packed. The bioreactor can be easily improved for 2 L of matrix to be packed. If 2 L of matrix is packed, 4000 times the EPO productivity of 175 cm^2 T-flask will be obtained.

5. CONCLUSION

In the radial flow bioreactor, over 1.3×10^8 cells/mL-matrix of high

334

cell density was attainable. Continuous production of EPO will be possible at the level of 4000 times higher productivity than that of 175 cm² T-flask.

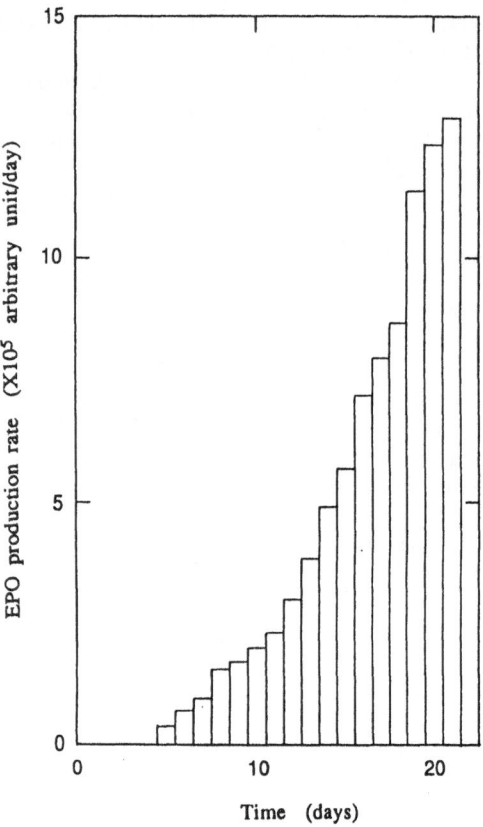

Fig. 5   Daily EPO production

6. REFERENCES

1. F.K.Lin, S.Suggs, C.H.Lin, J.K.Browne, R.Smalling, J.C. Egrie, K.K.Chen, G.M.Fox, F. Martin, Z. Stabinsky, S. M. Badrawi, P.H. Lain and E. Goldwasser (1985) "Cloning and expression of the human erythropoietin gene ", Proc. Natl. Acad. Sci. USA vol. 82, 7580-7584.
2. M.Takeuchi, N.Inoue, T.W. Strickland, M.Kubota, M.Wada, R.Shimizu, S.Hoshi, H. Kozutsumi S.Takasaki and A.Kobata (1989) " Relationship between sugar chain structure and biological activity of recombinant human erythro-poietin produced in Chinese hamster ovary cells "Proc.Natl. Acad.Sci.USA vol.86, 7819-7822.
3. H.Yoshida, S.Mizutani and H. Ikenaga " Radial flow bio-reactor for mammalian cell culture " Cytotechnology (in preparation).
4. N.Kurano, C.Leist, F.Messi, S.Kurano and A.Fiechter (1990) " Growth behavior of chinese hamster ovary cells in a compact loop bioreactor: 1. Effects of physical and chemical environments" J. Biotech. 15, 101-112.
5. J.I.Rearick, A.Chapman and S.Kornfeld (1981) "Glucose starvation alters lipid-linked oligosaccharide biosynthesis in chinese hamster ovary cells" J. Biol. Chem. 256, 6255-6261.

EFFECTS OF AUTOCRINE COMPONENTS ON GROWTH INHIBITION OF NH_4^+ AND ON GROWTH KINETICS OF HYBRIDOMA CELLS

Y. SHIRAI, K. HASHIMOTO, H. TAKAMATSU and T. YOSHIMI
Department of Chemical Engineering,
Faculty of Engineering
Kyoto University,
Sakyo-ku, Kyoto 606, Japan

ABSTRACT. We showed at the previous JAACT meeting that the growth inhibition of the hybridoma cells, 4C10B6, by NH_4^+ is attenuated by autocrine components that the 4C10B6 cells themselves produce. This paper deals with the characteristic of the autocrine components. The effects of autocrine components on the growth kinetics of the hybridoma cells are also discussed. The 4C10B6 hybridoma cells can proliferate well in conditioned medium in which NH_4^+ is included at the concentration of 10 mM, but they die in a fresh medium including 10 mM of NH_4^+, suggesting that any autocrine components produced by the cells would attenuate the growth inhibition by NH_4^+. However, the hybridoma cells did not proliferate even in the conditioned medium when the conditioned medium was treated with α-chymotrypsin. This suggests that the autocrine components are proteins. On the other hand, the cell growth yield for glucose increased at the late stage of batch culture or in high density culture 3 or 8 times as much as at the early stage of batch culture. Changes in the concentration of nutrients or waste products in the medium such as glucose or lactate little affected the growth yield. The growth yield also increased when the cells were cultivated in a conditioned medium. These facts suggests that the autocrine components would have something to do with the increase in the growth yield.

1. INTRODUCTION

Hybridoma cells are now widely used for producing monoclonal antibodies efficiently, which are useful not only for diagnostic applications, therapeutic treatments in tumors and in immunological diseases but also for ligands in affinity chromatography [1-3].

High density culture of hybridoma cells is necessary to produce monoclonal antibodies on an industrial scale. The effects of inhibitory components, typically NH_4^+ etc., on hybridoma growth and the growth kinetics in high density culture should be clarified to achieve the high density culture.

We have found that the inhibitory effects of NH_4^+ on hybridoma

R. Sasaki and K. Ikura (eds.), Animal Cell Culture and Production of Biologicals, 335–344.
© 1991 *Kluwer Academic Publishers.*

growth depends on the initial cell concentration, and that high molecular weight components produced by the cell are closely related to the attenuation of growth inhibition by NH_4^+ in our previous paper [4].

On the other hand, the growth kinetics of hybridoma cells in high density culture should be also clarified to use a medium with high efficiency for growing the cells and producing monoclonal antibodies in high density culture.

In this paper, it was confirmed if the attenuating factor(s) mentioned above would be proteins or not. Besides, the growth of hybridoma cells is compared between high density culture and low density culture.

2. MATERIALS AND METHODS

2.1. Materials

A mouse myeloma, P3/X63-Ag8-U1, was fused with mouse spleen cells to produce mouse-mouse hybridoma cells, 4C10B6. A non-serum medium developed by Murakami et al. [5] was used for cultivation of the cells: A mixture of PRMI1640 (Gibbco Co.,USA), F12 (Nissui Co., Japan) and DME (Nissui Co., Japan) media with a weight ratio of 2:1:1., in which only insulin, transferrin, monoethanolamine and sodium selenite were included as non-serum components. Triple-distilled water treated with potassium permanganate was used for preparing the medium. The detailed composition of the medium used is listed in our previous paper [4].

2.2 Conditioned Medium

A conditioned medium was obtained by separating the cells and the medium used after cultivation of the hybridoma cells for 1 day at initial cell concentrations between 1×10^5 cells/ml and 2×10^5 cells/ml to check the effects of autocrine components produced by the cells on the growth inhibition by NH_4^+ and on the growth kinetics of the cells. When the effects of autocrine components on the growth inhibition by NH_4^+ were investigated, the cells were cultivated in the medium containing 10 mM of NH_4^+ to obtain a conditioned medium with NH_4^+. The conditioned medium was treated with α-chymotrypsin (1mg/ml) for 10 hours at 37 ℃, and then it was treated with an inhibitor of the protease for 10 hours at 37 °C. Transferrin and insulin were added to the medium to compensate for those dissolved by the enzyme. This conditioned medium was used to confirm if the autocrine components that attenuate the growth inhibition by NH_4^+ are proteins or not. The conditioned medium treated by heat for 30 minutes at 60 °C was used for the same purpose.

2.3. Cultivation of the cells

The cells were cultivated statically in a flask as well as in a stirred fermentor shown in Fig. 1 in suspension. A perfusion system was used in the high density culture. The hybridoma cells were separated by a membrane sheet set at the bottom of the fermentor. The pore size of the membrane was 0.45 μm, and the volume of cell suspension in the fermentor was kept constant at 15 ml by adjusting the withdrawing rate and the supplying rate of the medium by two pumps. The perfusion rate was between 1 d^{-1} and 8 d^{-1}. The oxygen concentration in the medium was maintained at a level over half that saturated with air at an ordinary pressure. The cell suspension was mixed with stirring at the rate of 50 rpm and kept in a thermostatic chamber at 37 °C.

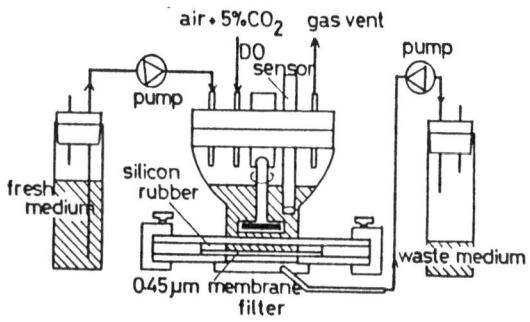

Fig. 1. Experimental apparatus for perfusion high density culture

Cells proliferating in an exponential growth phase were adopted as seed cells and cultivated one day at the cell concentration from 1 x 10^5 cells/ml to 3 x 10^5 cells/ml for a conditioning for all the culture experiments. The initial pH levels in the medium were adjusted by 0.1 M HCl solution before cultivation.

2.4. Assay

The concentration of glucose, glutamine, lactate and ammonium were measured using an enzyme reaction. Oxygen concentration was measured by a D. O. electrode. The oxygen consumption rate of the cells was measured in a small cell described in a previous paper [6], in which a D. O. electrode was inserted and the cell suspension was mixed. The measurement of the oxygen consumption rate of the cells cultured at high density was done in the medium used in the high density culture with a part of the cells removed from the cell suspension. This was done to avoid too rapid a change in oxygen concentration during the measurement owing to the oxygen uptake of many cells.

3. GROWTH KINETICS

The growth kinetics of the hybridoma cells was investigated based on the data obtained at the exponential growth phase in a batch culture. The growth kinetics were not compared with the steady-state stoichiometric coefficients which would be obtained in a chemostat culture. However, the values obtained in the batch culture do indicate general trends and are useful in this regard.

The following equations were used for calculating the growth kinetics of the hybridoma cells.

$$dX/dt = \mu X \qquad\qquad\qquad\qquad (1)$$

$$dS/dt = (-1/Y_x)(dX/dt) \qquad\qquad (2)$$

$$dP_i/dt = Y_i dX/dt \qquad i=1,2 \qquad (3)$$

where $\qquad\qquad Y_x = \Delta X/\Delta S \qquad\qquad\qquad (4)$

$$Y_i = \Delta P_i/\Delta X \qquad i=1,2 \qquad (5)$$

where μ is the specific growth rate and the values obtained experimentally from the slope of the exponential growth phase.

These equations were used for describing the growth kinetics of the hybridoma cells in high density culture. Material balances in the perfusion culture can be written as follows:

$$dX/dt = \mu X \qquad\qquad\qquad\qquad (6)$$

$$dS/dt = (-1/Y_x)(dX/dt)+(F/V)(S_0-S) \qquad (7)$$

$$dP_i/dt = Y_i(dX/dt)-(FP_i/V) \quad i=1,2 \qquad (8)$$

4. RESULTS AND DISCUSSION

4.1. On the Autocrine Components Attenuating Growth Inhibition by NH_4^+

In the previous JAACT meeting held at Tsukuba, we showed that high molecular weight components produced by the 4C10B6 cells attenuate the growth inhibition by NH_4^+. Here is discussion what the autocrine components are.

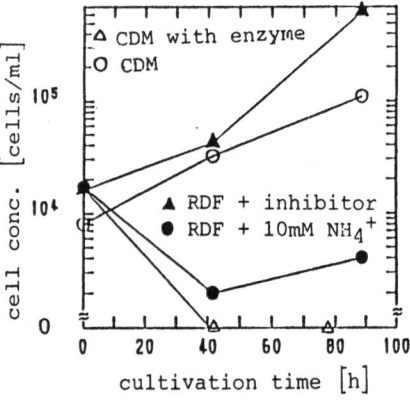

CDM: conditioned medium

Figure 2 shows a growth curve of the 4C10B6 cells cultivated in a conditioned medium with 10 mM of NH_4^+ treated with α-chymotrypsin. Other growth curves are also found in the figure for the cells cultivated in a conditioned medium with 10 mM of NH_4^+ but not treated with the enzyme, in a fresh medium with 10 mM of NH_4^+ and in a fresh medium without any supplement ammonium

Fig. 2. Growth curves of the 4C10B6 cells in the conditioned medium treated with α-chymotrypsin

ions treated with enzyme as well as an inhibitor. The 4C10B6 cells did not proliferate in the conditioned medium treated with the enzyme and the fresh medium with 10 mM of NH_4^+, while they did in the conditioned medium not treated at all and the fresh medium with the treatment. These facts indicate that the factors attenuating growth inhibition by NH_4^+ were damaged by the treatment with the protease, and that the treatment did not affect the cell growth at all, suggesting that the factors would be proteins.

A growth curve of the 4C10B6 cells in the conditioned medium treated by heat is shown in Fig.3, indicating that the effect of attenuation of the growth inhibition by NH_4^+ is reduced by a heat treatment. This also supports the above conclusion that the autocrine factors attenuating the growth inhibition by NH_4^+ are pro-

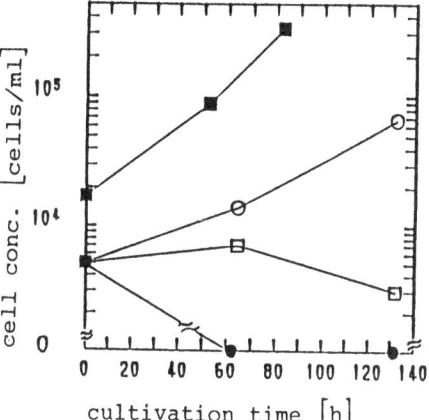

cultivation time [h]

O : CDM
□ : CDM treated by heat
■ : RDF treated with heat
● : RDF with 10 mM NH_4^+

Fig. 3. Growth curves of the 4C10B6 cells in the conditioned medium treated by heat

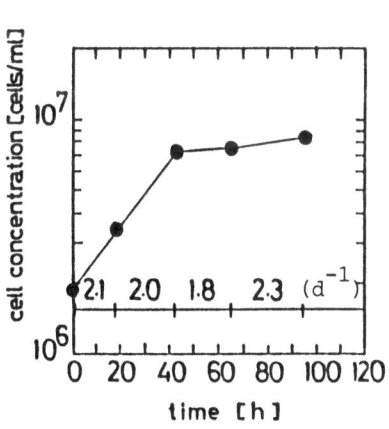

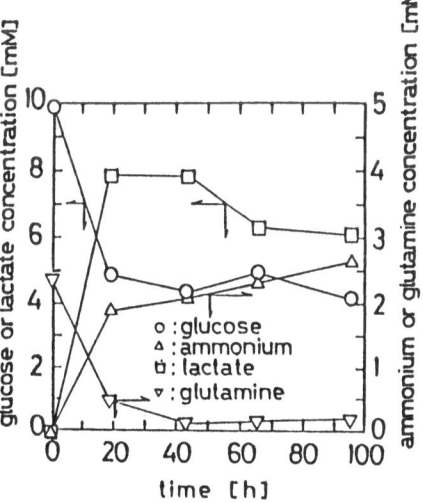

o : glucose
△ : ammonium
□ : lactate
▽ : glutamine

Fig. 4. The 4C10B6 growth and its substrate consumption and waste production in the perfusion system with lower perfusion rates

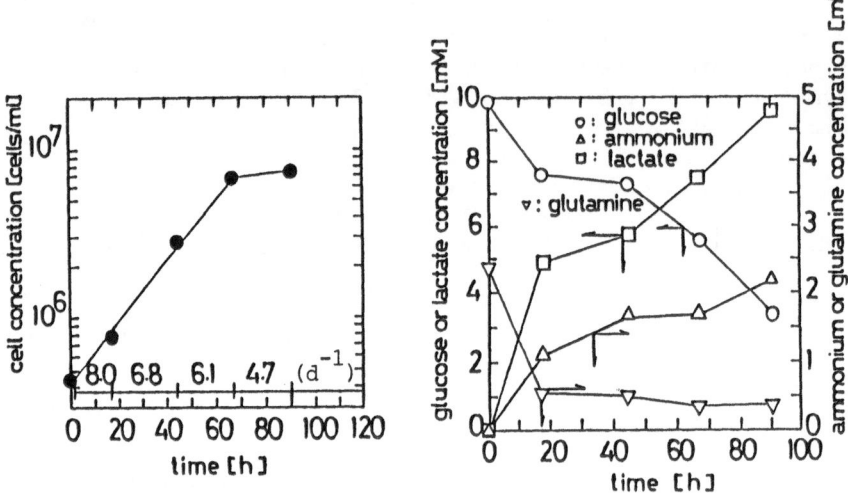

Fig. 5. The 4C10B6 growth and its
substrate consumption and waste
production in the perfusion sys-
teins.   tem with higher perfusion rates.

4.2. Growth Kinetics of the 4C10B6 Cells in High Density Culture

We often find that less medium is needed for proliferation of
hybridoma cells in high density culture than in a culture at cell
concentration around 1×10^5 cells/ml. This phenomenon was
investigated quantitatively. Figures 4 and 5 indicate the growth of 4C10B6 hybridoma cells and changes in substrates (glucose and glutamine) and products (lactate and ammonium) during two series of high density culture with lower perfusion rates (feed and withdrawal rate of the medium/volume of the medium in a fermentor)

Table 1-1. Growth kinetics in high density culture
with lower perfusion rates

time [h]	0 – 24	24 – 48	batch culture
F/V [1/d]	2.1	2.0	0
Y_x [cells/mmol]	1.9×10^8	4.7×10^8	5.9×10^7
μ/Y_x [mmol/cell h]	1.7×10^{-10}	0.7×10^{-10}	6.8×10^{-10}
μ/Y_{xg} [mmol/cell h]	0.7×10^{-10}	0.3×10^{-10}	2.0×10^{-10}
μ/Y_{xO_2} [mmol/cell h]	–	3.0×10^{-10}	2.1×10^{-10}
Y_1 [mmol/cell h]	2.7×10^{-10}	1.0×10^{-10}	1.8×10^{-9}
Y_2 [mmol/cell h]	0.7×10^{-10}	0.3×10^{-10}	1.9×10^{-10}

$\mu = 0.033 \text{ h}^{-1}$ in the exponential growth phase.

and higher perfusion rates, respectively. The numbers shown in the figures indicate the perfusion rates.

The growth kinetics were estimated based on the data obtained in the exponential growth phase by using Eqs. (6), (7) and (8). The growth kinetics in a culture at a cell concentration around 1×10^5 cells/ml were obtained by Eqs. (1) - (5) with experimental data from a batch culture of the 4C10B6 cells. The specific substrate consumption rates, the specific waste production rates and the growth yields for glucose and glutamine are summarized in Table 1. Each value of the consumption or production rates decreases with the cultivation time, but the growth yields increase, regardless of the perfusion rates.

Table 1-2. Growth kinetics in high density culture with higher perfusion rates

time (h)	0 - 24	24 - 48	48 - 72
F/V (1/d)	8.0	6.8	6.1
Y_x (cells/mmol)	5.4×10^7	1.4×10^8	2.2×10^8
μ/Y_x (mmol/cell h)	7.6×10^{-10}	2.9×10^{-10}	1.9×10^{-10}
μ/Y_{xg} (mmol/cell h)	2.1×10^{-10}	1.2×10^{-10}	0.7×10^{-10}
Y_1 (mmol/cell h)	1.7×10^{-9}	6.5×10^{-10}	3.4×10^{-10}
Y_2 (mmol/cell h)	3.7×10^{-10}	1.9×10^{-10}	0.8×10^{-10}

$\mu = 0.041 \ h^{-1}$ in the exponential growth phase.

The growth yields in the perfusion culture with a low perfusion rate are higher than that with a high perfusion rate. Table 1 indicates that the growth yields estimated from high density culture are more than eight times as high as that obtained from low density batch culture. In high density culture the specific growth rate did not decrease in spite of a decrease in the specific consumption and production rates, resulting in an increase in the growth yields. The specific oxygen consumption rate of the 4C10B6 hybridoma cells cultivated in high density culture were 8.3×10^{-14} mol/cell s, 1.5 times as high as that measured in low density culture [6]. Only the specific oxygen consumption rate increased in high density culture contrast to the other specific consumption rates.

The reason should be discovered why the growth kinetics of the 4C10B6 cells were changed so much in high density culture. In the perfusion cultures, cultivation conditions such as pH and substrate concentration will be changed during the cultivation. These factors were further investigated.

4.3. Effects of Substrate Concentration on the Hybridoma Growth

The Effects of glucose and glutamine concentrations on the growth of 4C10B6 hybridoma cells were investigated by adjusting the initial concentration of each component in the medium. Glucose concentration affects the cell growth rate. When less glucose was included in the medium, a lower growth rate was observed. Table 2 lists the specific growth rates, the specific glucose consumption rates and the growth yields for glucose at several initial glucose concentrations. The growth yields are close together at every initial glucose concentration.

On the other hand glutamine did not affect the cell growth rate unless glutamine was depleted completely.

Table 2. Growth yield and specific glucose consumption rate with change in initial glucose concentration

glucose conc. in the medium [mM]	specific growth rate μ [1/h]	Y_x [cells/mmol]	μ/Y_x [mmol/cell h]
2	0.025	1.06×10^8	2.35×10^{-10}
5	0.037	7.63×10^7	4.85×10^{-10}
9.9	0.042	6.13×10^7	6.84×10^{-10}

4.4. Effects of pH on the Growth Kinetics

McQueen and Bailey found that the growth yield for glucose in a hybridoma growth cultivated at pH 6.8 increases from three to ten times or more of the culture at pH 7.6, but that the growth yield for glutamine is close together at all pHs examined [7]. These trends have been also reported by Miller et al. [8].

Table 3. Effects of pH on the growth kinetics

	Run No. 1			2			3		
time [h]	pH [-]	ΔS [mM]	$Y_x \times 10^{-7}$ [cells/mmol]	pH	ΔS	$Y_x \times 10^{-7}$	pH	ΔS	$Y_x \times 10^{-7}$
0	7.4	-	-	7.0	-	-	6.6	-	-
24	7.0	4.5	3.7	6.6	2.8	6.1	6.6	2.4	4.6
48	6.6	2.6	22	6.4	2.3	23	6.6	2.1	19

Batch experiments with various initial pH values were done to confirm the effects of pH on the growth kinetics of 4C10B6 hybridoma cells. Table 3 summarizes the changes in cell concentration and glucose concentration as well as the growth yield estimated based on the changes in glucose concentration. A major change in the growth yield with the various pHs was not observed in the culture between pHs, but the same tendency reported by McQueen and Bailey was found in the culture at the initial pHs of 7.4 and 7.0. Above all, it is found in Table 3 that the growth yield increases with the cultivation time more than with the pH change.

4.5 Hybridoma Growth in a Conditioned Medium

The 4C10B6 hybridoma cells were cultivated in a conditioned medium in which the cells had been cultivated for 1 day at the initial cell concentration of 1×10^5 cells/ml. Glutamine was added to the conditioned medium to supply the glutamine consumed by the cells for one day. The culture condition was changed from that in an ordinary batch culture after 1 day by this treatment.

The growth yield for glucose was 1.5×10^8 cells/mmol, which was estimated from the culture with the conditioned medium. This value is 2.5 times as higher as that in an ordinary medium shown in Table 1.

4.6. Discussion

The substrate consumption and waste production rates decreased but the growth yields for glucose and glutamine increased in high density culture. This indicates that high density culture is more efficient for cultivating hybridoma cells than conventional batch culture with respect to energy.

Although enhancement of the growth yield which is ascribed to the slight pH change has been addressed by McQueen and Bailey [7], Table 3 indicates that the enhancement of the growth yields during the course of cultivation of the 4C10B6 hybridoma cells could not be explained only by changes in pH levels because the growth yield is quite different between the early 24 hours and the late 24 hours in the culture period in spite of the identical pH changes in each culture period.

The growth yield for glucose also increased to 2.5 times as high as that obtained in a conventional batch culture when the hybridoma cells were cultivated in the conditioned medium. The increase in the growth yield cannot be ascribed to the change in the initial glucose concentration, because a change in the initial glucose concentration in the medium affected the growth yield only slightly as shown in Table 2.

Autocrine components contribute to the animal cell growth [9-11]; for example, growth of fibroblast cells depends on the concentrations of the platelet-derived growth factor produced by the cells themselves [11] and here we showed that proteins from the 4C10B6 cells contribute to attenuate the growth inhibition by NH_4^+. If any autocrine components would contribute to the increase in the growth yield of the 4C10B6 cells, this would be explained because the autocrine components would be accumulated in the medium in the late period of culture or in the conditioned medium.

In high density culture the growth yields estimated in the lower perfusion rates were higher than those in the higher perfusion rates. The concentration of the autocrine components is higher in the system with the lower perfusion rates, resulting in the growth yields would thus increase.

A slight increase in oxygen consumption rate was observed in high density culture though glucose and glutamine were consumed less than in conventional batch culture, suggesting that the TCA cycle would be enhanced. The hybridoma cells would proliferate well in high density culture in spite of low substrate consumption rates owing to enhancement of the TCA cycle.

Based on the hypothesis that any autocrine components would contri-

344

bute to the change in the growth kinetics of the 4C10B6 cells, a pyruvate flux in the glycolytic pathway of the lactate production could be assumed to be changed to the TCA cycle by the autocrine components. Efficient amount of ATP reproduced through the TCA cycle hampers any glucose flux to the glycolytic pathway; the glucose consumption rate decreases at the constant growth rate, resulting in increase in the growth yield.

As mentioned in this paper the autocrine components produced by the 4C10B6 cells would work as important factors for their own growth.

Acknowledgment

The authors wish to thank Teijin Limited Co. Tokyo, Japan for kindly providing the 4C10B6 cells.

Nomenclature

F	= feeding rate of the fresh medium	$[cm^3/s]$
P_1	= concentration of lactate	$[mol/m^3]$
P_2	= concentration of ammonium	$[mol/m^3]$
S	= concentration of glucose	$[mol/m^3]$
V	= volume of cell suspension	$[m^3]$
Y_x	= growth yield for glucose	[cells/mol]
Y_{xg}	= growth yield for glutamine	[cells/mol]
Y_{xO2}	= growth yield for oxygen	[cells/mol]
Y_1	= lactate yield	[mol/cell]
Y_2	= ammonium yield	[mol/cell]
μ	= specific growth rate	[1/s]
μ_{max}	= maximum specific growth rate	[1/s]

References

1. Olsson, L. and Mathe, G., In Recent Results in Cancer Research, (1982) G. Mathe et al. Eds., 80, 334-337, Springer-Verlag, Berlin.
2. Ritz, J. and Schlossman, S. F., Blood, (1982) 59, 1-11.
3. Knight, P., Bio/Technol., (1989) 7, 243-249.
4. Shirai Y. et al., Trends in Animal Cell Culture Technology, (1990) H. Murakami Ed., 99-103.
5. Murakami, H., et al., Proc. Natl. Acad. Sci. USA, (1982) 79, 1158-1162.
6. Shirai, Y. et al., Appl. Microb. Biotechnol., (1988) 29, 113-118.
7. McQueen, A. and Bailey J. E., Biotechnol. Bioeng., (1990) 35, 1067-1077.
8. Miller, W. M. et al., Biotechnol. Bioeng., (1988) 32, 947-965.
9. Sporn, M. B. and Todaro, G. J., N. E. J. Med., (1980) 303, 878-880.
10. Huang, J. S., et al., Cell., (1984) 39, 79-87.
11. Lauffenberger, D. and Cozens, C., Biotechnol. Bioeng., (1989) 33, 1365-1378.

GLYCOSYLATION OF ERYTHROPOIETIN RECEPTOR

S. Masuda[1], Y. Hisada[1], M. Ueda[2] and R. Sasaki[1]
[1]Department of Food Science and Technology,
Faculty of Agriculture, Kyoto University, Kyoto 606 and
[2]Research Institute of Life Science, Snow Brand Milk
Products Co., Ltd, 519, Tochigi 329-05 (Japan)

ABSTRACT. The size of the murine erythropoietin (EPO) receptor
(EPO-R), 105 kDa, estimated from EPO·EPO-R cross-linked products,is
much higher than that, 53 kDa, predicted from EPO-R cDNA. The lectin-
binding properties of the solubilized EPO-R suggested that EPO-R was
$\underline{N}$-glycosylated. To find whether glycosylation of EPO-R accounted for
the difference in size, the decrease in size of $[^{125}I]$EPO·EPO-R cross-
linked products by enzymatic deglycosylation was examined. $\underline{N}$-glycosy-
lation of EPO-R was at most 10 kDa, not accounting for the difference
in size predicted from cDNA and from cross-linked experiments. The
possibilities that may cause the difference in receptor size are
discussed.

INTRODUCTION

Erythropoietin (EPO) acts on late erythroid precursor cells to
stimulate their growth and maturation. This action is mediated through
interaction of the hormone and its specific receptor on the cell surface [1]. EPO-R on murine [2-6] and human [7,8] cells was identified
by analyzing cross-linked complexes of the radiolabeled ligand and
EPO-R by SDS-PAGE; two cross-linked products of 140 and 120 kDa were
found under reducing conditions. Since EPO is a glycoprotein of 35
kDa [9], the molecular masses of EPO-R proteins are 105 and 85 kDa if
we assume that a single ligand molecule binds with EPO-R. Peptide
mapping of the two cross-linked products indicated that the primary
amino acid sequences of the two EPO-R proteins were similar and it was
proposed that the 85-kDa protein was produced by limited proteolysis
of the 105-kDa protein [10]. Analysis of murine EPO-R cDNA showed
that the matured EPO-R was a 483 amino acid polypeptide of 53 kDa
[11], much smaller than that estimated from the cross-linked product.
The amino acid sequence predicted from cDNA indicates that the EPO-R
protein has two potential sites of $\underline{N}$-linked glycosylation and also
many $\underline{O}$-linked glycosylation sites. We developed an assay of EPO-R
solubilized from murine erythroid precursor cells and found conditions
that yielded a solubilized EPO-R with similar properties to those of

R. Sasaki and K. Ikura (eds.), Animal Cell Culture and Production of Biologicals. 345–352.

EPO-R on the intact cells in terms of ligand-binding behavior and size
of the cross-linked products [12]. Here this paper describes binding
of the solubilized EPO-R with lectins, indicating that EPO-R is a
glycoprotein. We address the question whether glycosylation accounts
for the difference between the size of EPO-R predicted from cDNA and
the cross-linked product, using glycosidases that remove carbohydrates
from proteins.

RESULTS

Binding of the solubilized EPO-R to lectins——The cytoplasmic
membrane was prepared from spleen cells of mice to which mouse erythroleukemia cells bearing EPO-R were transplanted, and the membrane-

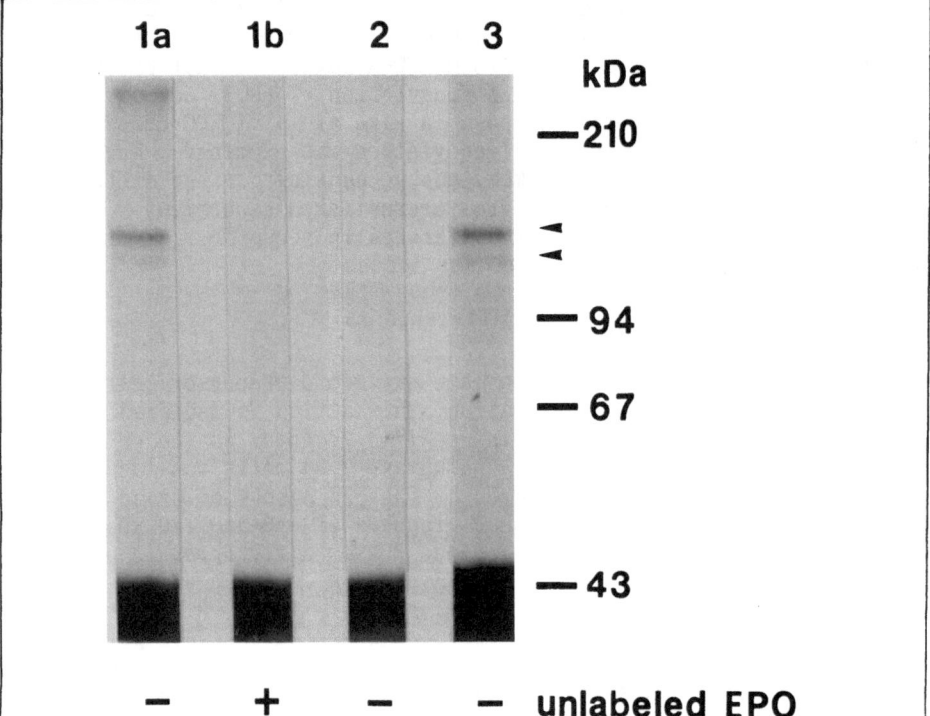

Fig. 1. Binding of the solubilized EPO-R to Con A. The cytoplasmic
membrane from murine erythroleukemia cells was solubilized in CHAPS
and was bound to Con A-agarose. EPO-R was detected as the cross-
linked products of EPO-R and [^{125}I]EPO on SDS-PAGE. Autoradiographs
show [^{125}I]EPO·EPO-R cross-linked products from the membrane solubilized by 0.5% CHAPS (lanes 1a and 1b), the through fraction appeared
without being adsorbed on Con A (lane 2), and the fraction eluted from
Con A by 0.5 M methyl-α-D-mannopyranoside (lane 3). The reaction
mixtures for binding of EPO to EPO-R in lanes 1a, 2, and 3 contained
[^{125}I]EPO and no unlabeled EPO but that in lane 1b contained 100-fold
unlabeled EPO.

bound receptor was solubilized with CHAPS as described previously [12]. The solubilized EPO-R was put on a column containing Con A-agarose to examine whether EPO-R was a glycoprotein. Ligand-binding activity of EPO-R was measured by detecting the specific cross-linked products of recombinant human [^{125}I]EPO and EPO-R on SDS-PAGE. Figure 1 shows autoradiographs of the cross-linked products. The solubilized EPO-R yielded two specific cross-linked products with molecular masses of 140 and 120 kDa (lane 1 in Fig. 1) but these products were not formed when the fraction eluted without being ad-sorbed (lane 2) and the washings (not shown) were used for the cross-linking experiment. EPO-R was eluted with methyl-α- D -mannopyranoside (lane 3), indicating that EPO-R is glycosylated.

A variety of lectins were used to further demonstrate that EPO-R contains sugars and also to find the class of sugars bound to EPO-R. EPO-R bound to Con A and lentil lectin, and the bound EPO-R was eluted with an appropriate compound, methyl-α-D-mannopyranoside. Both Con A and lentil lectin recognize carbohydrate residues that are present exclusively in N-linked sugars. The presence of a fucose residue at-tached to the Asn-linked N-acetylglucosamine residue is essential for high affinity binding to lentil lectin. It appears from these results that EPO-R contains N-linked sugars bearing fucose residues. Neither wheat germ lectin nor Ricinus communis$_{120}$ lectin binds to EPO-R, al-though these lectins interact with N-linked sugars. This result, however, is not contradictory with that of Con A and lentil lectin but rather suggests that EPO-R has N-linked sugars containing fucose residues and terminal sialic acids, since the presence of fucose residues in the N-linked sugars inhibits their interaction with wheat germ lectin and the presence of terminal sialic acids does so with Ricinus communis$_{120}$ lectin. Wheat germ lectin interacts not only with some of N-linked sugars but also with O-linked sugars having clustered sialic acids; unbinding of EPO-R to wheat germ lectin may indicate the absence of a muchin-type O-linked sugars in EPO-R. Peanut lectin binds to O-linked sugars containing no sialic acids with a high affin-ity and to those containing sialic acids with a decreased affinity. Although we have not done experiments with the desialylated EPO-R, unbinding of EPO-R to this lectin makes it very unlikely that EPO-R has a large amount of O-linked sugars.

Partial purification of EPO-R——We partially purified EPO-R by the use of binding properties of EPO-R to lectins. The solubilized membrane was put on a Con A column and EPO-R was eluted by 0.5 M methyl-α-D-mannopyranoside. The EPO-R fraction was put on a Ricinus communis$_{120}$ lectin column and EPO-R appeared in the flow through frac-tions without being adsorbed. By these purification procedures, EPO-R was purified 30-fold with an activity recovery of 50%, based on the intensities of autoradiographs of the cross-linked products.

Enzymatic deglycosylation of the cross-linked products of EPO and the solubilized receptor— To test if the difference in molecular size predicted from cDNA and estimated from the cross-linking experiment could be accounted for by glycosylation of EPO-R, the cross-linked complexes prepared using the partially purified EPO-R were digested with glycosidases and then analyzed on SDS-PAGE to estimate the decrease in the molecular size of the complexes by deglycosylation.

EPO (35 kDa) is made up by 18-kDa peptide and 17-kDa sugars. The completely deglycosylated EPO has the full activity [13] but is unstable in 0.5% CHAPS that is optimum for solubilization of EPO-R; we had to use the native EPO as a ligand when the solubilized EPO-R was used for cross-linking experiments. Thus, a size decrease of the cross-linked products by deglycosylation includes removal of sugars linked not only to EPO-R but also to EPO. First, decreases in EPO size by enzymatic deglycosylation were estimated (Fig. 2A). Since O-glycanase

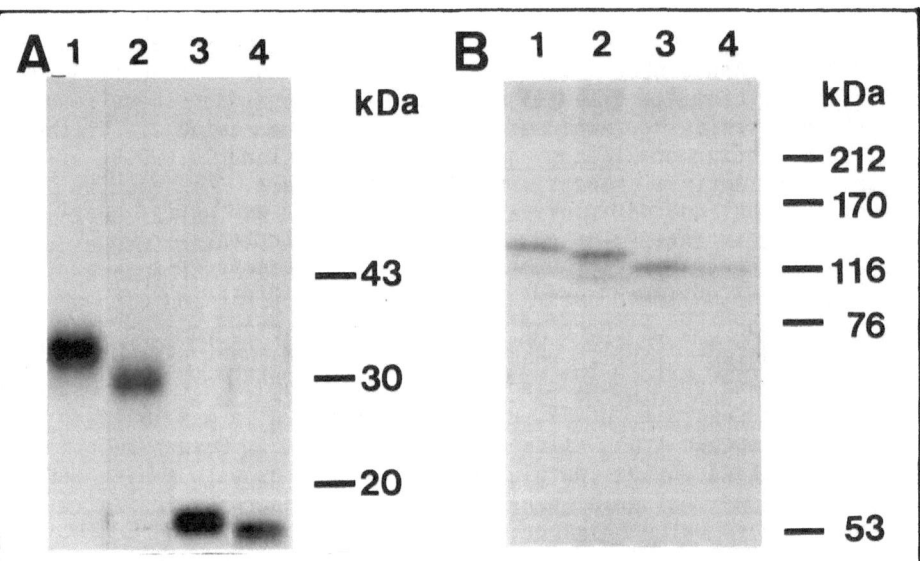

Fig. 2. Enzymatic deglycosylation of the cross-linked products between EPO and the solubilized receptor. Samples were separated by SDS-PAGE and the gels were autoradiographed. (A) shows deglycosylation of [^{125}I]EPO under the same conditions used for deglycosylation of the cross-linked products. Lane 1, [^{125}I]EPO treated in the absence of glycosidases; lane 2, treated with neuraminidase, and O-glycanase; lane 3, treated with N-glycanase; lane 4, treated with neuraminidase, O-glycanase, and N-glycanase. For electrophoresis, 12.5% gel was used. (B) shows deglycosylation of the cross-linked products between the partially purified EPO-R and [^{125}I]EPO. Lane 1, the cross-linked products treated in the absence of glycosidases; lane 2, treated with neuraminidase and O-glycanase; lane 3, treated with N-glycanase; lane 4, treated with neuraminidase, O-glycanase, and N-glycanase.

acted on the desialylated O-linked sugars, removal of O-linked sugars were done after desialylation with neuraminidase. Digestion with neuraminidase and the subsequent O-glycanase caused 5-kDa decrease in size of EPO (lane 2). It appears that this decrease is largely caused by desialylation from N-linked sugars because EPO has only one O-linked sugar chain [13] of 1 kDa (see lanes 3 and 4). Digestion by N-glycanase (lane 3) decreased the size of EPO to 19 kDa, which results from removal of N-linked sugars at three sites [13]. Complete deglycosylation by N-glycanase and O-glycanase yields EPO of 18 kDa (lane 4), consistent with the size predicted from EPO cDNA; contribution of O-linked sugar chain to the size of EPO is 1 kDa. These enzymatic deglycosylations were done under the same conditions that were used for deglycosylation of the cross-linked products described below. The presence of one distinct component in each lane after digestion indicates the completion of enzyme reactions and the conditions used here are appropriate for processing of the cross-linked products.

Results of enzymatic deglycosylation of the cross-linked products are shown in Fig. 2B. The presence of distinct bands again validates the completion of enzyme reactions. Digestion by neuraminidase together with O-glycanase converted the cross-linked products of 140 and 120 kDa to those of 135 and 115 kDa, respectively (compare lane 1 with 2). This reduction in size of the cross-linked products is equivalent to the 5-kDa decrease found when EPO was digested by neuraminidase together with O-glycanase (see lane 2 in Fig. 2A). Digestion of the cross-linked complexes by N-glycanase that cut off N-linked sugars yielded the products of 116 and 96 kDa (lane 3). Double digestion of the cross-linked products with O-glycanase and N-glycanase gave the same result as that of N-glycanase-digestion (lane 4). From these results, the total loss in size of the cross-linked complexes by deglycosylation was estimated to be 24 kDa. Since the sugars of EPO make up 17 kDa, the net reduction in size of EPO-R by deglycosylation

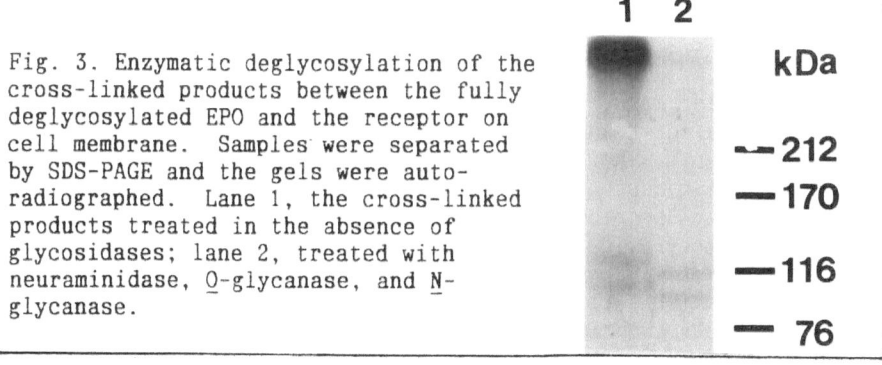

Fig. 3. Enzymatic deglycosylation of the cross-linked products between the fully deglycosylated EPO and the receptor on cell membrane. Samples were separated by SDS-PAGE and the gels were autoradiographed. Lane 1, the cross-linked products treated in the absence of glycosidases; lane 2, treated with neuraminidase, O-glycanase, and N-glycanase.

is 7 kDa. It appears from these results that glycosylation of EPO-R does not account for the large difference in size of EPO-R predicted from cDNA and estimated from the cross-linked complexes.

Taken together with lectin-binding properties of EPO-R, analyses of the cross-linked complexes deglycosylated by O-glycanase, N-glyca-nase, and both enzymes (lanes 2, 3, and 4 in Fig. 2B) indicate that EPO-R has N-linked sugars but little O-linked sugars. The two cross-linked products behaved similarly on glycosidase digestion, indicating that occurrence of a smaller receptor is not attributable to deglycosylation of the larger one.

Enzymatic deglycosylation of the cross-linked complexes between the fully deglycosylated EPO and the membrane receptor — We attempted to find the decrease in size by enzymatic deglycosylation of the cross-linked products between the fully deglycosylated [^{125}I]EPO and EPO-R. In this case, the decrease in size of the cross-linked products due to deglycosylation of EPO has been excluded and thereby interpretation of the experimental results would be easier. As described earlier, the fully deglycosylated EPO binds specifically to EPO-R on intact cells [13] but does not form the cross-linked complexes with EPO-R solubilized by 0.5% CHAPS. Here we did the experiments to detect the cross-linked products using the fully deglycosylated EPO and EPO-R on the cytoplasmic membrane preparation. Deglycosylated [^{125}I]EPO was bound to EPO-R on the cytoplasmic membrane from the erythroid precursor cells and the cross-linked products were formed. The cross-linked products were solubilized by CHAPS and deglycosylated. Figure 3 (lane 1) shows the cross-linked products before deglycosylation. Two products with molecular masses of 123 and 103 kDa were found; their sizes are consistent with those expected from the size of deglycosylated EPO, 18 kDa. Complete deglycosylation of the cross-linked products between deglycosylated EPO and EPO-R resulted in products of 116 and 96 kDa (lane 2). These results support the earlier conclusion that EPO-R is a glycoprotein but the contribution of sugars to the size of EPO-R molecule is 7 kDa.

DISCUSSION

The molecular mass, 53 kDa, of murine EPO-R p1predicted from cDNA [11] is much smaller than that, 105 kDa [2], estimated from cross-linking experiments. The amino acid sequence predicted from cDNA shows that EPO-R protein has potential sites for N- and O-glycosylation [11]. This paper describes the binding of EPO-R to lectins and the enzymatic deglycosylation of the [^{125}I]EPO·EPO-R cross-linked products to examine whether glycosylation of EPO-R accounts for the difference in size of EPO-R (105 kDa-53 kDa=52 kDa). The lectin binding properties of the solubilized EPO-R and analyses of the deglycosylated cross-linked complexes showed that EPO-R had N-linked sugars of 7 kDa but little O-linked sugars. Our findings appear rational from the fact that EPO-R has only two sites for N-glycosylation [11]. Thus the contribution of sugars to the size of EPO-R calculated from the cross-linking experiments was too small to make up for the size difference of EPO-R. Recently it was shown that lymphoid cell lines that have been transfered with the EPO-R cDNA produced 66-kDa EPO-R with complex-type sugars [14]. This result supports our conclusion of erythroid cells.

There are some ways to explain a large difference in size predicted from cDNA and from cross-linked product; (1) EPO-R is a dimeric form, (2) the cross-linked product contains other membrane components associated with EPO-R, or (3) two ligand molecules bind to EPO-R. Of these possibilities, dimerization of EPO-R or association with other membrane components is an interesting hypothesis. Dimer formation of EPO-R or interaction with an unidentified membrane component may be required to construct a high-affinity conformation. If the latter is the case, occurrence of the membrane component is not specific for erythroid cells because COS cells transformed with EPO-R cDNA express high-affinity sites [11].

REFERENCES

1. Krantz, S. B. and Goldwasser, E. (1984) 'Specific binding of erythropoietin to spleen cells infected with the anemia strain of Friend virus', Proc. Natl. Acad. Sci. USA 81, 7574-7578

2. Sasaki, R., Yanagawa, S., Hitomi, K. and Chiba, H. (1987) 'Characterization of erythropoietin receptor of murine erythroid cells', Eur. J. Biochem. 168, 43-48

3. Todokoro, K., Kanazawa, S., Amanuma, H. and Ikawa, Y. (1987) 'Specific binding of erythropoietin to its receptor on responsive erythroleukemia cells', Proc. Natl. Acad. Sci. USA 84, 4126-4134

4. Sawyer, S. T., Krantz, S. B. and Luma, J. (1987)'Identification of the receptor for erythropoietin by cross-linking to Friend virus-infected erythroid cells',Proc. Natl. Acad. Sci. USA 84, 3690-3694

5. Tojo, A., Fukamachi, H., Kasuga, M., Urabe, A. and Takaku, F. (1987)'Identification of erythropoietin receptors on fetal erythroid cells', Biochem. Biophys. Res. Commun. 148,443-448

6. Mayeux, P., Billat, C. and Jacquot, R. (1987) 'The erythropoietin receptor of rat erythroid progenitor cells', J. Biol. Chem. 262, 13985-13990

7. Hitomi, K., Fujita, K., Sasaki, R., Chiba, H., Okuno, Y., Ichiba, S., Takahashi, T. and Imura, H. (1988) 'Erythropoietin receptor of a human leukemic cell line with erythroid characteristics', Biochem. Biophys. Res. Commun. 154, 902-909

8. Broudy, V., Lin, N., Egrie, J., de Haen, C., Weiss, T., Papayannopoupolou, T. and Adamson, J. W. (1988) 'Identification of the receptor for erythropoietin on human and murine erythroleukemia cells and modulation by phorbol ester and dimethyl sufoxide', Proc. Natl. Acad. Sci. USA 85, 6513-6517

9. Goto, M., Akai, K., Murakami, A., Hashimoto, C., Tsuda, E., Ueda, M., Kawanishi, G., Takahashi, N., Ishimoto, A., Chiba, H. and Sasaki, R. (1988) 'Production of recombinant human erythropoietin in mammalian cells: host-cell dependency of the biological activity of the cloned glycoprotein', Bio/Technology 6, 67-71

10. Sawyer, S. T. (1989) 'The two proteins of the erythropoietin receptor are structurally similar', J. Biol. Chem. 264, 13343-13347

11. D'Andrea, A. D., Lodish, H. F. and Wong, G. G. (1989) 'Expression cloning of the murine erythropoietin receptor', Cell 57, 277-285

12. Hitomi, K., Masuda, S., Ito, K., Ueda, M. and Sasaki. R. (1989) 'Solubilization and characterization of erythropoietin receptor from transplantable mouse erythroblastic leukemic cells', Biochem. Biophys. Res. Commun. 160, 1140-1148

13. Tsuda, E., Kawanishi, G., Ueda, M., Masuda, S. and Sasaki, R. (1990) 'The role of carbohydrate in recombinant human erythropoietin', Eur. J. Biochem. 188, 405-411

14. Yoshimura, A., D'Andrea, A. D., and Lodish, H. (1990) 'Friend spleen focus-forming virus glycoprotein gp55 interacts with the erythropoietin receptor in the endoplasmic reticulum and affects receptor metabolism', Proc. Natl. Acad. Sci. USA 87, 4139-4143

EFFECTIVE PURIFICATION OF MONOCLONAL ANTIBODIES BY FAST FLOW AFFINITY CHROMATOGRAPHY

Saichi Yamada[2], Yoshihiro Kamiya[2], Nozomu Eto[1],
Koji Yamada[1], Hiroki Murakami[1], and Tsuyoshi Majima[2]
(1) Department of Food Science and Technology, Kyushu University
(2) Bio Research Section, NGK Insulators, Ltd ., 1 Maegata-
     cho Handa, 475 Japan

The aim of this study was to purify monoclonal antibodies (MAbs) from a large amount
of hybridoma supernatants using fast flow affinity chromatography (AFC) at high yield
and to investigate the optimal pH of MAb elution and antigen reactivity of Chuzan virus-specific MAbs at acidic pH . The fast flow affinity technique were efficient for pre-
parative-scale purification of M Ab, and some acidic buffers for M Ab elution increased
the reactivity of Chuzan virus-specific MAbs.

INTRODUCTION
High performance liquid chromatography (HPLC) methods have been used for analytical
and small-scale purification of MAbs from mouse ascites (1, 2, 3). However, in the case
of animal cell culture supernatants, it will be necessary to treat a large amount of the
supernatants since the M Ab concentration obtained in the cell culture is low in many
cases. Usually, the first step is enrichment of supernatants with ultrafiltration, which is
followed by ammonium sulfate precipitation and several chromatographic methods. In
some cases, multiple stages of purification reduce the antibody yield and activity. On
the other hand, affinity purification with the ligand of Protein A or G has been attempted to obtain M Ab at high purity and recovery, but it was impossible to operate at
high flow rates in soft adsorbents, such as agarose and cellulose. Further, excess amounts of affinity gels and a large column volume are required for process-scale purification,
and the concentration of products purified become lower due to M Ab elution from the a-
ffinity ligand at slow flow rates on larger AFC columns. In affinity purification of M Ab
from the hybridoma supernatant, the newly developed porous silica adsorbent, which
operates at high speed, have been used to overcome these problems. We report here this
fast flow affinity technique gives reduction of purification time, enriched products at high
concentration, and high MAb recovery.

Materials and Methods
Cell lines and cell culture.
Both CG1/1 and CG53/2/4 mouse hybridomas produce Chuzan virus-specific MAbs, and
the subclass of their MAbs are IgG1 and IgG2a respectively. They were grown in Dulbecco's Modified Eagle Medium and Ham's F-12 supplemented with 0.12% NaHCO3,
15 mM HEPES, 10 million units/l penicillin G sodium, 0.1 g titer/l streptomycin sulfate,
5 μg/ml insulin, 20 μg/ml transferrin, 20 μM ethanolamine, and 25 nM sodium selenite.

R. Sasaki and K. Ikura (eds.), Animal Cell Culture and Production of Biologicals, 353–357.

MAb purification by fast flow AFC.

Process-scale affinity purification of MAb was done on a Chromatop protein A or G column (250 mm × 4.6 mm I.D.) (ChromatoChem . Inc. , America). The column was equilibrated with the adsorption buffer, such as phosphate buffer solution or 1.5 M glycine plus 3 M Na Cl (pH 8.9). A cell culture suspension was centrifuged at 3,000 rpm for 10 minutes to remove animal cells. After filtration with a 0.45 μm Millipore filter or centrifugation at 10, 000 rpm for 20 minutes, the hybridoma supernatant was pumped into the column with a HPLC pump at high flow rates. The column was washed with the adsorption buffer at a faster flow rate. After the absorbance at 280 nm returned to the baseline, MAb was eluted with various acidic buffers.

Analysis of MAb recovery.

After purification with fast flow AFC, MAb yield was measured by an enzyme-linked immunosorbent assay (ELISA).

Preparation of antibody solutions for immunogloblin reactivity.

First, serum-free hybridoma supernatants containing Chuzan virus-specific MAbs were treated with ammonium sulfate precipitation, and then dialyzed against PBS. One-half ml of antibody solution was mixed into test tubes with 2 ml of elution buffer, and left for 5 minutes. Next, they were neutralized with alkali solution (0.1 N phosphate, 2 M Tris), and dialyzed against PBS for 4 hours.

Analysis of immunogloblin reactivity.

Chuzan virus-specific immunogloblin reactivity was measured by E L I S A with Chuzan virus coated immuno plate and peroxidase modified anti-mouse IgG.

Results and Discussion

Rapid purification of MAb on a preparative scale with fast flow AFC.

To purify M Ab from a large amount of the hybridoma supernatant with high recovery, one-step purification has been done on a Cromatop A F C column. As shown in Fig. 1, one liter of the culture supernatant was treated at a high flow rate of 10 ml / min. for approximately two hours using a Chromatop protein A column. It has been reported that mouse IgG1 binds to Protein A only weakly and IgG1 recovery is very low (4). However , IgG1 recovery was increased by the pretreatment of supernatants with high pH and salt concentration (e.g. 1.5 M glycine plus 3 M NaCl (pH 8.9)) greatly. In this case , high recovery of 92.3 % was obtained without enrichment of supernatant .

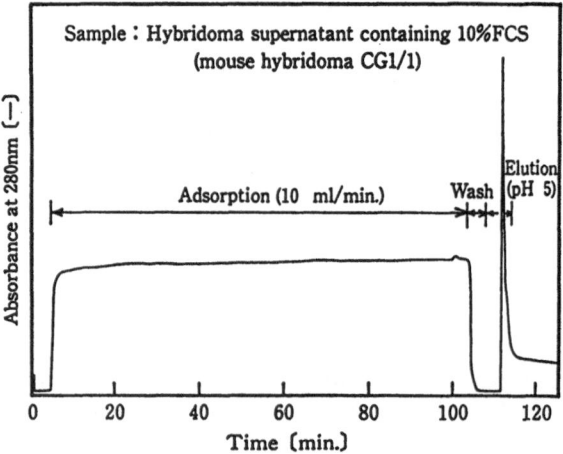

Figure 1. Rapid purification of M Ab on a preparative scale using a Chromatop Protein A Column.

Also, the ligand of Protein G has been used to purify CG 1 / 1 M Ab (IgG1) from the serum-free hybridoma supernatant directly. This result is shown in Fig. 2. For this run, 442 ml of the supernatant was put onto the column at a flow rate of 4 ml / min. MAb ' recovery on fraction number 3-5 was 82 %. Moreover, the peak of MAb elution was very sharp , so a purified product was obtained at a high concentration of about 0.1 mg/ml.

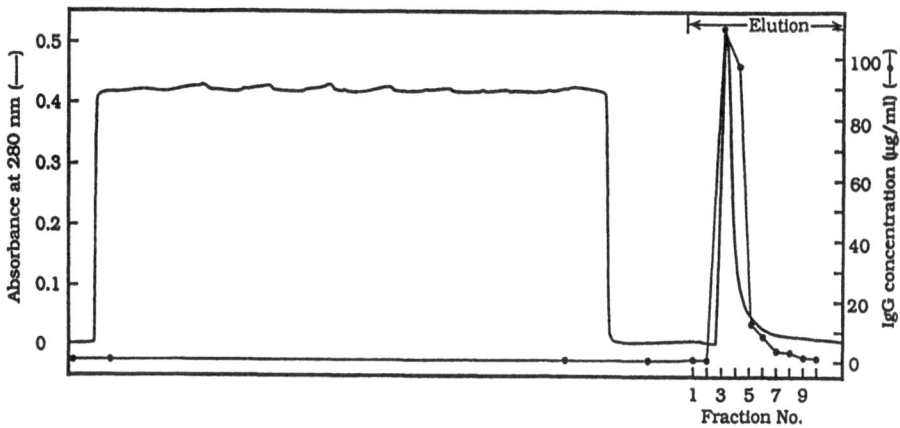

Figure 2.   Purification of mouse MAb in serum-free CG1/1 culture supernatant on Chromatop Protein G Column.

Effects of eluted pH on activity of antigen-binding site.
The low pH used to elute immunogloblins from Protein A may be deleterious to antibody activity. To measure the drop of antibody activity caused by lower pH , the MAb activity for antigen-binding site was measured by ELISA . Anti-human albumin M Ab was purified from mouse ascites on Chromatop Protein A at each pH. After neutralization with 1 M Tris-Cl (pH 8.0), acidic eluates were dialyzed against PBS. As shown in Fig. 3, MAb activity for antigen binding site decreased with the drop of pH eluted. Comparing the pH necessary for Ig G 1 elution between Protein A and G , pH 5 for Protein A is milder than pH 3 for Protein G. This result suggests that the Protein A column is more useful than the Protein G column in purification of mouse subclass IgG1 because of high activity of products purified.

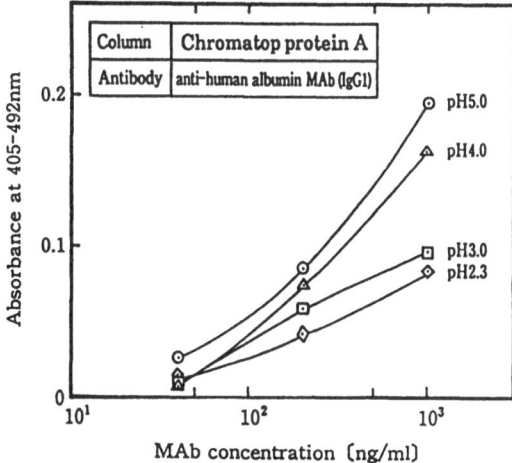

Figure . 3 Effects of pH eluted on activity of antigen- binding site.

Increase of MAb reactivity with affinity purification.

It is necessary to expose MAbs into acidic conditions in affinity purification of mouse MA bs. To investigate stability of MAbs at low pH, various kinds of elution buffers that have buffer ability at around pH 3.0 were used. For CG53/2/4, reactivity for antigen advanced due to the use of sodium acetate at the concentration of more than 0.1 mol/l. Also, potassium hydrogen phthalate-HCl brought out a slight change in reactivity, but the other buffers have no effect on MAb reactivity (shown in Table 1.). Control test was done with PBS.

TABLE 1. Effects of various buffers (pH 3.0) on reactivity of mouse MAbs

Buffer	Reactivity (A 405-492 nm)	
	CG53/2/4	CG1/1
0.1 M Citric acid-Na citrate	0.041	0.115
3.5 M Acetic acid-Na acetate	0.296	0.089
0.1 M Acetic acid-Na acetate	0.229	0.126
0.1 M Glycine-HCl	0.026	-
0.1 M Glycine/2% Acetic acid	0.025	-
0.15 M NaCl/2% Acetic acid	0.029	-
0.1 M Potassium hydrogen phthalate-HCl	0.075	-
Phosphate buffered saline	0.036	0.128

These results suggest that mouse MAbs will be unstable in acetic acid buffer, and MAb reactivity for antigen decline with the kinds of buffer. It is desirable that MAb reactivity not be affected with elution buffer, but there may be utility value in elution buffers that advance M Ab reactivity greatly. On the basis of this result, CG53 /2/4 were purified on Chromatop protein G column, and MAb reactivities were compared between CG 53/2/4 purified and unpurified. In Fig. 4, MAbs with increased reactivity for Chuzan virus were obtained with MAb elution of 3.5 M sodium acetate. Measuring the M Ab concentration necessary for absorbance value of 0.8 between 3.5 M sodium acetate and culture supernatant, they are approximately 2.6 and 45 µg/ml, respectively. Thus, MAb eluted with 3.5 M sodium acetate had 17 times the reactivity of MAb unpurified. On the other hand, MA b eluted with other buffers, such as 0.01 M and 0.1 M citric acid-Na citrate, and 0.1 M potassium citrate indicated the similar reactivity as MAb in culture supernatants

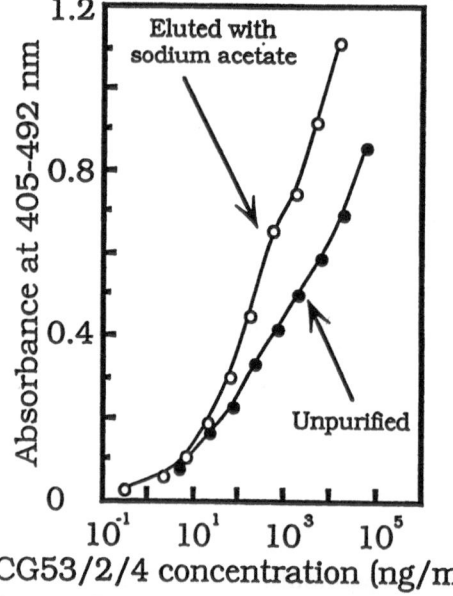

Figure 4. Dose response curves of MAb with or without Chromatop purification

Acknowledgement
We are grateful to Drs. K. Yamada and H. Murakami (Kyushu University) for their supply of mouse hybridoma and helpful advice.

References

1. Boudan, P., Ulla, J., Christina, N., and Anders, W. (1986) ' Rapid purification of monoclonal antibodies by high-performance liquid chromatography ', J.Chrom. 359, 449-460

2. Bruch, C., Portetelle ,D., Glineur,C., and Bollen,A. (1982) 'One-Step Purification of Mouse Monoclonal Antibodies from Ascitic Fluid by DEAE Affi-Gel Blue Chromatography', J.Immunol.Methods 53, 313-319

3. NeohH,S.,Gordon, C.,Potter, A.,and Zola, H. (1986) 'The purification of mouse monoclonal antibodies from ascitic fluid',J.Immunol.Methods 91, 231-235

4. Ey, P.L.,Prowse, S.J.,and Jenkin, C.R. (1978) ' Isolation of pure IgG1, IgG2a, and IgG 2b immunogloblins from mouse serum using protein A-Sepharose ', Immunochemistry 15, 429-436

DETECTION OF BOVINE VIRUSES BY MONOCLONAL ANTIBODIES

N. ETO,(1) S. SHIRAHATA,(2) K. YAMADA,(1) A. KOGA,(1) and
H. MURAKAMI(2)
1) *Department of Food Science and Technology*; 2) *Graduate
School of Genetic Resources Technology, Faculty of
Agriculture, Kyushu University 46-09, 6-10-1 Hakozaki,
Higashi-ku, Fukuoka 812, Japan*

ABSTRACT. We established sixteen mouse monoclonal antibodies (MAbs)
reactive to Chuzan virus K-47 strain using P3-X63-Ag8-U1 cells as
fusion partners. Among them, CG53/2/4 recognized a 100K structural
protein of the virus. This 100K antigen is a glycoprotein which is
essential for the infection of Chuzan virus. The other anti-Chuzan
virus MAbs reacted with a 41K antigen of the virus. Especially CG1/1
showed the highest reactivity to the virus. Forward step sandwich
assay using CG1/1 and biotinylated CG53/2/4 could detect Chuzan virus
at 10 $TCID_{50}$/ml. Furthermore we established fourteen MAbs reactive to
bovine ephemeral fever virus (BEFV) YHL strain. Among them, YM4/9
reacted specifically with a 43K antigen of BEFV, corresponding to the
matrix protein 1. The other MAbs reacted not only with the 43K
antigen, but also with unknown 23K and 21K antigens. By a simultaneous
two-site method using YM4/9 and YG3/4, it was possible to detect
$10^{4.10} TCID_{50}$/ml of BEFV in the presence of bovine serum. Therefore,
these MAbs can eventually predict the virus infection of cattle before
sideration.

INTRODUCTION

Domestic animal hygiene is important to the livestock business.
However, many kinds of epizootic diseases occurred in Japan, inducing
great damage in productivities of milk and meat. To prevent severe
outbreaks of epizootic diseases, rapid and accurate virus detecting
systems are desirable. An epizootic of congenital abnormalities of
calves characterized by hydranencephaly-cerebellar hypoplasia (HCH)
syndrome is caused by Chuzan virus belonging to the Palyam subgroup of
genus *Orbivirus* (Miura *et al.*, 1988a). The main clinical signs of the
disease are impairment of mobility and signs of impairment of the
nervous system. The main macroscopic pathologic changes are HCH (Goto
et al., 1988). The pathogenic virus was isolated from the blood of a
calf (Miura *et al.*, 1988a) and was named strain K-47 virus.
  On the other hand, bovine ephemeral fever is caused by bovine
ephemeral fever virus (BEFV), a rhabdovirus (Tanaka *et al.*, 1969).
This disease is essentially characterized by respiratory symptoms with

R. Sasaki and K. Ikura (eds.), Animal Cell Culture and Production of Biologicals, 359–366.
© 1991 *Kluwer Academic Publishers.*

increased respiration, a temporary dyspnea, nasooral-pharyngeal secretions, and lacrimation. Anorexia, decreased lactation, joint pain, and muscle tremor are also common symptoms of the disease. Though the disease has low mortality, it causes great economic damage due to decrease or stop in milk production and decrease in the quality and production of meat.

In addition to therapeutic work on the diseases, the development of rapid and simple detection of the virus is also important for effective uses of stockbreeding resources. The purpose of our study is to produce MAbs to bovine viruses and apply them as diagnostic or therapeutic reagents. We report here the establishment of MAbs reactive to Chuzan virus or BEFV and the characterization of the MAbs to develop a diagnostic system for the viruses.

MATERIALS AND METHODS

Viruses and host cells. Four arboviruses known to infect cattle were used in this study. They included 2 orbiviruses, Chuzan virus K-47 strain (Miura *et al.*, 1988a) of the Palyam subgroup and Ibaraki virus No. 2 strain (Omori *et al.*, 1969) of the epizootic hemorrhagic disease of deer subgroup; A Bunya virus, Akabane virus OBE-1 strain (Kurogi *et al.*, 1976) of the Simbu virus group; and a lyssa virus, bovine ephemeral fever virus YHL strain. These viruses were prepared in serum-free condition (Eto *et al.*, 1991) using hamster lung cells (HmLu-1) as a host cell line. HmLu-1 cells were grown in a DF medium (Barnes & Sato, 1980). The virus titer was expressed as 50% tissue culture infectious dose (TCID$_{50}$), by the method of Kärber (1931).

Immunization and hybridization. BALB/c male mice were injected intraperitoneally with virus preparation five times (100 μg protein in each injection) every 10 days. On the third day after the final immunization, splenocytes were isolated and fused with a HAT-sensitive mouse myeloma cell line, P3-X63-Ag8-U1 cells, by the PEG method (Galfrè & Milstein, 1981). Hybridomas producing MAb reactive to the virus were cloned twice by the limited dilution method (Goding, 1980). Established hybridomas were cultured in DF medium supplemented with 10 μg/ml insulin, 35 μg/ml of transferrin, 20 μM ethanolamine, and 25 nM selenium (ITES) (Murakami *et al.*, 1982).

Purification and biotinylation of MAbs. IgG type MAbs were purified using a Chromatop Protein G column (0.46× 25 cm, NGK Insulators, Ltd, Nagoya, Japan). Then CG53/2/4 was biotinylated using a Biotinylation kit (Amersham International plc., Amersham, UK). An IgM type MAbs were purified using a Shim-pack HAC column (0.75× 5 cm, Shimadzu Corporation, Kyoto, Japan).

Reactivity of MAbs to the virus. The reactivity of MAbs to viruses was measured by ELISA using peroxidase-labeled anti-mouse IgG or IgM (Organon Teknika N.V.-Cappel Products, Westchester, USA) as a 2nd antibody. The molecular weight of viral antigen was measured by transferred blotting according to the method of Towbin *et al.* (1979).

Periodate treatment of antigens. α-Glycol in glycosidic moiety of the antigens was cleaved by mild periodate oxidation at an acidic pH, by the method of Woodward *et al.*(1985). Briefly, viral antigen on blotted strips were incubated with 10 mM sodium periodate dissolved in 50 mM sodium acetate (pH 4.5) for 1 hr, and aldehyde groups were blocked by exposing to 50 mM sodium borohydride dissolved in PBS for 30 min. Then the strips were used for the immunoassay described above.

Virus neutralization test. The virus solution with a titer of 100 $TCID_{50}$ and MAb solutions were mixed together and incubated for 1 hr at 37℃. HmLu-1 cells in 96 well-tissue culture plates at 1×10^5 cells/ml were infected with the mixture of virus and MAb for 1 hr, and then cultured in the fresh DF medium for 2 days to measure the ND_{50} titer, by the method of Reed & Muench (1938).

RESULTS

Specificity of MAbs
After two fusion experiments, sixteen hybridomas producing MAb reactive to Chuzan virus K-47 were isolated. As shown in Table 1, these MAbs had no cross-reactivity with Ibaraki virus No.2, BEFV-YHL and Akabane virus OBE-1 strains, and with their host cell line, HmLu-1 cells.

In the case of BEFV, fourteen hybridomas producing MAb reactive to BEFV were isolated by two fusion experiments. Their characteristics are summarized in Table 2. These MAbs showed no cross-reactivity with Ibaraki virus No.2 and Akabane virus OBE-1 strains, or their host cells.

Viral Antigens recognized by MAbs
Tables 1 and 2 also summarize the antigen specificities of the MAbs, each of which showed characteristic reactivity to the virus. In the case of anti-Chuzan virus MAbs, CG53/2/4 reacted with a 100K antigen of the virus. The reactivity was diminished after periodate treatment of this antigen, suggesting that the 100K viral antigen is a glycoprotein. The other anti-Chuzan virus MAbs reacted only with the 41K viral antigen of the virus. Since the reactivity of the 41K antigen to the MAbs was not changed after the periodate treatment, the epitope of 41K viral antigen recognized by CG1/1 may not contain glycosides.

In the case of anti-BEFV MAbs, YM4/9 reacted only with a 43K antigen, corresponding to the matrix protein 1 (M1 protein) of BEFV (Table 2). On the other hand, the other MAbs reacted most strongly with the 43K antigen but also reacted with unknown 23 and 21K antigens.

Table 1. List of anti-Chuzan virus MAbs

MAb	Isotype	Light chain	Cross reactivity [a]					Molecule size of antigen (kDa)	Sensitivity of periodate oxidation	Neutralizing value
			HmLu-1 cells	Chuzan K-47	Ibaraki No. 2	BEFV[b] YHL	Akabane OBE-1			ND50 (ng)
CG1/1	G1	κ	0.011	1.559	0.010	0.016	0.015	41	-	>2.7x10³
CG2/2D	G1	κ	0.017	1.396	0.020	0.020	0.032	41	-	N.T.[c]
CG12	G1	κ	0.016	0.086	N.T.	N.T.	N.T.	41	-	N.T.
CG7	G2a	κ	0.011	0.244	N.T.	N.T.	N.T.	41	-	N.T.
CG13/14/2	G2a	κ	0.008	0.188	0.011	0.016	0.014	41	-	N.T.
CG31	G2a	κ	0.018	0.133	N.T.	N.T.	N.T.	41	-	N.T.
CG41/44/11	G2a	κ	0.012	0.140	0.012	0.013	0.012	41	-	N.T.
CG49	G2a	κ, λ	0.011	0.076	0.010	0.015	0.015	41	-	N.T.
CG53/2/4	G2a	κ	0.017	0.256	0.012	0.014	0.015	100	+	2.2x10²
CG3/12/26	G2b	κ	0.013	0.065	0.011	0.013	0.014	41	-	N.T.
CG14/3/1	G2b	κ	0.010	0.142	0.010	0.015	0.015	41	-	N.T.
CM1/33	M	κ	0.018	0.036	N.T.	N.T.	N.T.	41	-	N.T.
CM3/5/3	M	κ	0.029	0.193	N.T.	N.T.	N.T.	41	-	N.T.
CM5/57	M	κ	0.025	0.050	N.T.	N.T.	N.T.	41	-	N.T.
CM11/33	M	κ	0.016	0.477	0.017	0.013	0.014	41	-	>5.5x10³
CM23/24	M	κ	0.023	0.063	N.T.	N.T.	N.T.	41	-	N.T.

a. Data of cross reactivity are shown by O.D. in ELISA.
b. BEFV=bovine ephemeral fever virus
c. N.T.=not tested

Table 2. List of anti-bovine ephemeral fever virus MAbs

MAb	Isotype	Light chain	Cross reactivitiy [a]				Molecule size of antigen (kDa)	Neutralizing value
			BEFV[b] YHL	Ibaraki No.2	Akabane OBE-1	HmLu-1 cell		ND50 (ng)
YG1/51	G1	κ	0.134	0.000	0.006	0.003	43, 23, 21	N.T.[c]
YG2/3/2	G1	κ	0.244	0.002	0.005	0.007	43, 23, 21	N.T.
YG3/4	G1	κ	0.271	0.002	0.007	0.005	43, 23, 21	>3.1x10²
YG3A/3	G1	κ	0.318	-0.001	0.001	0.005	43, 23, 21	N.T.
YG4	G1	κ	0.072	N.T.	N.T.	0.004	43, 23, 21	N.T.
YG5/8	G1	κ	0.021	-0.004	0.001	0.003	43, 23, 21	>2.7x10²
YG6/7	G1	κ	0.029	0.002	0.006	0.008	43, 23, 21	>3.2x10²
YM2/6	M	κ	0.157	0.006	0.021	0.008	43, 23, 21	>6.3x10²
YM2B/13	M	κ	0.153	0.014	0.019	0.022	43, 23, 21	N.T.
YM3/19	M	κ	0.035	0.004	0.005	0.000	43, 23, 21	N.T.
YM4/9	M	κ	0.726	0.007	0.005	0.000	43	>1.2x10³
YM4C/9	M	κ	0.650	0.010	0.004	0.007	43, 23, 21	N.T.
YM5/2	M	κ	0.079	0.010	0.006	0.007	43, 23, 21	N.T.
YM6/8	M	κ	0.079	0.014	0.010	0.012	43, 23, 21	>4.8x10²

a. Data of cross reactivity are shown by O.D. in ELISA.
b. BEFV=bovine ephemeral fever virus
c. N.T.=not tested

Neutralizing activity of MAbs

In the case of anti-Chuzan virus MAbs, only CG53/2/4 neutralized the virus of which the titer was 100 $TCID_{50}$. In this condition, the ND_{50} was 220 ng. The other MAbs did not show any neutralizing activity even at the elevated amount of more than 2 μg (Table 1). None of anti-BEFV MAbs had neutralizing activity against the BEFV.

Detection of Chuzan virus K-47 by the forward step sandwich method using CG1/1 and biotinylated CG53/2/4

A set of IgGs, CG1/1 and CG53/2/4, which recognized different epitopes of Chuzan virus, was used to design an efficient Chuzan virus detection system. Immunoplates were coated with 10 μg/ml of the purified CG1/1. Then the virus and 2.5 μg/ml of biotinylated CG53/2/4 were added successively to the immunoplate. The amount of CG1/1-Chuzan virus-biotinylated CG53/2/4 complex was enzymatically measured using peroxidase-labeled streptavidin (Amersham International plc.). As shown in Fig 1A, the forward step sandwich method was able to detect Chuzan virus at virus titers over $10^{1.0} TCID_{50}/ml$.

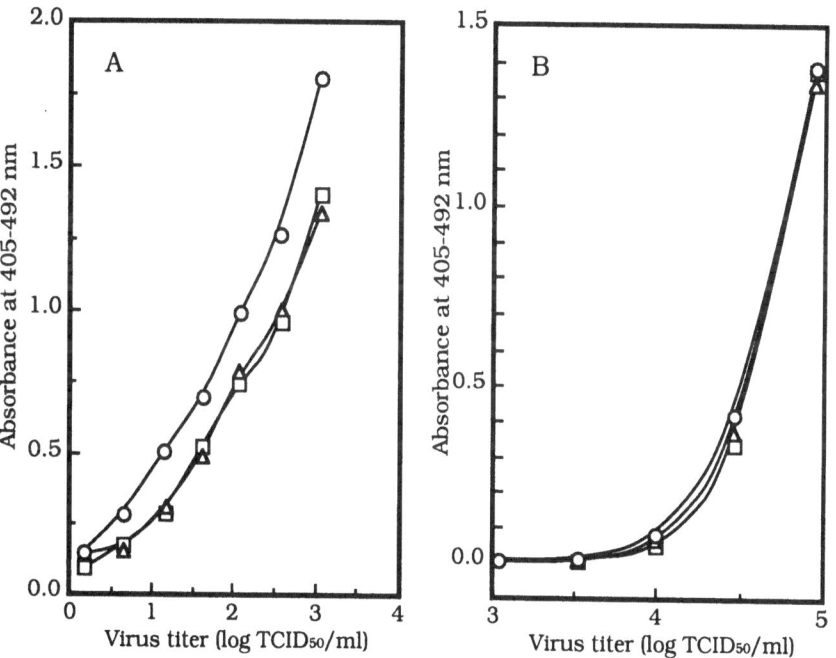

Fig. 1. Detection of Chuzan virus K-47 by the forward step sandwich method using CG1/1 and biotinylated CG53/2/4 (A) and detection of BEFV by the simultaneous two-site method using YM4/9 and YG3/4 (B). Calf serum was added to the virus samples at the concentrations of 0%(O), 25%(▲), and 50%(□), to estimate the interference of serum components in the system.

Detection of BEFV by the simultaneous two-site method using YG3/4 and YM4/9

To detect the virus antigen, a simultaneous two-site assay (Endo *et al.*, 1982) was done using a set of IgG (YG3/4) and IgM (YM4/9), which gave the highest reactivity in each class. Purified YM4/9 (50 μg/ml) was coated on an immunoplate and then the mixture of crude virus and YG3/4 (140 μg/ml) was added to the immunoplate. The amount of virus-YG3/4 complex bound to the virus-specific YM4/9 was measured using peroxidase-labeled anti-mouse IgG and the substrate solution described above. As shown in Fig.1B, the simultaneous two-site method was able to detect the virus at virus titers over $10^{4.10}$ TCID$_{50}$/ml. Bovine serum did not interfere the detection of the virus using the simple sandwich ELISA method.

MAbs react to other viruses

We established moreover eight anti-Ibaraki virus and seven anti-Akabane virus MAbs. Two of eight anti-Ibaraki virus MAbs, IG6 and IG7, were cross-reactive to BEFV (data not shown). This suggests BEFV and Ibaraki virus have an identical epitope, although they are classified differently in serology.

DISCUSSION

To clarify the functions of virus components and to develop a diagnostic system for the virus, we obtained sixteen anti-Chuzan virus and fourteen anti-BEFV MAbs. All of the MAbs reacted specifically with their immunogens.

These sixteen anti-Chuzan virus MAbs were divided into two groups, according to their reactivity to Chuzan virus K-47 components. An IgG2a type MAb, CG53/2/4 reacted with the 100K glycoprotein of Chuzan virus. The other MAbs reacted with the 41K antigen of the virus. Since CG53/2/4 reactive to the 100K glycoprotein neutralized the viral infectivity, the epitope for CG53/2/4 may be the receptor-binding site on 100K antigen. So 100K glycoprotein may be an essential viral component protein for Chuzan virus infection. Therefore, CG53/2/4 may be useful not only for diagnosis but also for therapy of the virus disease. In addition, the 100K glycoprotein can be used as a component vaccine against the virus.

We adopted CG1/1, which had the highest affinity for the virus, as solid phase of sandwich ELISA. CG53/2/4 which recognized different antigen from CG1/1 was adopted as biotin labelled MAb. This combination of MAbs could detect the virus at 10 TCID$_{50}$/ml. Since cattle that have the virus below $10^{3.0}$ TCID$_{50}$/ml in blood do not fall ill (Miura *et al.*, 1988b), the ELISA system enabled us to detect the virus even in apparently healthy cattle that had only a little virus. Conventional quantitative analysis of virus using the cytopathic effect observation method requires 5-7 days, while these ELISA methods using CG1/1 and biotinylated CG53/2/4 require only a few hours. These MAbs will be pertinent to construct a rapid and accurate diagnosis system for detecting Chuzan virus.

Among anti-BEFV MAbs, only an IgM type MAb, YM4/9, reacted only with the 43K antigen of BEFV. On the other hand, the other MAbs reacted not only with the 43K antigen strongly, but also with 23 and 21K antigens. The 43K antigen recognized by these MAbs may correspond to the M1 protein of the virus. The differences in the reactivity of YM4/9 and the other MAbs to BEFV antigens suggested that these anti-BEFV MAbs recognized different epitopes on the M1 protein. Since YG3/4 and YM4/9 showed high affinities, these MAbs were used for the development of the virus detection system. The virus could be quantitatively detected by the simultaneous two-site method using the two MAbs, even in the presence of serum. Since the method is simpler and less time-consuming than conventional forward step sandwich ELISA methods, it would be suitable for pathological examinations.

Establishments of MAbs to pathogenic viruses as described here may contribute to effective use of stockbreeding resources through the protection against outbreak of epizootic diseases. Moreover, quality control of bovine serum is especially important in production bio-material by animal cell culture. MAbs that react to bovine viruses may be used for serum quality control.

ACKNOWLEDGMENTS

The authors are grateful to the National Institute of Animal Health, Ministry of Agriculture, Forestry, and Fisheries of Japan for supply of HmLu-1 cells and the virus strains.

REFERENCES

Barnes, D. and Sato, G. (1980) 'Method for growth of cultured cells in serum-free medium', Anal. Biochem. 102, 255-270.
Endo, Y., Ohtaki, S., Kan, H., Yoshitake S., and Ishikawa, E. (1982) 'A simultaneous enzyme-linked immunoassay for human thyroglobulin' Anal. Lett. 15, 1301-1315.
Eto, N., Yamada, K., Nagamine, K., Haramaki, K., Shirahata, S. and Murakami, H. (1991) 'Multiplication of bovine viruses in hamster lung HmLu-1 cells cultured in protein-free medium' Agric. Biol. Chem. 55, (in press).
Galfrè, G. and Milstein, C. (1981) 'Preparation of monoclonal antibodies: strategies and procedures' Methods Enzymol. 73, 3-46.
Goding, J. W. (1980) 'Antibody production by hybridomas' J. Immunol. Meth. 39, 285-308.
Goto, Y., Miura, Y. and Kono, Y. (1988) 'Serologic evidence for the etiologic role of Chuzan virus in an epizootic of congenital abnormalities with hydranencephaly-cerebellar hypoplasia syndrome of calves in Japan' Am. J. Vet. Res. 49, 2026-2029.
Kurogi, H., Inaba, Y., Takahashi, E., Sato, K., Omori, T., Miura, Y., Goto, Y., Fujiwara, Y., Hatano, Y., Kodama, K., Fukuyama, S., Sasaki, N. and Matumoto, M. (1976) 'Epizootic congenital arthrogryposis-hydranencephaly syndrome in cattle: Isolation of Akabane virus from affected fetuses' Arch. Virol. 51, 67-74.

Kärber, G. (1931) 'Beitrag zur kollektiven Behandlung pharmakologischer Reihenversuche' Naunyn-schmiedeber's Archiv fur Experimentelle Pathologie und Pharmakologie 162, 480-483.

Miura, Y., Goto, Y., Kubo, M. and Kono, Y. (1988a) 'Isolation of Chuzan virus, a new member of the Palyam subgroup of the genus *Orbivirus*, from cattle and *Culicoides oxystoma* in Japan' Am. J. Vet. Res. 49, 2022-2025.

Miura, Y., Goto, Y., Kubo, M. and Kono, Y. (1988b) 'Pathogenicity of Chuzan virus, a new member of the Palyam subgroup of genus *Orbivirus* for cattle' Jpn. J. Vet. Sci. 50, 632-637.

Murakami, H., Masui, H., Sato, G. H., Sueoka, N., Chow, T. P. and Kano-Sueoka, T. (1982) 'Growth of hybridoma cells in serum-free medium: ethanolamine is an essential component' Proc. Natl. Acad. Sci. U.S.A. 79, 1158-1162.

Omori, T., Inaba, Y., Morimoto, T., Tanaka, Y., Kono, M., Kurogi, H. and Matumoto, M. (1969) 'Ibaraki virus, an agent of epizootic disease of cattle resembling bluetongue II. Isolation of the virus in bovine cell culture' Japan. J. Microbiol. 13, 159-168.

Reed, L. J. and Muench, H. (1938) 'A simple method of estimating fifty per cent endpoints' Am. J. Hyg. 27, 493-497.

Tanaka, Y., Inaba, Y., Sato, K., Ito, H., Omori T. and Matumoto, M. (1969) 'Bovine epizootic fever. II. Physicochemical properties of the virus' Japan. J. Microbiol. 13, 169-176.

Towbin, H., Staehelin, T. and Gordon, J. (1979) 'Electrophoretic transfer of proteins from polyacrylamide gels to nitrocellulose sheets: Procedure and some applications' Proc. Natl. Acad. Sci. U.S.A. 76, 4350-4354.

Woodward, M. P., Young, W. W. Jr and Bloodgood, R. A. (1985) 'Detection of monoclonal antibodies specific for carbohydrate epitopes using periodate oxidation' J. Immunol. Methods. 78, 143-153.

Interaction between human lactoferrin and B lymphocytes.

Kaoru Sato, Hiroshi Shinmoto, Morimasa Tanimoto, Yoshihiro Kawasaki, Shun'ichi Dosako, and Shin'ichi Taneya
Technical Research Institute, Snow Brand Milk Products Co., Ltd., 1-1-2 Minamidai , Kawagoe 350, Japan

Abstract
Our previous studies demonstrated that B lymphocytes incorporated human lactoferrin (LF) and then released it to an extracellular medium. We did competition experiments for better understanding of the interaction between LF and B lymphocytes, using a reverse hemolytic plaque forming assay.   Competition between human and bovine LF reduced the number of plaque forming cells (PFC), but human transferrin (TF) showed no inhibitory effect on the PFC counts. These suggested that the receptor on B lymphocytes to human LF did not share with TF but bovine LF. The competition between a synthetic N-terminal nonapeptide of human LF (Gly-Arg-Arg-Arg-Arg-Ser-Val-Gln-Trp) and human LF demonstrated reduced number of PFC dose-dependently. Mild tryptic human LF also reduced the human LF-PFC counts, although removal of the C-terminal region did not affect the PFC counts. From these results, we concluded that highly cationic areas of human LF, such as N-terminal, accounted for the interaction with B lymphocytes.

Introduction
Lactoferrin (LF) is an iron-binding glycoprotein that is found widely in external secretions such as milk, saliva, tears, sperm, and in neutrophil granules. One of the major roles of LF is effects on cellular functions such as suppressing the production of colony-stimulating activity (CSA)(1),   growth stimulation for certain cell lines in serum-free medium (2,3) and enhancing iron absorption in the intestinal tract (4). These are mediated by the interaction between LF and cells; many reports have shown the binding of LF to certain cells (5) and the presence of LF receptor on the cells (6). We have also demonstrated that B lymphocytes incorporate LF and then release it to the extracellular medium (7). Because this receptor-mediated endocytosis is very similar to that of transferrin (TF), we hypothesized that human LF (hLF) and human TF (hTF) shared the same receptor. If this is true, the cellular functions of LF within the cell may be similar to those of TF. The interaction between LF and cells is thus of particular interest to understand the role of LF within a cell. According to Rochard et al.(8), recognition sites of the LF molecule by lymphocyte

R. Sasaki and K. Ikura (eds.), Animal Cell Culture and Production of Biologicals, 367–372.

receptors are located in the N-terminal domain 1 (residues 4-90 and/or 258-281). Since LF is a basic protein and binds to heparin (9) and DNA (10) electrostatically through its cationic residues, we focused our attention on the highly cationic areas of human LF, such as the N-terminal, to elucidate the interaction with B lymphocytes.

Materials and methods
1. Sample preparation
hLF or bovine LF (bLF) were isolated from defatted human or bovine milk by affinity chromatography using Affigel-10 (Bio-Rad) immobilized with mouse anti-hLF or anti-bLF monoclonal antibodies (11). hTF was purchased from Sigma. A nonapeptide (Gly-Arg-Arg-Arg-Arg-Ser-Val-Gln-Trp) was synthesized by the conventional solution method.

2. Enzymatic hydrolysis of hLF
hLF (1mg/ml) was hydrolyzed mildly with 10 U of trypsin at 37°C for 1 hr . The digestion was stoppted with trypsin inhibitor (Sigma). The hydrolysates were fractionated through affinity chromatography on heparin-Sepharose CL-6B (Pharmacia) and chromatofocusing on Mono Q (Pharmacia). The mild tryptic hLF showed single band on SDS-polyacrylamide gel electrophoresis (SDS-PAGE) and almost similar molecular weight to that of intact hLF (Fig.1).

Removal of C-terminal residues was done with carboxypeptidase Y (Sigma) at 37°C for 4 hr at the E/S ratio of 1:4 (w/w).

3. Reverse hemolytic plaque assay
Peripheral blood lymphocytes (PBL) from healthy donors were used in this assay without separating T lymphocytes, because hLF preferably interacts with B lymphocytes rather than T lymphocytes (7). PBLs were cultured with 100 µg/ml of hLF or bLF for four days and washed twice with Hank's medium. The number of cells secreting LF were measured with a reverse hemolytic plaque assay using *Staphylococcus aureus* protein A-coated SRBC and a polyclonal anti-hLF (Dako) or anti-bLF (prepared in our laboratory) serum by the method of Gronowicz et al. (12).

Results and Discussion
1. Receptor on B lymphocytes to hLF does not share with hTF but bLF.
It is not surprising that B lymphocytes incorporate hLF and re-secrete it into the extracellular medium because of its high homology in the primary structure with hTF(13) whose receptor-mediated recycling(14) is well understood. This reminded us to carry out the competition assays between hLF and hTF or bLF to confirm the hypothesis that these proteins share the same receptor. However, the competition assays demonstrated that the hLF-receptor on B lymphocytes did not share with hTF but bLF (Fig.2). This suggested that hLF and bLF functioned similarly within B lymphocytes; contrarily their intracellular functions differed considerably from those of hTF.

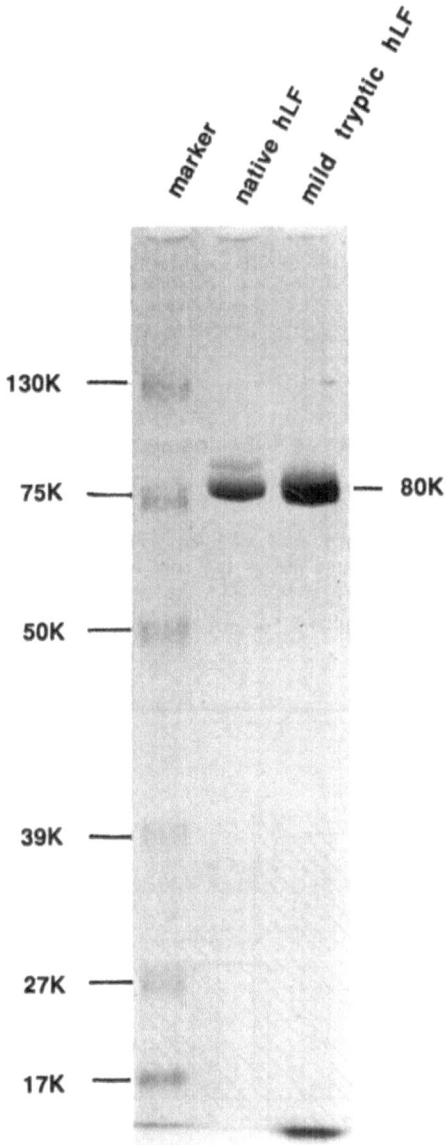

Fig.1　SDS-PAGE patterns of hLF and mild tryptic hLF

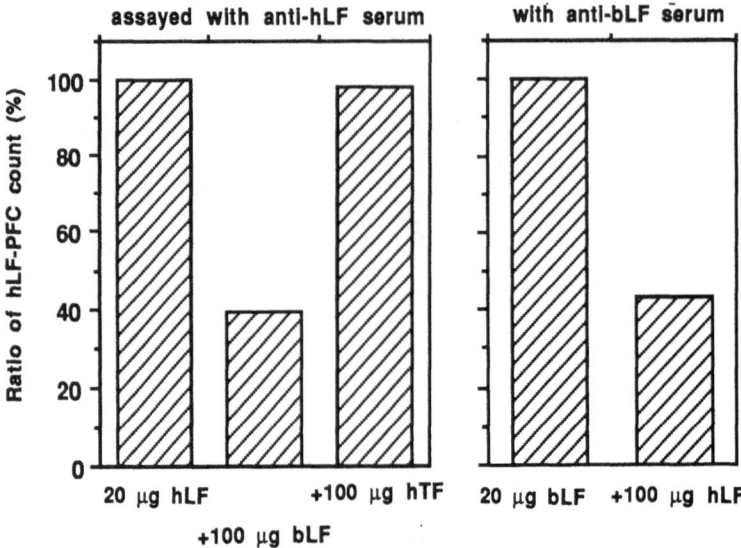

Fig.2 The competition between hLF and bLF or hTF
The cells were cultured with 20 μg hLF or bLF and 100 μg of either bLF, hTF, or hTF at 37 °C for 4 days. The re-secretion of hLF from the cells was assayed by the reverse hemolytic plaque forming assay with anti-hLF serum using protein A coated SRBC as an indicator.

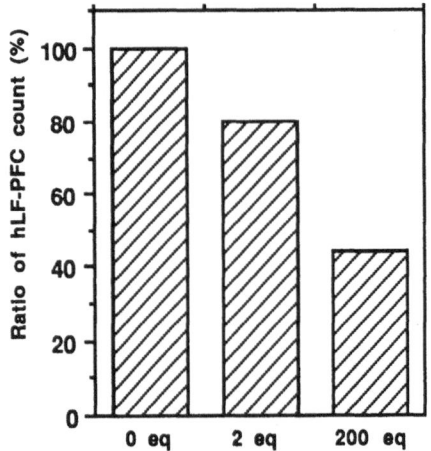

Equivalent of synthetic nonapeptide to 100 μg hLF

Fig.3 The competition between synthetic N-terminal nonapeptide of hLF and native hLF
The cells were cultured with 100 μg hLF and the nonapeptide (Gly-Arg-Arg-Arg-Arg-Ser-Val-Gln-Trp) at 37 °C for 4 days.

2. Highly cationic regions of hLF are responsible for the interaction with B lymphocyte.

Since hLF binds to heparin (9) and DNA (10) electrostatically through its highly cationic residues, it is possible that hLF interacts with its receptor on B lymphocyte electrostatically. Taking the highly positive residues at the N-terminal of hLF into account, the competition assay was done between hLF and the synthetic nonapeptide. As shown in Fig. 3, hLF-plaque forming cell (PFC) counts reduced dose-dependently with the addition of the synthetic peptide. Additionally removal of N-terminal peptide from hLF showed lowered hLF-PFC to about 30% of that of intact hLF (Fig.4). The results implied the positive residues at the N-terminal of hLF were responsible for the interaction with B lymphocytes. However, since the hLF-PFC of the mild tryptic hLF still remained to some extent, not only the N-terminal peptide but also other cationic regions seemed to be recognized by the receptor. Unlike the N-terminal, the C-terminal region was irrelevant to the interaction (Fig. 4). This is in agreement with Rochard et al.(8) who have indicated the importance of the N-terminal domain 1 (residues 4-90 and/or 258-281) as the recognition sites. However, it should be pointed out that the tripeptide (residues 1-3) plays a key role in the interaction, because the hydrolysis with trypsin mildly removes the tripeptide from the N-terminal of hLF (15). Although the intracellular function of hLF within B lymphocytes still remains unsolved, the receptor-mediated endocytosis of hLF and bLF would be promising to deliver certain substances of biologically importance into B lymphocytes, since we have developed a technology to isolate bLF from bovine milk in a large scale.

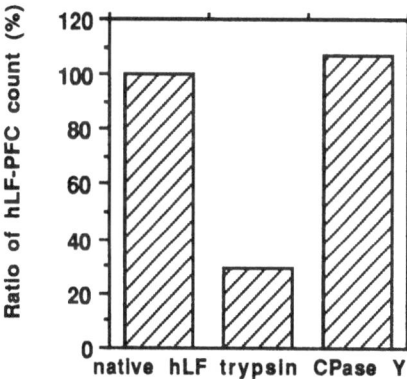

Fig.4 hLF-PFC counts of native hLF, hLF treated with trypsin and hLF treated with carboxypeptidase Y (CPase Y).

Ten U of trypsin were added to 1 mg/ml of hLF and indicated at 37 °C for 1 hr. Then trypsin inhibitor was added to stop the digestion. The mild tryptic hLF was purified with affinity chromatography on heparin-Sepharose CL-6B and chormatofocusing on Mono P.

One mg/ml solution of hLF in PBS was digested with carboxypeptidase Y (enzyme/substrate ratio; 1: 40 by weight) at 37 °C for 4 hr.

References

(1) Bagby, G. C., McCall, Jr. E. and Layman, D. L. (1989)' Regulation of colony-stimulating activity production.' J. Clin. Invest. 71, 340-344.

(2) Hashizume, S., Kuroda, K. and Murakami, H. (1983) ' Identification of lactoferrin as an essential growth factor for human lymphocytic cell lines in serum-free medium.' Biochim. Biophys. Acta, 763, 377-382.

(3) Azuma, N., Mori, N., Kaminogawa, S. and Yamauchi, K. (1989) ' Stimulatory effect of human lactoferrin on DNA synthesis in BALB/c 3T3 cells. ' Agric. Biol. Chem., 53, 31-35.

(4) Kawakami, H., Hiratsuka, M. and Dosako S. (1988) ' Effect of iron-saturated lactoferrin on iron absorption.' Agric. Biol. Chem., 52, 903-908.

(5) Bennett, R. and Davis, J. (1981) ' Lactoferrin binding to human peripheral blood cells: an interaction with a B-enriched population of lymphocytes and a subpopulation of adherent mononuclear cells.' J. Immunol., 127, 1211-1216.

(6) Mazurier, J., Legrand, D., Hu, W.-L., Montreuil, J. and Spik, G. (1989) ' Expression of human lactotransferrin receptors in phytohemagglutinin-stimulated human peripheral blood lymphocytes.' Eur. J. Biochem., 179, 481-487.

(7) Sato, K., Shinmoto, H., Tanimoto, M., Dosako, S. and Nakajima, I. (1990) ' Uptake and re-secretion of human lactoferrin by B lymphocytes.' Agric. Biol. Chem., 54, 1275-1279.

(8) Rochard, E., Legrand, D., Mazurier, J., Montreuil, J. and Spik, G. (1989) ' The N-terminal domain 1 of human lactotransferrin binds specifically to phytohemagglutinin-stimulated peripheral blood human lymphocyte receptors.' FEBS Lett., 255, 201-204.

(9) Finkelstein, M. B. and Finkelstein, R. A. (1982) ' Sequential purification of lactoferrin, lysozyme and secretory immunoglobulin A from human milk.' FEBS Lett., 144, 1-5.

(10) Bennett, R., Davis, J., Campbell, S. and Portnoff, S. (1983) ' Lactoferrin binds to cell membrane DNA.' J. Clin. Invest., 71, 611-618.

(11) Kawakami, H., Shinmoto, H., Dosako, S. and Sogo, Y. (1987) ' One-step isolation of lactoferrin using immobilized monoclonal antibodies.' J. Dairy Sci., 70, 752-759.

(12) Gronowicz, E., Coutinho, A. and Melchers, F. (1976) ' A plaque assay for all cells secreting Ig of a given type or class.' Eur. J. Immunol., 6, 588-590.

(13) Metz-Boutigue, M.-H., Jolles, J., Mazurier, J., Schoentgen, F., Legrand, D., Spik, G., Montreuil, J. and Jolles, P. (1984) ' Human lactotransferrin: amino acid sequence and structural comparisons with other transferrins.' Eur. J. Biochem., 145, 659-676.

(14) Hunt, R. C. and Carlson, L. M. (1986) ' Internalization and recycling of transferrin and its receptor.' J. Biol. Chem., 261, 3681-3686.

(15) Legrand, D., Mazurier, J., Metz-Boutigue, M.-H., Jolles, J., Jolles, P., Montreuil, J. and Spik, G. (1984) ' Characterization and localization of an iron-binding 18-kDa glycopeptide isolated from the N-terminal half of human lactotransferrin.' Biochim. Biophys. Acta, 787, 90-96.

COMPARISON OF CYTOTOXICITY CAUSED BY DIFFERENT HYDROPEROXIDES

Tsutomu Nakayama, Kenzo Hori, Kaoru Terazawa, and Shunro Kawakishi. Department of Food Science and Technology, Faculty of Agriculture, Nagoya University, Chikusa-ku, Nagoya 464-01, Japan

ABSTRACT. We have established a simple method for comparing the cyto-toxicity of hydrogen peroxide (H_2O_2), tertiary butyl hydroperoxide (t-BHP), and methyl linoleate hydroperoxide (MLHP) on V79 cells, using a colony-formation assay. In all cases L-buthionine-(S,R)-sulfoximine enhanced the cytotoxicity, and Quin 2 inhibited it. Nordihydroguaiaretic acid (NDGA) and o-phenanthroline suppressed the cytotoxicity of H_2O_2 and t-BHP, but they had no effect on the cytotoxicity of MLHP. These results suggest that the biological effects of t-BHP are similar to those of H_2O_2 and not to those of lipid hydroperoxides. In the course of the experiments, we found that this assay system is useful for screening of antioxidants that prevent cytotoxic effects of these hydroperoxides.

INTRODUCTION

Cytotoxic effects of free radicals and active oxygen species, i.e. H_2O_2, superoxide, hydroxyl radical, singlet oxygen, lipid hydro-peroxides and their secondary degradation products, have attracted the interest of many researchers during the last decade. These active species are formed in cells, foods, and cigarette smoke and have been supposed to cause many human diseases, e.g. ischemia, atherosclerosis, aging, and cancer. Since these species are very unstable, researchers in this field often use hydroperoxides as precursors of these free radicals. There are many reports describing the effects of hydroperoxides on mammalian cells or their components. Various parameters, e.g. viability, DNA lesions, mutation, or chromosomal aberrations have been measured in cells treated with hydrogen peroxide (H_2O_2), tertiary butyl hydroperoxide (t-BHP), or lipid hydroperoxide [1-6]. Since there is no report comparing the cytotoxicity of these hydroperoxides under the same conditions, we tried to establish a simple and reliable system, using a colony-formation assay, one of the most reliable methods of assessing cytotoxic effects [7].

R. Sasaki and K. Ikura (eds.), Animal Cell Culture and Production of Biologicals, 373–380.

MATERIALS AND METHODS

Reagents

Methyl linoleate hydroperoxide (MLHP) was prepared by allowing methyl linoleate to stand in an incubator at 40°C for seven days. Crude MLHP was obtained by silica column chromatography. It was further purified by high performance liquid chromatography immediately before use to avoid formation of secondary degradation products. Methyl linoleate hydroxide (MLH) was prepared by reduction of MLHP with sodium borohydride and purified by thin layer chromatography. The structure of MLH was identified by mass spectrometry. MLHP and MLH, which consist of 4 stereoisomers, were sonicated in 1% Tween 20 and filtered through a cellulose nitrate filter (pore size: 0.2 μm). Then their concentrations in the sterilized solutions were measured by absorbance at 233 nm. TMNDGA (tetramethoxy NDGA) was prepared by methylation of NDGA (nordihydroguaiaretic acid, which was purchased from Sigma) with diazomethane and identified by mass spectrometry and NMR.

Cell culture

We adopted V79 cells from Chinese hamster lung fibroblasts, because this cell line has been often used in tests of cytotoxicity, mutagenicity, and clastogenicity. The cells were grown in minimum essential medium (MEM) supplemented with 10% heat-inactivated fetal bovine serum (FBS).

Standard colony formation assay

The cells were seeded in 200/60-mm petri dishes and incubated in MEM supplemented with 10% FBS, which is necessary for the cells to attach to the bottom of dishes. After 4 hours, the cells were washed with HEPES buffered saline (HBS) and incubated in HBS with each hydroperoxide. (We added each antioxidant and hydroperoxide to MEM without serum and HBS without serum, respectively, because the serum probably adsorbs to or reacts with each reagent. We also incubated each hydroperoxide and antioxidant at separate times to avoid direct reaction of the two compounds. Consequently, we were able to assess the effects of antioxidant taken in the cells on the cytotoxicity of hydroperoxides.) After culture in MEM supplemented with 10% FBS for 5 days, the colonies were fixed with methanol and stained with Giemsa's solution. The number of colonies consisting of more than 50 cells was counted. The relative survival fraction was calculated by dividing the number of colonies on treated petri dishes by the number of colonies on untreated petri dishes. Results are expressed as the means and S.D. of four separately treated cultures unless specified otherwise. The data were analyzed statistically using Student's t-test, comparing control vs. treatment groups.

RESULTS AND DISCUSSION

Each hydroperoxide was added to HBS instead of MEM or MEM supplemented with 10% serum to avoid direct reaction of each hydroperoxide with components of the medium or the serum. Since incubation of the cells in HBS for more than 90 min had cytotoxic effects (the relative survival was 98% for 60 min of incubation and 76% for 90 of min incubation), the cells were incubated with relatively concentrated hydroperoxides for a short time. Figure 1 shows that the toxicity of each hydroperoxide similarly increased with incubation time up to 45 min. This enables us to compare the cytotoxicity at the same incubation time. Figure 2 shows typical dose effect curves of the relative survivals. Although it was difficult to keep a constant dose dependency in each experiment, MLHP was always most toxic, followed by H_2O_2 and t-BHP. Therefore, the cytotoxicity of each hydroperoxide is independent of its lipophilicity, because MLHP is most lipophilic, followed by t-BHP and H_2O_2. The hydroperoxide structure seems essential for cytotoxicity, since MLH and t-butyl alcohol were much less toxic than the corresponding hydroperoxides. Considering these results, we fixed the incubation time at 30 min and arranged the concentration of each hydroperoxide to keep the relative survivals at 20-40% in the following experiments; 2-10 μM of MLHP, 100-300 μM of t-BHP, and 30-50 μM of H_2O_2 were added to HBS to adjust their toxic potency. When confluent cells were first treated with each hydroperoxide and 200 cells were seeded for the colony formation assay, the sequence of the toxic potency of three hydroperoxides was the same as in the experiments described above (figure 3), but more concentrated solutions were necessary to give similar survival fractions to those of figure 2. We did not use this assay protocol in the following experiments, because the data might be less reproducible owing to the variation of numbers of the seeded cells.

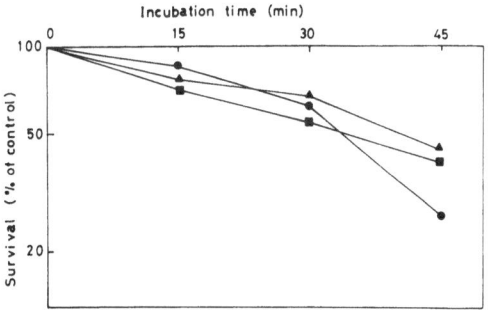

Figure 1
Time dependency of the survival of cells treated with different hydroperoxides. V79 cells were treated with hydroperoxides in HBS at 37°C. Each concentration was adjusted to give similar cytotoxic potency. The results are expressed as the mean of triplicate experiments. (●) 7 uM MLHP, (▲) 30 uM H_2O_2, (■) 300 uM t-BHP.

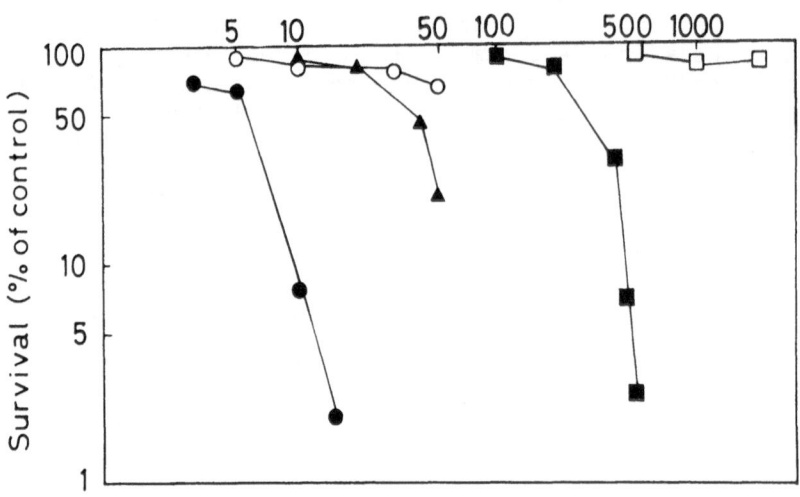

Figure 2
Typical dose effect curves of the relative survivals. V79 cells were treated
with the hydroperoxides and the corresponding hydroxides in HBS for 30 min at
37°C. The number of colonies formed after 5 days was counted. The results
are expressed as the mean of triplicate experiments. (●) MLHP,(○) MLH,
(▲) H_2O_2, (■) t-BHP, (□) tertiary butyl alcohol.

Effects of BSO, Quin-2-AM, o-phenanthroline, and NDGA

BSO is an inhibitor of glutathione synthesis in cells and it is
already known that BSO enhances the cytotoxicity of t-BHP[2]. We
added BSO to the medium after seeding and also to HBS at the time
of treatment with hydroperoxide. After treatment with hydroperoxide, the cells were incubated in the medium without BSO. Figure
4(a) shows that BSO enhanced the cytotoxic effects of all types of
hydroperoxides used in our experiment. This suggests that intracellular glutathione reduced these hydroperoxides in the presence
of glutathione peroxidase, consistent with the view that the
hydroperoxide structure is essential to the cytotoxicity.

When the lipophilic Quin-2-AM is added to cells, it diffuses
across the plasma membrane and is hydrolyzed to Quin-2 by
cytosolic esterases. Then, Quin-2 tightly binds Ca^{2+} in the cytosol [8]. It is reported that Quin 2 reduces the cellular toxicity
caused by H_2O_2 [9]. We added 25 μl of Quin-2-AM dimethylsulfoxide
solution to HBS (final 10 nM) 5 min before treatment with hydroperoxide. The effects of Quin-2-AM on the relative survival fraction were quite similar (Fig. 4(b)). Since Quin-2-AM reduced the
toxic effect of all three hydroperoxides, Ca^{2+} seems an important
factor in cytotoxicity induced by any hydroperoxides.

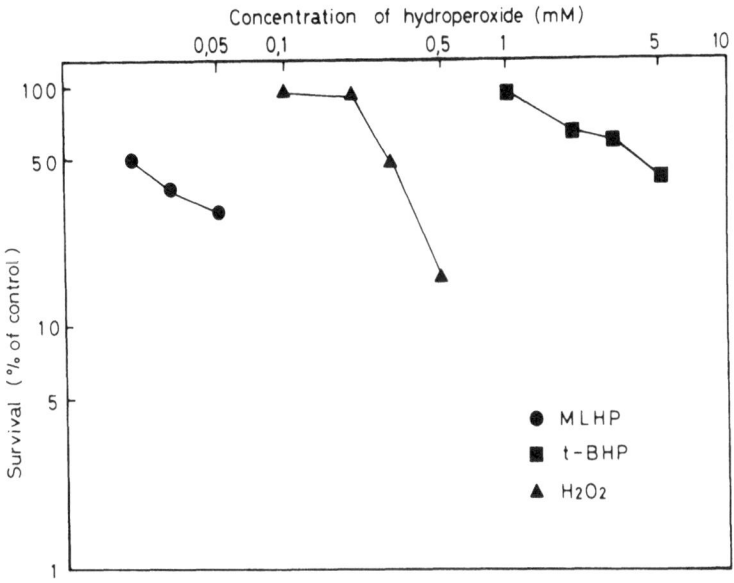

Figure 3

Dose effect curves of the relative survivals, when cells
were first treated with hydroperoxide, then trypsinized and
seeded for colony formation assay. The results are expressed
in the same manner as in figure 2.

We added o-phenanthroline (final 25 μM), a potent iron chelator
rendering the metal incapable of generating hydroxyl radicals [1],
to HBS and incubated for 10 min before treatment with hydroperoxides. Figure 4 (c) shows that o-phenanthroline inhibited the
cytotoxicity caused by H_2O_2 and t-BHP, confirming previous findings
[2,10] On the other hand, no inhibitory effect of o-phenanthroline
on the cytotoxicity by MLHP was observed, suggesting that iron in
the cells is not necessary for the toxic mechanism of MLHP.
    NDGA, a lipid-soluble antioxidant, is known to trap free
radicals and inhibit lipoxygenases from a wide variety of sources
[11]. In our experiments we avoided direct reaction of NDGA with
each hydroperoxide added to HBS and the lipid fraction of serum.
We seeded 200 cells and incubated them in MEM supplemented with
10% FBS for 2 hours. After changing the medium to MEM free of FBS,
we added 25 μl of 1 mM NDGA ethanol solution (final 5 μM) and
incubated the cells for 4 hours. After washing the cells with HBS,
the cells were treated with each hydroperoxide. Since NDGA had
potent inhibitory effects on the cytotoxicity caused by t-BHP and
H_2O_2, we changed the concentration of these hydroperoxides so that
the survival of the cell without treatment with NDGA became nearly

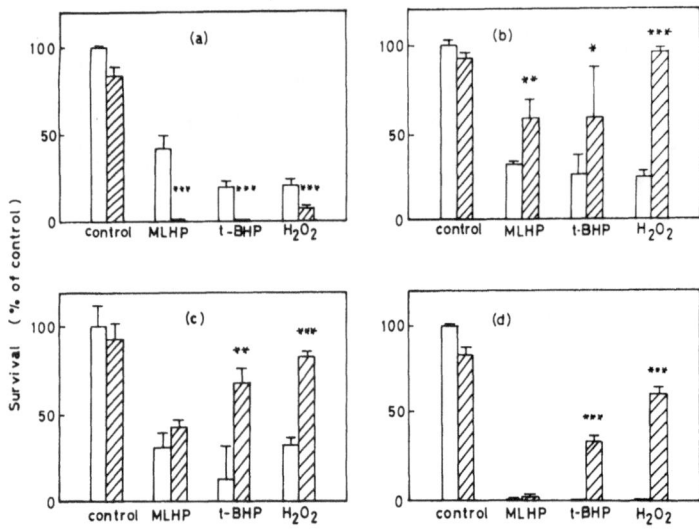

Figure 4
Effects of BSO, Quin-2, o-phenanthroline, and NDGA on the cytotoxicity of hydroperoxides. V79 cells were exposed to hydroperoxide without (□) or with (▨) each reagent for 30 min at 37 C: (a) 50 μM of BSO, (b) 10 nM of Quin-2-AM, (c) 25 μM of o-phenanthroline. In experiment (d),V79 cells pretreated with (▨) or without (□) 5 μM of NDGA were exposed to hydroperoxide for 30 min at 37°C. The results are expressed as the mean ± S.D. of four separately treated cultures. * P<0.05, ** P<0.01, ***P<0.001.

0%. Even in this concentration, NDGA also showed potent inhibitory effects (Fig. 4(d)). On the other hand, the cytotoxicity of MLHP was not reduced. When the survival fraction of the cells treated with MLHP was adjusted to 30-40%, no inhibitory effect of NDGA was observed (data not shown). Since NDGA prevented the toxic effects of H_2O_2, we tried to clarify the structure-activity relationship. As shown in figure 5, incubation with 50 μM H_2O_2 made the survival fraction nearly 0%. Under these severe conditions, pretreatment of the cells with NDGA inhibited the cytotoxicity of H_2O_2, but TMNDGA had no inhibitory effects. Therefore, the ortho-dihydroxy on the benzene ring (catechol structure) of NDGA seems necessary for the inhibitory effects.

Considering the effects of the chelators and the inhibitors, the cytotoxic mechanism of t-BHP seems similar to that of H_2O_2. On the other hand, the absence of an effect of NDGA or of o-phenanthroline on the cytotoxicity of MLHP suggests that the cytotoxic effects of lipid hydroperoxides are different from those of t-BHP. Previously, there was no report on any antioxidants that protected mammalian cells from cytotoxic effects of hydroperoxides. Using our system, we found that tocopherol, BHT, ascorbic acid, catechol, catechin, and ferulic acid did not have any protective effects on the cytotoxicity of the three hydroperoxides (data not shown).

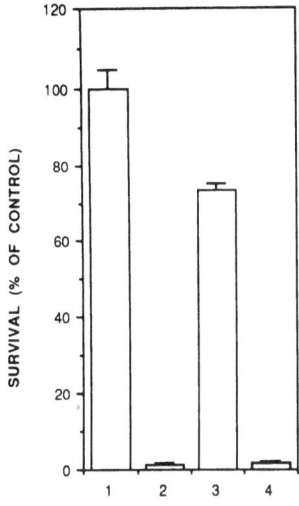

Figure 5
Effects of NDGA and TMNDGA on the cytotoxicity of H_2O_2. V79 cells were pre-treated with 5 μM of NDGA or TMNDGA in MEM without serum for four hours. After the medium was changed to HBS, cells were exposed 50 μM of H_2O_2 for 30 min at 37°C. 1. control, 2. 50 μM of H_2O_2, 3. 5 μM of NDGA + 50 μM of H_2O_2, 4. 5 μM of TMNDGA + 50 μM of H_2O_2. The results are expressed as the mean of quadruplicate experiments.

But since a potent inhibitory action of NDGA on the cytotoxicity of H_2O_2 was found for the first time, our assay system will be useful for screening other antioxidants having similar effects.

REFERENCES

1. Meneghini, R. (1988) 'Genotoxicity of active oxygen species in mammalian cells', Mutation Research, **195**, 215-230.
2. Ochi, T., Miyaura, S. (1989) 'Cytotoxicity of an organic hydroperoxide and cellular antioxidant defense system against hydroperoxides in cultured mammalian cells', Toxicology, **55**, 69-82.
3. Nakayama, T., Kaneko, M., Kodama, M. (1986) 'Detection of DNA damage in cultured human fibroblasts induced by methyl linoleate hydroperoxide', Agricultural Biological Chemistry, **50**, 261-262.
4. Nakayama, T., Kaneko, M., Kodama, M., Nagata, C. (1985) 'Cigarette smoke induces DNA single-strand breaks in human cells', Nature, **314**, 462-464.
5. Masaki, N., Kyle, M.E., Serroni, A., Farber, J.L. (1989) 'Mitochondrial damage as a mechanism of cell injury in the killing of cultured hepatocytes by tert-butyl hydroperoxide', Archives of Biochemistry and Biophysics, **270**, 672-680.
6. Kaneko, T. Honda, S., Nakano, S, Matsuo, M. (1987) 'Lethal effects of a linoleic acid hydroperoxide and its autoxidation products, unsaturated aliphatic aldehydes, on human diploid fibroblasts', Chemico-Biological Interactions, **63**, 127-137.
7. Cook, J.A., Mitchell, J.B. (1989) 'Viability measurements in mammalian cell systems', Analytical Biochemistry, **179**, 1-7.
8. Tsien, R.Y, Pozzan, T., Rink, T.J. (1982) 'Calcium homeostasis in intact lymphocytes: cytoplasmic free calcium monitored with a new, intracellularly trapped fluorescent indicator', The Journal of Cell Biology, **94**, 325-334.
9. Cantoni, O.Sestili, P., Cattabeni, F., Bellomo, G., Pou, S., Cohen, M., Cerutti, P. (1989) 'Calcium chelator Quin 2 prevents hydrogen-peroxide-induced DNA breakage and cytotoxicity', European Journal of Biochemistry, **182**, 209-212.
10. Mello Filho, A.C., Hoffmann, M.E., Meneghini, R.(1984) 'Cell killing and DNA damage by hydrogen peroxide are mediated by intracellular iron', The Biochemical Journal, **218**, 273-275.
11. Kemal, C., Louis-Flamberg, P., Krupinski-Olsen, R., Shorter, A.L.(1987) 'Reductive inactivation of soybean lipoxygenase 1 by catechols: A possible mechanism for regulation of lipoxygenase activity', Biochemistry, **26**, 7064-7072.

EFFECTS OF SPINACH ON GROWTH OF HUMAN-DERIVED NORMAL AND CANCER CELLS

Z-L. Kong,(1) H. Murakami,(2) and K. Shinohara(3)
(1) Institute of Food Chemistry, Faculty of Agriculture, Kyushu University, Hakozaki, Fukuoka 812, Japan
(2) Graduate School of Genetic Resources Technology, Faculty of Agriculture, Kyushu University, Hakozaki, Fukuoka 812, Japan
(3) National Food Research Institute, The Ministry of Agriculture, Forestry, and Fisheries, Kannondai, Tsukuba, Ibaraki 305, Japan

Cultured animal cells are useful for evaluation of physiological functions of food components at the cellular level. Our previous studies have demonstrated that algal phycocyanins had a growth-promoting activity on some serum-free cultured human-derived cell lines (1). These suggest that the components of plants such as vegetables and fruits would also affect the growth of animal cells. We have found that extracts of vegetables and fruits had growth-promoting and -inhibiting effects on some human-derived normal and cancer cells (2). In this study, isolation of active principles of an aqueous dialyzate of spinach was done and the effects of each fraction on the growth of human-derived normal and cancer cell lines were examined.

Materials and Methods

Cell lines and medium
The human-derived cell lines used in this study were from a histiocytic lymphoma (U-937), hybridomas (HB4C5 and SI102), a normal bladder fetus (FHs738Bl), a normal colon fibroblast (CCD-18Co), a normal whole embryo (FHs173We), a normal embryonic lung (WI-38), a breast adenocarcinoma (MCF-7), a epidermoid (A431), a differentiated hepatoma (HuH-7), a lung adenocarcinoma (PC-8), a lung squamous carcinoma (QG-56), a lung anaplastic carcinoma (QG-90), a bladder cancer cell (KU-1), a stomach adenocarcinoma (MKN-28) and a melanoma (Bowes). These cell lines were obtained from the American Type Tissue Collection and the Japanese Cancer Research Resources Bank (Tokyo). U-937, HB4C5, and SI102 cells were usually maintained in enriched RDF (eRDF, Kyokuto Pharmaceutical Kogyo Co. Tokyo, Japan) supplemented with 1.7

381

R. Sasaki and K. Ikura (eds.), Animal Cell Culture and Production of Biologicals, 381–386.

μM insulin (I; Novo Pharmaceutical Co., U.S.A.), 0.4 μM iron-free human transferrin (T; Wako Pure Chemical Industries Ltd., Osaka, Japan), 20 μM ethanolamine (E; Wako Pure Chemical Industries Ltd., Osaka, Japan), 2.5 x 10^{-8} M selenite (S; Wako Pure Chemical Industries Ltd., Osaka, Japan) (eRDF-ITES medium). All other cell lines were maintained in eRDF medium containing 10% fetal calf serum (FCS; General Scientific Lab., U.S.A.).

Viability and growth-promoting assay

The effects of fractions of dialyzate on the viability of cells were estimated by an MTT assay (2). In the MTT assay, the formazan formation from MTT was measured in the cells incubated with the samples for 12 hr. The growth-promoting activity of samples toward U-937, HB4C5, and SI102 cells was estimated by counting the cells mwith a Sysmex microcell counter after culturing the cells with the samples at 37°C for 4 days in a humidified 5% CO2-95% air atmosphere. In the cases when adhesive cells were used, the viability was estimated by a dye-exclusion method after trypsinization with 0.1% trypsin containing 0.005% EDTA.

Gel filtration and DEAE-cellulose ion exchange chromatography of dialyzate

The dialyzate was filtered on Sephadex G-100 and G-25 gel columns (1.5 x 50 cm). Elution was done with distilled water and monitored by measuring the absorbancy at 280 nm. In a DEAE ion exchange chromatography, the elution was done with a gradient of 0.05-1.0 M NaCl and monitored by measuring the absorbancy at 280 nm.

Results

Effects of the gel-filtered fractions of dialyzate of spinach on the growth of SI102, HB4C5 and U-937 cells in eRDF-ITES medium

The aqueous dialyzate of spinach (50 mg/ml) was first separated with a Sephadex G-100 gel filtration. The dialyzate was mainly separated into two fractions of SPW1 and SPW2, as shown in Fig. 1.

The growth-promoting activity on HB4C5 cells was detected in the SPW2 fraction. The SPW2 fraction was gel-filtered into four fractions of SPW2-1, SPW2-2, SPW2-3, and SPW2-4 with a Sephadex G-25 column as shown in Fig. 2.

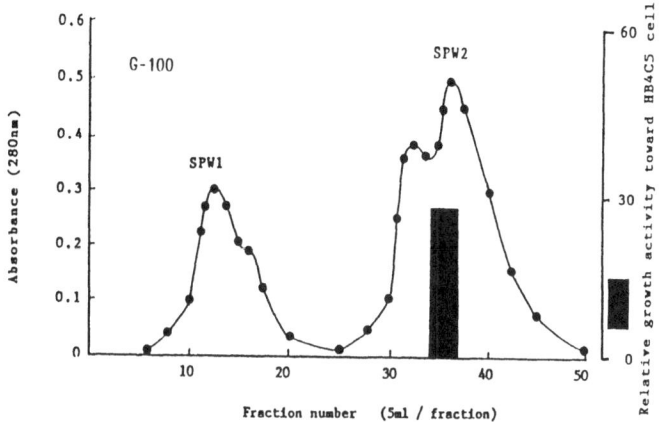

Fig. 1. Sephadex G-100 Gel-filtration of Aqueous Dialyzate of Spinach and Effect of the Fractions on Growth of HB4C5 Cells.

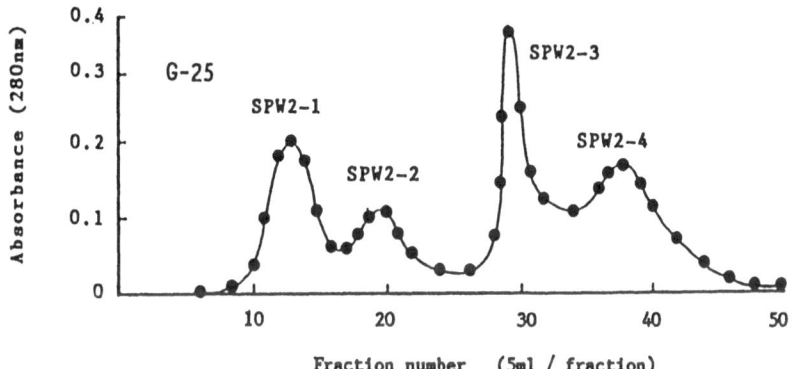

Fig. 2. Sephadex G-25 Gel-filtration of SPW2 Fraction.

A MTT assay of these fractions on SI102 cells indicated that the fractions SPW2, SPW2-1, SPW2-2, SPW2-3 and SPW2-4 gave the higher absorbancy at 550 nm at the low concentrations than the control did, suggesting that these fractions may have the growth-promoting activity on SI102 cells. The original dialyzate and SPW1 fraction reduced the viability of SI102 cells markedly.

The effects of Sephadex G-25 fractions on the growth of U-937, SI102 and HB4C5 cells in eRDF-ITES medium were shown in Fig..3 These fractions, especially, SPW2-1, SPW2-2, and SPW2-4, promoted the growth of the 3 kinds of cells. These accelerated the growth of U-937, SI102 and HB4C5 cells in eRDF-ITES medium by about 35-45%, although the optimal concentrations for the growth-promoting activity varied with the fractions. The growth-promoting activity of SPW2-2 was lower than that of the other fractions.

384

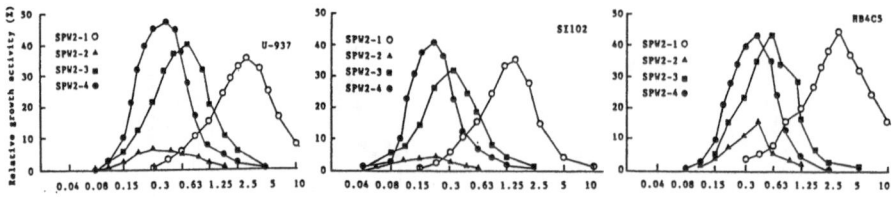

Concentration (ug/ml)

Fig. 3. Effects of the Gel-filtered Fractions of Dialyzate
of Spinach on Growth of U-937, SI102 and HB4C5 Cells.

Effect of Sephadex G-25 fractions on the growth of normal
and tumor cell lines

The effects of Sephadex G-25 fractions on the viability
of the breast adenocarcinoma cell line, MCF-7 were then examined.   Among the fractions, SPW2-3 and SPW2-4 evidently
reduced the viability of MCF-7 dose-dependently, while the
activity of SPW2-1 and SPW2-2 was low.   The effects of SPW2-
3 on the viability of various kinds of normal and cancer
cell lines was examined.   As shown in Table I, a tendency
for SPW2-3 to reduce the viability of cancer cell lines more
significantly than that of normal cell lines was observed.
Among the cancer cell lines, the viability of breast
adenocarcinoma (MCF-7), differentiated hepatoma (HuH-7),
lung carcinomas(PC-8, QG-56 and QG-90) and stomach adenocarcinoma (MKN-28) was more markedly inhibited by SPW2-3.

Table 1. Effect of SPW2-3 Fraction on the Growth of Several
Human-derived Cultured Cell Line

	Cell line	Viability (%)		
		Dose (µg/ml)		
		0	0.5	1	2
Normal cells	FHs738B1	100	100	98	97
	CCD-18Co	100	100	98	95
	FHs173We	100	100	94	85
	WI-38	100	100	92	82
Cancer cells	MCF-7	100	90	88	68
	A431	100	96	95	80
	HuH-7	100	86	75	58
	PC-8	100	93	86	71
	QG-56	100	95	92	77
	QG-90	100	90	84	67
	KU-1	100	98	97	89
	MKN-28	100	89	83	77
	Bowes	100	97	91	85

MCF-7 cells without treatment with SPW2-3 attached to the
dish and grew, while the MCF-7 cells treated with SPW2-3 for
24 hr were detached from a dish, leading to the death. This
phenomenon resembled the case when a antitumoric reagent,
actinomycin D(1 µg/ml), was used. The proliferation of normal cell lines such as FHs173We and CCD-18Co cells was not
affected by the presence of the fraction.

DEAE-cellulose ion chromatography of SPW2-3 fraction and effect of the fractions on the growth of HB4C5 and MCF-7 cells

A fraction of SPW2-3 was then fractionated by DEAE-
cellulose ion chromatography. The fractionation pattern,
the growth-promoting activity on HB4C5 cells and the inhibiting activity on the viability on MCF-7 cells are shown
in Fig. 4. Among the fractions, the high growth-promoting
activity on HB4C5 cells in eRDF-ITES medium was detected
in the fraction SPWD3, while the high inhibiting activity on

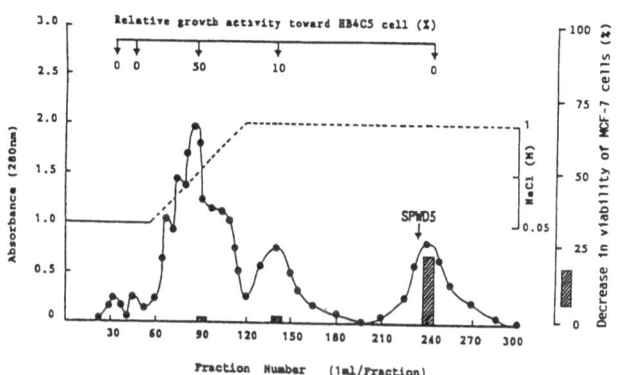

Fig. 4. DEAE-cellulose Chromatography of SPW2-3 Fraction and
Effects of the Fractions on the Growth of HB4C5 or MCF-7
Cells.

the viability on MCF-7 cells in eRDF medium containing 5%
FCS in the fraction SPWD5. A SDS-electrophoretic study of
the SPWD5 fraction suggested that an active principle might
be a glycoprotein having a molecular mass of 16 kdalton.

Effects of SPW2-3 and SPWD5 on the morphological alteration of U-937 cells

A human histiocytic lymphoma cell line, U-937, is known
to be susceptible to morphological alteration in the
presence of phorbol esters and vitamin D3 derivatives to
differentiate into macrophage-like cells which attach to the
culture dishes(3). The U-937 cells (1 x 10^6 cells/ml) were

cultured with SPW2-3 (1 µg/ml) or SPWD5 (0.5 µg/ml)
fraction in eRDF medium containing 10% FCS for 18 hr in CO_2
incubator, and the morphological alteration of U-937 cells
was examined by the viability test with dye exclusion and by
counting the cells attached to the dishes. As shown in
Table II, the fraction SPW2-3 and SPWD5 were found to cause
morphological alteration of U-937 cells and increase the
adhesion of U-937 cells. The adhesion percentage with SPWD5
was higher than that with SPW2-3, suggesting that the
specific activity on the adhesive effect of SPWD5 fraction
on U-937 cells was increased by about 8fold, compared with
the SPW2-3 fraction.

Table II. Morphological Alteration of U-937 cells with
SPW2-3 and SPWD5 Fractions

Treatment	Viability(%)	Adhesion(%)
Control (PBS)	98	0
SPW2-3 (1 µg/ml)	96	>5
SPWD5 (0.5 µg/ml)	95	>20

Discussion

This study has found that in the aqueous dialyzate of
spinach extracts, two different cell physiological activators existed; one having a growth-promoting activity on
hybridoma and lymphoma cells and another reducing the
viability of human cancer cells. A growth-inhibiting constituent was suggested to be a glycoprotein Mr of 16kdalton.
Partially purified fractions of the dialyzate also change
the morphology of a histiocytic lymphoma cell, U-937.

References

1)Shinohara, K., Okura, Y., Koyano, T., Murakami, H. and
  Omura, H. (1988)'Algal phycocyanins promote growth of
  human cells in culture', In Vitro Cellular & Developmental
  Biology, 24, 1057-1061.
2)Mosmann, T.(1983)'Rapid colorimetric assay for cellular
  growth and survival', J. Immunol. Methods, 65, 55-63.
3)Dodd, R. C., Cohn, M.S., Newman, S.L. and Gray, T.K.
  (1983)'Vitamin D metabolites change the phenotype of
  monoblastic U-937 cells', Proc. Natl. Acad. Sci. USA, 80,
  7538-7541.

ASSAY SYSTEMS USING CULTURED HEPATOCYTES FOR PHARMACOLOGICAL AND NUTRITIONAL ACTIONS OF AMINO ACIDS AND RELATED COMPOUNDS

K. YAGASAKI, K. ISHIHARA, N. MORISAKI, A. MIURA, AND R. FUNABIKI
Department of Applied Biological Science, Tokyo Noko University
Fuchu, Tokyo 183, Japan

ABSTRACT. Actions of amino acids and related compounds on lipid and protein metabolism were studied in two kinds of rat hepatocytes. In primary monolayer cultures of hepatocytes from rats bearing hepatomas, L-cysteine was found to stimulate bile acid production. A cystine-enriched diet reduced *in vivo* a hepatoma-induced hypercholesterolemia with an enhancement of fecal bile acid excretion. In an established cell line of RLC-16 hepatocytes, L-leucine and its ketoacid derivative were found to stimulate protein synthesis. From inhibitor studies, their stimulatory action on protein synthesis in RLC-16 cells was suggested to be mediated by protein kinase C activation. The cultured hepatocytes used here may be effective for screening the actions of amino acids, and for studying the modes of their actions.

1. Introduction

The liver is important in lipid and protein metabolism. The conversion of cholesterol to bile acids is the major pathway for removal of cholesterol from the body and takes place exclusively in the liver. The ability for bile acid synthesis from cholesterol in the liver is therefore thought to be one of factors inducing or improving hypercholesterolemia in various nutritional, hormonal, and pathological states [1]. Rats bearing hepatomas have been found to show hypercholesterolemia characterized by an enormous elevation of very-low-density lipoprotein + low-density lipoprotein (VLDL+LDL)-cholesterol (Ch) and a reduction of high-density lipoprotein (HDL)-Ch in the serum [2]. Fecal bile acid excretion is found to be reduced in hepatoma-bearing rats when compared to tumor-free (normal) rats [3] and this reduction to be improved by concomitant addition of methionine and glycine to a basal diet [3]. Primary monolayer cultures of adult rat hepatocytes have been shown to be suitable for studies of cholesterol and bile acid metabolism [4]. In this study, we investigated the effects of cyst(e)ine on bile acid synthesis from cholesterol in primary cultured hepatocytes and on fecal bile acid excretion using hepatoma-bearing rats, to examine a correlation between *in vitro* and *in vivo* studies.

R. Sasaki and K. Ikura (eds.), Animal Cell Culture and Production of Biologicals, 387–393.
© 1991 *Kluwer Academic Publishers.*

Leucine, one of the branched chain amino acids, has been reported to inhibit protein degradation in the liver [5,6] and muscles [7], and to stimulate protein synthesis in isolated rat muscles [7] and cultured myotubes [8]. The regulation of leucine catabolism in the liver is different from that in the muscles: leucine is catabolized rapidly in the muscles [7], while it is not oxidized rapidly in the liver because of low activity of transaminase [5,9]. In this study, we also studied the effects of leucine on protein synthesis using an established cell line of RLC-16 rat hepatocytes [10], to find if leucine stimulates protein synthesis in hepatocytes like muscle cells despite the presence of the disparity in leucine catabolism between the two cells.

2. Materials and Methods

2.1. MEASUREMENT OF BILE ACID SYNTHESIS IN PRIMARY CULTURED HEPATOCYTES

An ascites hepatoma line of AH109A cells (5×10^5 cells/rat, provided by SRL, Tokyo, Japan) were subcutaneously implanted in the back of male Donryu rats (5-6 weeks of age, Nippon Rat Co., Saitama, Japan) as described previously [2]. The rats were then fed on a semisynthetic diet containing 20% casein [3]. Hepatocytes were isolated from hepatoma-bearing rats around 2 weeks after the implantation, essentially by the method described by Seglen [11] using collagenase (Type S-1, Iwaki Glass, Tokyo, Japan). Hepatocytes (2×10^6 cells) were attached for 4 hr to collagen-coated Corning dishes (60 mm) in 3 ml of Williams' E medium (WE) containing 10% fetal calf serum (FCS, Hazleton, KS, U.S.A.). The hepatocytes were thereafter kept for 20 hr in serum-free medium [4] (WE supplemented with 1 μM corticosterone, 10 μM dexamethasone, and 1 μM testosterone). The cells then received 0.27 μCi of [4-^{14}C]cholesterol (57.5 mCi/mmol, New England Nuclear) mixed with normal rat serum (final concentration of 3.3%). The conversion of [^{14}C]cholesterol to bile acids was measured 48 hr later after extraction of bile acids from the medium and cells [3] and removal of contaminating [^{14}C]cholesterol from bile acid fraction using digitonin [12].

2.2. ESTIMATION OF FECAL BILE ACID EXCRETION

Male Donryu rats implanted with AH109A cells were fed for 2 weeks on either the 20% casein diet or the diet supplemented with 1.2% L-cystine. Tumor-free rats were fed on the 20% casein diet as the normal group. Feces were individually collected on days 12-14, and fecal neutral and acidic steroids were extracted and measured by enzymatic methods as described previously [3].

2.3. MEASUREMENT OF PROTEIN SYNTHESIS IN RLC-16 HEPATOCYTES

RLC-16 hepatocytes [10] were provided by the RIKEN Cell Bank (RCB069, Tsukuba, Japan). Stock cultures of RLC-16 cells were maintained in DM-160 medium supplemented with 10% FCS [10% FCS/DM-160]. The cells (1.2×10^5 cells/well) were subcultured into Nunc 24-place multiwell plates and grown for 7 days in 10% FCS/DM-160, and then kept for 12 hr in 0.2% FCS/ leucine-free DM-160. After the leucine depletion, hepatocytes were exposed to serum-free, experimental media for 30 min. The cells then received 0.5 μ Ci of L-[ring-3,5-^{3}H]tyrosine (50 Ci/mmol, American Radiolabeled Chemicals Inc.). The 1-hr incorporation of [^{3}H]tyrosine into total cellular protein was measured essentially the same method as described previously [13]. Results were expressed as dpm/hr/mg protein.

2.4. STATISTICAL METHOD

Statistical analysis was done using Student's t test, and a p value of < 0.05 was considered significant.

3. Results and Discussion

3.1. ACTIONS OF CYST(E)INE ON BILE ACID METABOLISM

The effects of L-cysteine on the conversion of [4-^{14}C]cholesterol to bile acids were examined in primary monolayer cultures of hepatocytes

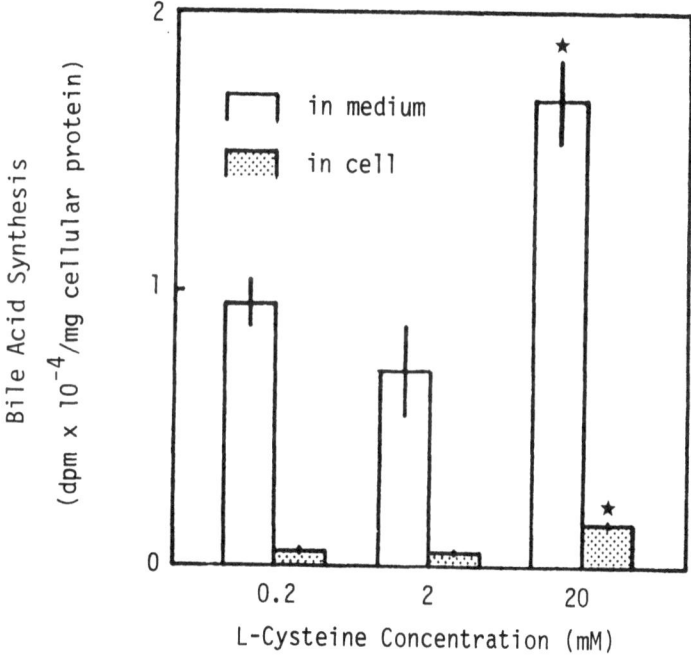

Figure 1. Effects of L-cysteine on the conversion of [4-^{14}C]cholesterol into bile acids in monolayer cultures of hepatocytes from hepatoma-bearing rats. Each value represents the mean ± standard error of three to four assays. ★Significantly different from the corresponding control (0.2 mM) groups at p < 0.05.

390

from hepatoma-bearing rats in the serum-free medium (Figure 1). Cysteine significantly stimulated bile acid production at a high concentration of 20 mM. More than 90% of bile acids synthesized by hepatocytes existed in the medium.

Since the stimulation of bile acid production in the liver is thought to be one of hypocholesterolemic mechanisms for nutrients and drugs, the effects of dietary supplemented L-cystine on fecal bile acid excretion were tested *in vivo* in hepatoma-bearing rats which have been found to show endogenous hypercholesterolemia [2]. As shown in Figure 2, fecal bile acid excretion tended to be lower in hepatoma-bearing rats than in normal rats (normal vs. control). Dietary supplemented cystine (1.2%) restored the hepatoma-induced decrease in fecal bile acid excretion to the normal level (control vs. cystine). Neutral steroids were also decreased by hepatoma bearing, but cystine failed to restore the decrease in neutral steroids. In hepatoma-bearing rats, an enormous elevation in (VLDL+LDL)-Ch and a notable decrease in HDL-Ch were observed when compared to normal rats, and both the hepatoma-induced elevation in (VLDL+LDL)-Ch and decrease in HDL-Ch were suppressed by dietary supplementation of 1.2% cystine (data not shown). From *in vitro* and *in vivo* studies, L-cyst(e)ine seems to stimulate bile acid synthesis via cholesterol 7α-hydroxylase activation [14].

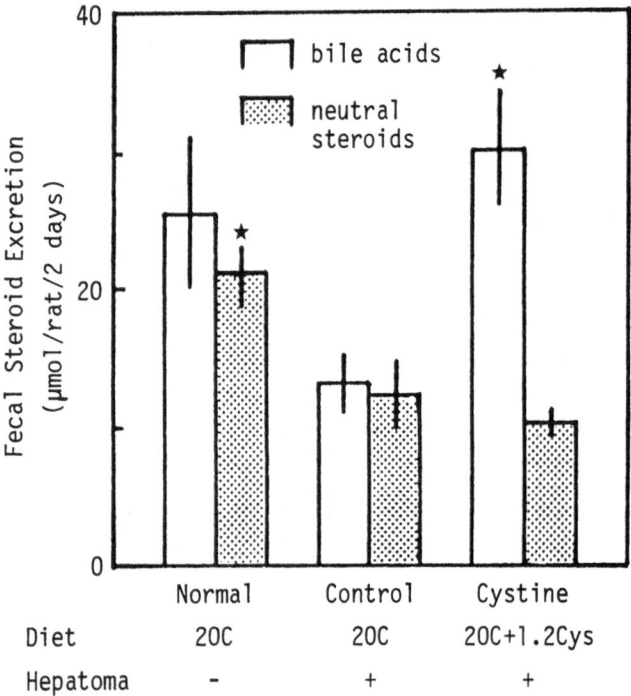

Figure 2. Effects of dietary addition of L-cystine on fecal steroid excretion in hepatoma-bearing rats. Each value represents the mean ± standard error of four to five rats. ★Significantly different from the corresponding control groups at $p < 0.05$.

These results indicate that primary cultured hepatocytes of hepatoma-bearing rats are useful for estimating *in vivo* effects of amino acids and probably other compounds on hypercholesterolemia in, at least, the hepatoma-bearing rats.

3.2. ACTIONS OF LEUCINE AND RELATED COMPOUNDS ON PROTEIN SYNTHESIS

Effects of L-leucine and its related compounds on protein synthesis were examined in RLC-16 hepatocytes. In preliminary experiments, L-leucine was found to stimulate protein synthesis at 1.5 mM. A deamination product of leucine, α-ketoisocaproic acid (KIC), also stimulated protein synthesis at 1.5 mM like L-leucine, while D-leucine (1.5 mM) showed no influence (data not shown). To examine the mode of stimulatory action of L-leucine, effects of known inhibitors of arachidonate and inositolphospholipid metabolism were studied. The results are summarized in Figure 3. The stimulatory action of L-leucine on protein synthesis was significantly interrupted by mepacrine (phospholipase A_2 and C inhibitor) and 1-0-hexadecyl-2-0-methylglycerol (diacylglycerol-dependent protein kinase C inhibitor) at concentrations where these inhibitors had no significant effect on basal protein synthesis, while indomethacin (cyclooxygenase inhibitor) and caffeic acid (lipoxygenase inhibitor) did not interrupt the stimulatory action

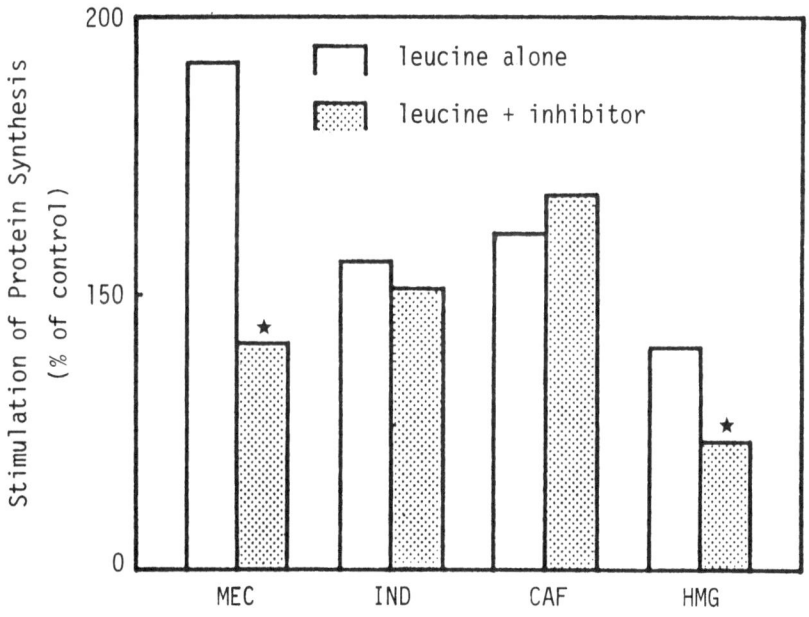

Figure 3. Influences of mepacrine (MEC), indomethacin (IND), caffeic acid (CAF), and 1-0-hexadecyl-2-0-methylglycerol (HMG) on L-leucine-stimulated protein synthesis in RLC-16 hepatocytes. Each group consisted of five to six assays. MEC, 20 μM; IND, 50 μM; CAF, 200 μM; HMG, 1000 μM. *Significantly different from the corresponding leucine groups at $p < 0.05$.

of L-leucine. Responses of KIC to the inhibitors were similar to those of L-leucine (data not shown). These results suggest a partial involvement of protein kinase C in the stimulatory actions of L-leucine and KIC.

In conclusion, cultured hepatocytes employed here may be effective for screening the actions of amino acids and related compounds on lipid and protein metabolism, and for studying their modes of actions.

4. References

[1] Yagasaki, K. (1989) 'Endogenous hypercholesterolemia and dietary amino acids: Sulfur amino acids and glycine', in M. Friedman (ed.), Absorption and Utilization of Amino Acids, Vol. II, CRC Press, Florida, pp. 275-287.

[2] Irikura, T., Takagi, K., Okada, K., and Yagasaki, K. (1985) 'Effect of KCD-232, a new hypolipidemic agent, on serum lipoprotein changes in hepatoma-bearing rats', Lipids, 20, 420-424.

[3] Yagasaki, K., Machida-Takehana, M., and Funabiki, R. (1990) 'Effects of dietary methionine and glycine on serum lipoprotein profiles and fecal sterol excretion in normal and hepatoma-bearing rats', J. Nutr. Sci. Vitaminol., 36, 45-54.

[4] Hylemon, P. B., Gurley, E. C., Kubaska, W. M., Whitehead, T. R., Guzelian, P. S., and Vlahcevic, Z. R. (1985) 'Suitability of primary monolayer cultures of adult rat hepatocytes for studies of cholesterol and bile acid metabolism', J. Biol. Chem., 260, 1015-1019.

[5] Pösö, A. R., Wert, J. J., Jr., and Mortimore, G. E. (1982) 'Multifunctional control by amino acids of deprivation-induced proteolysis in liver', J. Biol. Chem., 257, 12114-12120.

[6] Mortimore, G. E., Pösö, A. R., Kadowaki, M., and Wert, J. J., Jr. (1987) 'Multiphasic control of hepatic protein degradation by regulatory amino acids', J. Biol. Chem., 262, 16322-16327.

[7] Tischler, M. E., Desautels, M., and Goldberg, A. L. (1982) 'Does leucine, leucyl-tRNA, or some metabolite of leucine regulate protein synthesis and degradation in skeletal and cardiac muscle?', J. Biol. Chem., 257, 1613-1621.

[8] Yagasaki, K. (1990) 'Signaling in L-leucine-induced stimulatory action on protein synthesis of cultured muscle cells', Rep. Res. Comm. Essen. Amino Acids, 126, 1-5.

[9] Crabb, D. W., and Harris, R. A. (1978) 'Studies on the regulation of leucine catabolism in the liver', J. Biol. Chem., 253, 1481-1487.

[10] Takaoka, T., Yasumoto, S., and Katsuta, H. (1975) 'A simple method for the cultivation of rat liver cells', Japan. J. Exp. Med., 45, 317-326.

[11] Seglen, P. O. (1976) 'Preparation of isolated rat liver cells', Methods Cell Biol., 13, 29-83.

[12] Okada, K., Yagasaki, K., Mochizuki, T., Takagi, K., and Irikura, T. (1985) 'Effect of 4-(4'-chlorobenzyloxy)benzyl nicotinate (KCD-232) on cholesterol metabolism in rats', Biochem. Pharmacol., 34, 3361-

3367.

[13] Yagasaki, K., Shinonaga, M., and Funabiki, R. (1986) 'Effect of isopentenyladenine, a cytokinin, on proliferation and protein synthesis in cultured myoblasts', Agric. Biol. Chem., 50, 2791-2794.

[14] Stephan, Z. F., Lindsey, S., and Hayes, K. C. (1987) 'Taurine enhances low density lipoprotein binding', J. Biol. Chem., 262, 6069-6073.

PREPARATION AND CHARACTERISTICS OF MONOCLONAL ANTIBODIES REACTIVE TO SATSUMA DWARF VIRUS.

Keita Hirashima,[*1] Yasuhiro Noguchi,[*1] and Hiroki Murakami[*2]
[*1] The Branch of Fruit Tree Saplings, Fukuoka Agricultural Research Center, 17-3 Ishigaki, Tanushimaru-town, Ukiha-gun, Fukuoka, 839-12
[*2] Graduate School of Genetic Resources Technology, Faculty of Agriculture, Kyushu University 46-09, 6-10-1 Hakozaki, Higashi-ku, Fukuoka 812, Japan

Satsuma Dwarf Virus (SDV) is the first virus disease reported(3) on citrus in Japan. It is now distributed widely in almost all citrus-growing districts in Japan. This virus is a soil-transmissible plant pathogen(1). It is difficult to grow citrus trees on contaminated fields. The detection of the virus before planting is effective in preventing the spread of the disease. Enzyme-linked immunosorbent assay (ELISA) first came into used in 1980(2) for SDV detection. However, it is diffic ult to obtain reproducible results using polyclonal antibodies. We attem ped to produce monoclonal antibodies to SDV.

MATELIALS AND METHODS

VIRUSES AND IMMUNIZATION.
The virus was propagated in <u>Physalis floridana</u>, and purified from infected leaves. The purified SDV was electrophoresed on SDS-polyacrylamide gel to separate viral components. Purified SDV emulsified in Freund's complete adjuvant was injected into the peritoneum of mice on days 0, 15, and 30. Finally virus was administered intravenously 3 days before cell fusion.

PRODUCTION OF HYBRIDOMAS.
Immune spleen cells from immunized mice were fused with P3-X63-Ag8-U1 (P3U1) myeloma cells at a ratio of 10 : 1, in the presence of 50% polyethylene glycol 1,000 (Wako) in serum-free ERDF medium. Fusion products were suspended in HAT selective medium (ERDF supplemented with 10^{-4} M hypoxanthine, $4x10^{-7}$ M aminopterin, $1.6x10^{-5}$ M thymidine, and 15% fetal bovine serum) to a density of $5x10^5$ spleen cells/ml, and were inoculated into 96-well tissue culture plates (0.1 ml / well) and incubated in 5% CO^2 atmosphere in a humidified incubator at 37℃. Approximately 10 days after fusion, hybridomas were screened for antibody production by an indirect ELISA method. Hybridoma cells producing SDV-reactive MABs were cloned twice by limiting dilution. Cloned cells were grown in rat thymocytes-conditioned HT medium (HAT medium with aminopterin removed).

R. Sasaki and K. Ikura (eds.), Animal Cell Culture and Production of Biologicals, 395–399.

Infected leaves of <u>Physalis floridana</u>
├─ Homogenize in 0.1 M citrate buffer, pH 6.5,
│ containing 0.1% mercaptoacetic acid
├─ filter through two layers of cheesecloth
├─ Add Mg-bentonite suspension (about 40 mg/ml
│ bentonite, 1 ml/10 g tissues) and stir
├─ Centrifuge at 10,000 rpm for 10 min
Supernatant
├─ Add ammonium sulfate (25 g/100 ml supernatant)
├─ Centrifuge at 10,000 rpm for 15 min
Pellet
├─ Resuspend in 5 mM borate buffer, pH 8.6,
│ containing 1 mM EDTA
├─ Emulsify briskly with carbon tetrachloride
│ (1 ml/5 ml supernatant)
├─ Centrifuge at 3,000 rpm for 15 min
Aqueous phase
├─ Ultracentrifuge at 40,000 rpm for 90 min
Pellet
├─ Resuspend in 5 mM borate buffer, pH 8.6,
│ containing 1 mM EDTA
├─ Centrifuge at 10,000 rpm for 15 min
Supernatant
├─ Ultracentrifuge at 24,000 rpm for 120 min
│ in 10~40% sucrose density gradient
Viral fraction

Fig. 1. Procedure for purification of SDV

PRODUCTION AND PURIFICATION OF MABs.

After 4 days of cultivation of hybridomas in serum-free ITES-ERDF medium (containing 0.01 mg/ml insulin, 0.02 mg/ml transferrin, 0.02 mM ethanol amine, and 2.5×10^{-8} M sodium selenite), the culture supernatant was harvested and concentrated by salting out with 50% saturation ammonium sulfate. The concentrated culture supernatant were used for purification of MABs by hydroxyapatite column chromatography.

ASSAY OF MABs

The wells of microplates were incubated with purified SDV in the buffer (50 mM carbonate buffer, pH 9.6) overnight at 4°C and then blocked by incubation for 2 hr with 0.5% bovine serum albumin (BSA) in PBS. After washing with PBS containing 0.05% Tween 20 (PBST), hybridoma cultured supernatant was added and incubated for 1 hr at 37°C. After washed, plates were incubated for 1 hr at 37°C with peroxidase labeled affinity purified goat antibodies to mouse immunoglobulins (Zymed) in PBS. Plates were washed a final time and enzyme substrate (ABTS) was added and incubated for 30 min at 37°C. Reactions were stopped by adding 1.5% oxalic acid to each well.

REACTIVITY OF MABs
Viral components separated on SDS-polyacrylamide gel electrophoresis
were electrophoretically transferred to a nitrocellulose membrane, and
then blocked by incubation for 1 hr with PBS containing 3% BSA. After
being washed, purified MABs of cultured supernatant were added and
incubated for 1 hr at room temperature. After being washed, they were
incubated 1 hr with biotinylated anti-mouse immunoglobulins (Bio-Rad) in
PBS at 37°C. After being washed, they were incubated for 1 hr with
peroxidase-labeled avidin in PBS at 37°C. They were washed a final time
and incubated with substrate solution (4-chloro-1-naphtnol) for 30 min
at 37°C.

RESULTS AND DISCUSSION
The result of the SDS-polyacrylamide gel electrophoresis indicated that
the purified SDV was composed of 4 protein subunits. Molecular sizes of
these proteins were 23, 46, 59, and 63K daltons.

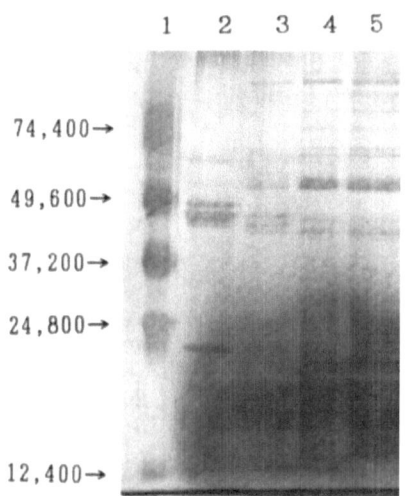

Fig.2. SDS-polyacrylamide gel electrophoresis of SDV from various
      purification steps.
      1, Molecular weight standards (12,400; 24,800; 37,200;
         49,600; 74,400).
      2, Viral fraction from using sucrose density gradient
         ultracentrifugation.
      3, Pellet from the ultracentrifugation.
      4, Before ultracentrifugation.
      5, Pellet from ammonium sulfate precipitation.

We obtained the hybridoma clones which showed positive reaction in
indirect ELISA. Their characteristics are summaraized in Table 1.
Among them, 3D2 hybridoma showed the most MAB productivity, and it
produced MAB with high reactivity to SDV.

Table 1. Characteristics of SDV-reactive MABs

Cell lines	Doubling time[a]	MAB		
		Class	Ig conc.[b]	Titer to SDV[c]
	(hr)		(μg/ml)	(ng/ml)
1F4	19.2	IgG	0.56	100
2A10	20.0	IgM	n.t[d]	10
3D2	26.4	IgM	2.75	10
3D3	22.0	IgG	n.t	n.t
3E1	15.2	IgM	2.30	100
3H9	22.0	IgM	n.t	10
4F10	20.4	IgM	1.71	100
4G6	28.8	IgM	1.96	100
5D7	18.0	IgG	n.t	n.t
5G10	27.3	IgM	n.t	100

a) Cultured in serum-free ITES-ERDF medium.
b) 1×10^5 cells/24 hr.
c) Tested by indirect ELISA.
d) Not tested.

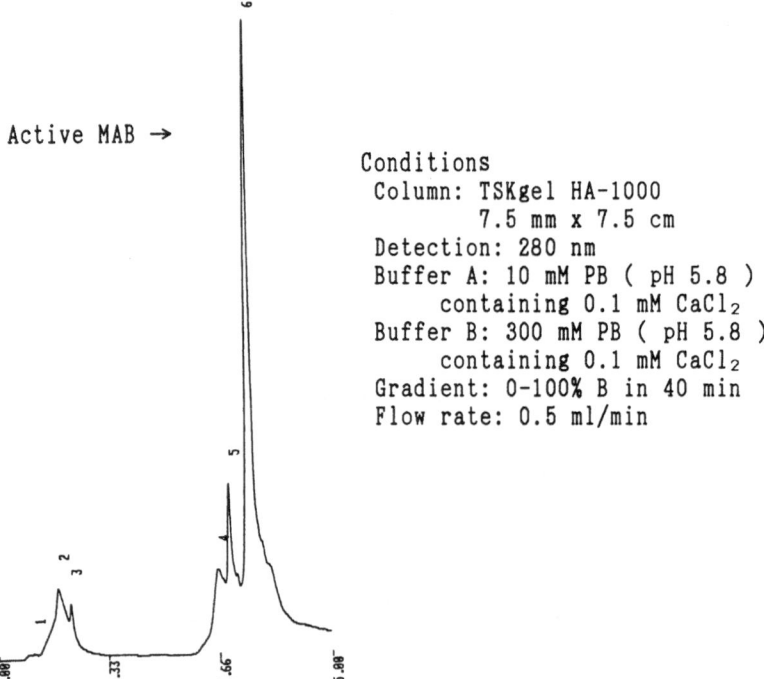

Active MAB →

Conditions
  Column: TSKgel HA-1000
         7.5 mm x 7.5 cm
  Detection: 280 nm
  Buffer A: 10 mM PB (pH 5.8)
       containing 0.1 mM CaCl$_2$
  Buffer B: 300 mM PB (pH 5.8)
       containing 0.1 mM CaCl$_2$
  Gradient: 0-100% B in 40 min
  Flow rate: 0.5 ml/min

Fig.3. Purification of MABs in cultured supernatant.

The result of hydroxyapatite column chromatography are shown in Fig.3. All MABs reacted with the 23KD antigen on purified SDV immunoblotted membranes. This indicates that the 23KD component contains the epitope to the MABs, and the 23K component was a typical antigen on the coat protein of SDV.

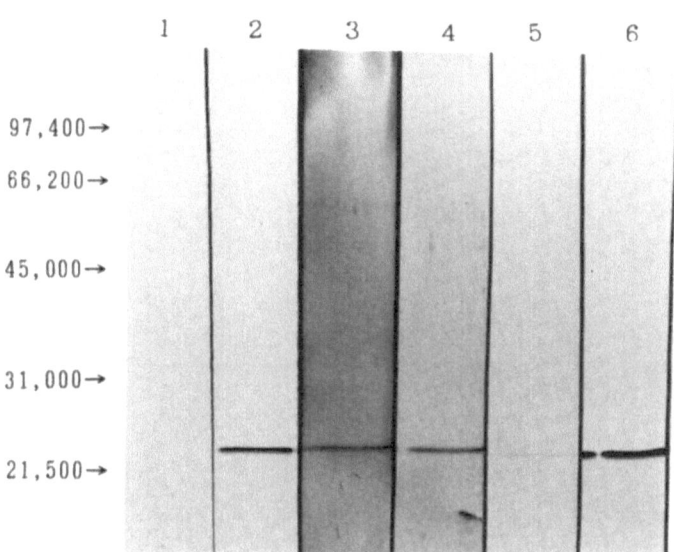

Fig.5. The reactivity of the monoclonal antibodies
to SDV components on immunoblotted membranes.
1, Culture supernatant of myeloma cells;
2, 1F4; 3, 5G10; 4, 3H9; 5, 4F10; 6, 3D2;

LITERATURE CITED
1. Izawa, H. 1966. Investigations on withering disease of Citrus unshu Marcov. in Gamagori District, Aichi Prefecture. Aichi Hort. Exp. Sta. 5:1-9.
2. Usugi, T. and Saito, Y. 1980. Detection of Satsuma Dwarf Virus by Enzyme-Linked Immunosorubent Assay. Abstr. Ann. Phytopath. soc. Japan 46: 106.
3. Yamada, S., and K. Sawamura. 1950. Satsuma dwarf. Abstr. Proc. Cong. Hort. Soc. Autumun 1950. 36-37.

RELATIVE ADVANTAGES OF CONTINUOUS VERSUS BATCH PROCESSES

JB GRIFFITHS
PHLS CAMR, Porton Down
Salisbury, Wiltshire SP4 0JG
UK

1. INTRODUCTION

New process technology is often developed to increase unit productivity which should give manufacturers real cost-saving benefits and increased efficiency. However for this new technology to become accepted as a production process, manufacturers need the assurance that it is as reliable as existing systems, and that regulatory requirements can be met. Thus this discussion of whether the future of animal cell biotechnology is with a batch or continuous process examines both the technical feasibility and the problem of what additional control data are necessary to ensure regulatory approval for a continuous process.

2. CURRENT SITUATION

Batch is the preferred production process for the historical reason that it is well proven and because substantial investments have been made in plant. It is also an ideal package for licencing control. Although amplified heterologous gene expression and medium developments have increased unit productivity of animal cell systems it is still 1-2 factors below an equivalent bacterial system. To reduce this productivity gap continuous processes with enhanced cell density have been developed. However the present situation with regard to the manufacture of biologicals is that, for large quantity products, high volume (to 10,000L) low cell density (below 3×10^6/ml) batch systems are favoured. The high density (10^8/ml) continuous systems, eg hollow fibres, are restricted to low volume operation (under 1L) and are therefore only suited to low quantity products (eg diagnostic MAB's). There is some evidence of movement towards converting existing batch

401

R. Sasaki and K. Ikura (eds.), Animal Cell Culture and Production of Biologicals, 401–410.
© 1991 Kluwer Academic Publishers.

technology into medium cell density (10^7/ml) longer-term processes by the use of spin-filters etc. The question is whether this can be sustained into a large volume, high density continuous system with the properties needed to be a successful production technology (ie simple design and operation, controllable, scaleable, flexible and licensable).

3. CONTINUOUS PROCESSES - ADVANTAGES

3.1 Unit productivity

In Table 1 the main types of culture process are listed with theoretical cell densities, and an extrapolation of product yield per week and per month based on an average monoclonal antibody production rate.

TABLE I. Relative productivities in different culture systems.

CULTURE TYPE	CELL NO.	PRODUCT YIELD PER LITRE		LENGTH
	(Millions)	mg./week*	mg./month	Days
1. BATCH	3	100	200	7
2. SEMI-CONTINUOUS BATCH	3	200	600	21
3. FED BATCH	6	200	500	14
4. CONTINUOUS PERFUSION	30+	3000	12000	>30
5. CONTINUOUS FLOW	2	300	1200	>100

(* values allow for turn-round time of non-continuous cultures)

The advantages of a continuous process are obviously reduced 'down-time' and, because of the higher cell density possible with perfusion, higher unit titres. A comparison of continuous perfusion with other systems shows a significantly higher productivity, but can this be translated to a significant reduction in costs per product unit?

3.2. Unit Cost

Various cost analyses have been made which show a range of differences from less than 10% to 75% lower unit cost with a continuous process. The variations come about by what is included in the formula, and the scale of operation.
In Table 2 the analysis is based on a 1000L batch process and takes into account equipment and plant investment, running costs of staff and consumables and then relates these costs to a unit product cost.

TABLE 2. Production costs of continuous processes as a ratio of a 1000L batch culture.

	1000L BATCH £K	100L SPIN FILTER	CONTINUOUS PROCESSES+ (eg FIBRES/OPTICELL)			FLUID.BED *100L
			1X1L	2X1L	5X1L	
CAPITAL						
- EQUIPMENT	1000	0.5	0.1	0.2	0.5	1.0
- PLANT	3000	0.5	0.24	0.24	0.5	1.0
PER ANNUM						
- CAPITAL (5YR)	800	0.5	0.2	0.23	0.48	1.0
- STAFF	200	0.75	0.4	0.5	0.6	1.0
- MEDIA	150	0.55	0.2	0.4	1.0	8.0
YIELD (gm)	2500	0.6	0.2	0.4	1.0	8.0
COST (gm)	0.46	0.92	1.2	0.75	0.57	0.24

* System not yet available at this scale

+ NB Extra development, quality assurance and licencing costs

It can be seen that either a multiple continuous process or a large-scale (100L) continuous process equates on total production figures and brings about significant savings. With respect to the 100L fluidised bed quoted in Table 2 currently this system has only been scaled to 24L (1) thus these data can be seen as target rather than actual. The question is can such a system be developed and accepted?

4. TECHNICAL FEASIBILITY FOR A LARGE SCALE CONTINUOUS PROCESS

Development in cell reactors has been aimed at overcoming limiting factors in scale-up ie oxygen, shear sensitivity, nutrient exhaustion/metabolic toxicity, and surface area (2). This led to the use of perfusion, which in turn led to the possibility of maintaining cell densities at 20-100 fold higher than in conventional bioreactors (3). In order to have efficient perfusion suspension cells have to be retained (immobilised) in the bioreactor so that fast flow rates can be used without cell washout. Means of retaining cells are summarised in Table 3. Separation by spin-filters etc are ideal for microcarriers (4) but have limited application for suspension cells, although in-line centrifugation techniques are becoming a real possibility for 100-1000L scales. (5, 6,) Immurement techniques are very effective for maintaining densities of over 10^8 cells/ml but are almost impossible to scale-up as a unit beyond the litre scale. Thus they are very popular for the production of small quantity products but have no scale-up potential. The entrapment technique, especially those based on dynamic matrices, do have a scale-up potential and are the best means of achieving the target process identified in Table 2.

TABLE 3 Methods for retaining cells in bioreactors to enable efficient perfusion and the development of continuous processes.

STRATEGY	DEVICE	REFERENCE
1. SEPARATION	Spin Filters	4
	Loop Reactors	7
	In-Line Centrifugation	5, 6
2. IMMUREMENT	Hollow Fibre Bioreactors	8
	Encapsulation	9
	Membrane Bioreactors	10
3. ENTRAPMENT	Ceramic (Opticell)	11
	Aggregation in fibres	12
	Flocculation	13
	Sponges/Foam matrices	14
	Porous Microcarriers	1, 15, 16

5.   POROUS MICROCARRIERS

Solid microcarriers (eg Cytodex, Dormacell etc) are the most
successful large-scale process for anchorage-dependent cells
and have proven operation at 4,000L.   Their limitation has
been on how the unit process intensity can be increased.
This was achieved by the fabrication of microcarriers with
continuous pores constituting over 90% of the void volume.
This process was pioneered by the Verax Corporation (1) with
collagen.   Other carriers now available are Cultispher
(gelatin), and Siran (glass)(15).   Not only do these
microcarriers provide a huge increase in surface area for
attached cells but they entrap suspension cells at high
internal densities.   Examples of porous microcarriers are
shown in Fig. 1.

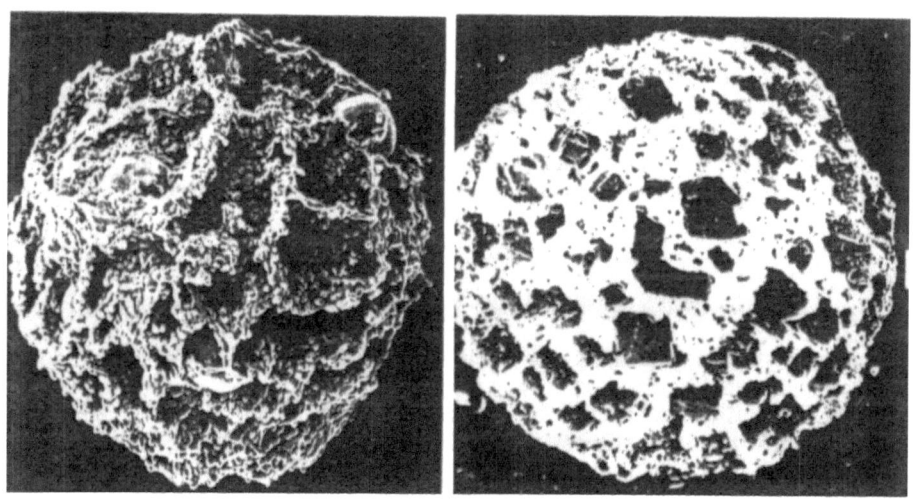

Fig. 1 (A) Verax microsphere   (B) Siran (Schott Glaswerke)
porous glass bead.

These porous microcarriers can be used in fluidised, fixed (Siran), or stirred (Cultispher) bioreactors and have many process advantages (listed in Table 4). Currently the Verax system has been sealed to 24 litres which by controlled process optimisation and definition is equal in productivity to 2-3,000 litre conventional stirred bioreactor. The capability to scale-up to 100 litres is attainable - it only requires a product needed in such high quantities - thus it is technically feasible to have the systems targeted in Table 2.

TABLE 4   Porous carriers - advantages compared to solid carriers

---

A: Unit cell density 20 - 50 fold higher
B: Support both attached and suspension cells
C: Immobilisation in 3D configuration easily
   derivitised
D: Short diffusion paths into a sphere
E: Suitable for stirred, fluidised or fixed bed
   reactors
F: Good scale-up potential by comparison with
   analogous systems (eg. microcarrier at
   4,000L)
G: Cells protected from shear
H: Capable of long term continuous culture

---

6. LICENCING REQUIREMENTS OF A CONTINUOUS PROCESS

The change from a batch to a continuous process needs a more careful definition of a production 'lot', and extra validation that a cell is stable, and neither it nor the product changes in character during a long culture process. Differences in batch and continuous processes are summarised in Table 5. With regard to product licencing the factors listed under cells and product need the most consideration. The proposed strategy is to compare the master working cell bank with a post-production (end) cell bank for; consistent cell number, viability and productivity; reproducible metabolic parameters (eg glucose utilisation rate); copy number, DNA sequence and restriction map of inserted genetic material. Extra process parameters such as media exchange rate, fluidisation or mixing rates will have to be monitored and controlled.

TABLE 5   Comparison of batch and continuous processes

	FACTOR	BATCH	CONTINUOUS
EQUIPMENT	Bioreactor	Simple	Complex
	Reliability	Proven	Less than batch
	Volume		
	(bioreactor)	High	Low
	(medium tanks)	High	High
PROCESS	Seed scale-up	Multiple steps	Limited steps
	Manipulations	Limited	Numerous
	Develop. /Validation	Short	Long
	Downtime		
	(between runs)	Short	Long
	(annually)	Long	Short
	Production facilities	Large plant	Laboratory/ small plant
CELLS	Age and stability	Limited significance	Highly significant
PRODUCT	Residence time	Long	Short
	Expression stability	Limited significance	Very important
	Identity	Between runs	Between harvests & runs
ECONOMICS	Productivity	Low	High
	Unit cost	High	Low
LICENCING	Batch definition	1 (2) harvests	Multiple harvests
	Validation	Defined	Many extra parameters

Product yield, purity, identity, potency etc is a standard requirement met by such tests as SDS Page, Isoelectric focusing, HPLC, DNA fingerprint, peptide mapping etc. The extra requirement is that this may have to be compared between multiple harvests within a production run, as well as between runs. If a continuous harvest is pooled then product stability needs careful validation. Finally plant has to remain reliable over extra long culture periods thus the life expectancy of parts that wear (O-rings, filters etc) need more thorough validation.

These factors show that a product from a continuous process can be validated with enough assurance to be licensed. However care must be taken in the selection of criteria so that the time and cost of carrying out the quality assurance tests do not negate the economic advantages of a continuous process.

7. SUMMARY

In this paper the advantages of a continuous high density process can be clearly seen. The technical feasibility, economic advantage, and ability to licence a product from such a process are also demonstrated. It also serves as a model to show that developments in process technology must be made with an awareness of the total process so that it remains simple and reliable when scaled-up, and in particular will meet all the requirements of the regulatory agencies. With respect to whether the future is with a batch or continuous process the answer is that both will fulfil vital roles in animal cell biotechnology. It is reassuring for manufacturers that if a product comes along which needs to be made in over 100 Kg amounts there is a production process waiting to meet such demands.

REFERENCES

1.   Runstadler, P.W. and Cernek, S.R. (1988). Large-
      scale fluidized bed, immobilized cultivation of
      animal cells at high densities. In: Spier, R.E.
      and Griffiths, J.B. (Eds.), Animal Cell
      Biotechnology 3. Academic Press, London. pp 305-
      320.

2.   Griffiths, J.B. (1990a). Animal Cells - the
      breakthrough to a dominant technology.
      Cytotechnology 3, 109-116.

3.   Griffiths, J.B. (1990b). Perfusion systems for cell
     cultivation. In: Lubiniecki, A.S. (Ed), Large-
     Scale Mammalian Cell Culture Technology, Marcel
     Dekker Inc., NY. pp 217-250

4.   Griffiths, J.B., Cameron, D.R. and Looby, D. (1987).
     A comparison of unit process system for anchorage
     dependent cells. Dev. Biol. Stand. 66, 331-338.

5.   Gronvik, K.O., Frieburg, H. and Malmstrom, U.
     (1989). Centritech Cell - a new separation device
     for mammalian cells. In: Spier, R.E., Griffiths,
     J.B., Stephenne, J. and Crooy, J.J. (Eds.), Advances
     in Animal Cell Biology and Technology for
     Bioprocesses, Butterworths, Guildford, pp 336-344.

6.   Tokashiki, M., Arai, T., Hamamoto, K., and Ishimaru,
     K. (1990). High density culture of hybridoma cells
     using a perfusion culture vessel with an external
     centrifuge. Cytotechnology 3, 239-244.

7.   Emery, A.A., Lavery, M., Williams, B., and Handa,
     A., (1987). Large scale hybridoma culture. In:
     Webb, C. and Marituna, F. (Eds.); Plant and Animal
     Cells: Process Possibilities, Ellis Horwood. pp
     137-146.

8.   Knight, P. (1989). Hollow fiber bioreactors for
     mammalian cell culture. Bio/Technology 7, 459-461.

9.   Duff, R.G. (1989). Microencapsulation technology: a
     novel method of monoclonal antibody production.
     Trends Biotechnol. 3, 167-170.

10. Scheirer, W. (1988). High density growth of animal
     cells within cell retention fermenters equipped with
     membranes. In: Spier, R.E. and Griffiths, J.B.
     (Eds.) Animal Cell Biotechnology 3, Academic Press,
     London. pp 263-281.

11. Berg, G.J. and Bodeker, B.G.D. (1988). Employing a
     ceramic matrix for the immobilisation of mammalian
     cells in culture. In: Spier, R.E. and Griffiths,
     J.B. (Eds.) Animal Cell Biotechnology 3, Academic
     Press, London. pp 321-325.

12. Larsson, B. and Litwin, J. (1987). The growth of
     polio virus in human diploid fibroblasts grown with
     cellulose microcarriers in suspension. Dev. Biol.
     Stand. 66, 385-390.

13.   Aunins, J.G. and Wang, I.C. (1989) Induced
      flocculation of animal cells in suspension culture.
      Biotech. Bioeng. 34, 629-638.

14.   Murdin, A.D., Thorpe, J.S., Groves, D.J. and Spier,
      R.E. (1989). Growth and metabolism of hybridomas
      immobilized in packed beds; comparison with static
      and suspension cultures. Enz. Microb. Technol. 11,
      341-346.

15.   Looby, D. and Griffiths, J.B. (1988). Fixed bed
      porous glass sphere (porosphere) bioreactors for
      animal cells. Cytotechnology 1, 339-346.

16.   Looby, D. and Griffiths, J.B. (1990).
      Immobilization of animal cells in porous carrier
      culture. TIBTECH 8, 204-209.

PRE-CLINICAL REQUIREMENTS FOR BIOTECHNOLOGY DERIVED PRODUCTS

SEIICHI ENOMOTO
Nippon Hazleton
4-7-10 Nihonbashi Honcho, Chuo-ku,
Tokyo 103 Japan

ABSTRACT. A major concern about continuous cell lines is that tumor-inducing factors such as oncogenes or oncogenic viruses derived from these cell cultures might be imported with the product and lead to formation of a tumor in the recipient. Therefore, governmental agencies in many countries provided guidelines to manufacturers of cell culture products. These guidelines define appropriate testing to address the crucial issues involved in the production of biologicals in cell lines. These include general cell characterization, *in vitro* and *in vivo* tests for rodent and non-rodent viruses, retroviruses assays, tests for bacterial and mycoplasmal contaminants, final and bulk-final product tests and validation of the purification process for inactivation and / or removal of contaminants. This presentation will offer an overview of the current guidelines.

1. Introduction

Biotechnology has become an important method to produce drugs. The characteristic of this method is to use *E. coli*, yeast, or mammalian cells to produce new drugs. Compared with chemically synthesized drugs,the production process is very complicated. The most commonly used cell lines are Chinese hamster ovary cells, BHK-21 cells, mouse cells, and human cell lines. In many cases, the history of the culture is vague, the cells have been cultured for considerable periods, and they have passed to a number of different laboratories. Because of this, very strict quality control is required by the regulatory authorities.

2. Potential Risk Factors

Table 1 shows potential risk factors when cell lines are used for drug production. The first potential risk is the cell itself. As you know, we have used animal cells for vaccine production. But this experience has been restricted to normal cell lines. It was not until 1987 that FDA approved the use of continuous cell lines as host cells. Most of these host cells are tumor cells. Therefore, if they are present in the final product, there is the risk that tumor formation may be initiated. For that reason, we have to eliminate contamination by host cells. The next factor is exogenous DNA. If a high level of exogenous DNA is in the product, there is the risk in the patient that DNA expression may occur. This is very important in the case of oncogene contamination. Host cell-derived proteins are also risky for patients. Exogenous proteins may cause an immune response.

R. Sasaki and K. Ikura (eds.), Animal Cell Culture and Production of Biologicals, 411–418.

Table 1. Potential Risk Factors Associated with Cells

Factor	Potential Risk
• Cell	• Tumor
• DNA	• Expression of Abnormal Gene
• Proteins	• Immune Response
• Endogenous Viruses	• Transformation

The last one is endogenous viruses. The mammalian cells currently used originate from rodents such as hamsters or mice. These rodents may have endogenous viruses. Many viruses are known to be infectious to humans so that contamination with these viruses must be eliminated.

3. Data Requirements for Biotechnology-derived Drugs in Japan

To eliminate these potential risks, two notifications of data requirements for biotechnology-derived drugs were issued by the Pharmaceutical Affairs Bureau of the Ministry of Health and Welfare (MHW) in Japan. The first one was issued in March, 1984 and is concerned with recombinant DNA technology. When this notification was issued, the main host system used was prokaryotic, that is *E. coli* and the risks I previously mentioned were less than those at present. The second notification was issued in 1988 resulting because mammalian cell culture technology became popular. Most of the testing which I will present was developed to meet this second notification as well as FDA and EC guidelines for cell line testing.

4. Key Elements

There are some key elements that should be considered especially for mammalian cell derived drugs. First of all, of course, cell lines that are used for drug production must be tested. In this case, appropriate testing is conducted at each sequential cell preparation step, that is : cell seed, master cell bank, working cell bank, and cells at the post production level. The final product should also be tested. Also the methods of product separation and purification should be evaluated to validate that inactivation or elimination of virus or other contamination is assured.

5. Transformed Cell Substrate Testing

5. 1. General Characterization

Table 2 shows the summary of requirements for mammalian cell derived drugs. The testing is classified into five groups. The first one is general cell characterization. **Tumorigenicity** testing is required by the Japanese authorities. Positive tumorigenicity does not prohibit the use of the cell line. The purpose of this testing is to understand the existence or the degree of tumorigenicity. We usually inject cells into athymic nude mice and observe the mice to monitor tumor formation for 28 days. If no tumor formation occurs, a further five month observation period is necessary.

Karyotyping is conducted to confirm the uniformity of the cell line and to identify any inadvertent exposure of the cell line to contamination. This is done by optical microscope.

The observation with **electron microscopy** gives us much useful information. Master cell bank and cells above the production level are examined by transmission electron microscopy for the presence of viruses or other contaminants. Treatment with an inducing agent such as bromodeoxyuridine to enhance the detection of retroviruses is also recommended in Japan.

Table 2. Transformed Cell Substrate Testing

Test Type	MCB	>PL	FP
General Characterization			
Tumorigenicity	+(1)	+(1)	
Karyotyping	+	+	
Electron Microscopy	+	+	
Specific Contaminant Testing			
Mycoplasma	+	+	
Mouse Antibody Production (MAP) Test	+	+	
in vitro Virus Assay	+	+	
in vivo Virus Assay	+	+	
EBV	+(2)	+(2)	
Cytomegalovirus	+(2)	+(2)	
Hepatitis B Virus	+(2)	+(2)	
Bovine Viruses	+(2)	+(2)	
Sterility	+	+	
Retrovirus testing			
XC plaque	+	+	
S+L- mink cell focus	+	+	
Cocultivation		+(3)	
Reverse transcriptase (Double template)	+	+	
Induction with chemical inducer and assay for retrovirus particles		+(1)	
HIV and HTLV assay	+(2)	+(2)	
Evaluation of Purification Process			
Final Product Evaluation			
Mycoplasma			+
General Safety			+
DNA			+
Sterility			+
Pyrogen			+

MCB: Master Cell Bank >PL: Cells above the Production Level FP: Final Product
+: Required. (1): May not always be required. (2): Depending on origin of cells or biological materials. (3): Depending on results of retrovirus screening in MCB.

5. 2. Specific Contaminant testing

The following tests belong to the Specific Contaminant Testing group. Tests should be done for the presence of both cultivable and non-cultivable **mycoplasma**.

Mouse antibody production tests for murine viruses can be easily conducted and have relatively high sensitivity. Culture supernatant is inoculated into mice and 28 days later serum antibodies are measured. Normally about 10 to 20 mouse viruses are checked. It is important to include a challenge for Lymphocytic Choriomeningitis Virus as well as testing for Thymic viruses, EDIM virus, and mouse Cytomegalovirus.

Cell banks should be tested in an **in vitro virus assay**. This assay is done using indicator cell lines. Indicator cell lines, like a primary human cell line and a monkey kidney cell line, are co-cultured with host cells for a minimum of 14 days and then examined for morphology and for hemadsorption and hemagglutination.

An **in vivo virus assay** is recommended because some viruses can be detected by testing in animals. We use adult and suckling mice , embryonated hen's eggs, and guinea pigs. Cells and culture fluids are inoculated into these animals and the animals are observed for 28 days.

Tests for specific **human viruses** are recommended when the host cell originates from human biological materials. These cells include human-human or human-mouse hybridomas and lymphoblastoid cell lines. Table 3 shows some of the assays for human viruses. **Human Immunodeficiency Virus** is detected by co-cultivation with human peripheral blood mononuclear cells for 6 weeks followed by antigen-capture ELISA or reverse transcriptase assay. Assay for **Herpes virus** is done in almost the same way. **Hepatitis B** screening is conducted with commercially available third-generation kits. **Cytomegalovirus** testing is required by a 6-week human-embryo culture assay with a blind passage between the third and fourth weeks. The detection is done by immunofluorescence. **Epstein-Barr virus** may be detected by viral DNA hybridization, nuclear antigen fluorescent staining, or immortalization of umbilical cord blood lymphocytes. PCR technique can aid in the detection of EBV. The detection of **papilloma virus** is done by RNA hybridization.

Table 3. Human Viruses

Virus	Test Method
HIV-1	Co-cultivation, observation for CPE and sequential subpassage, analysis by antigen capture ELISA or RT
HIV-2	Co-cultivation, observation for CPE and sequential subpassage, analysis by RT
Herpes virus VI	Co-cultivation, sequential subpassage and analysis by immunofluorescence with MAb
Hepatitis B	Third generation antigen detection assay
CMV	Co-cultivation, observation for CPE and confirmation by immunofluorescence
EBV	PCR / DNA hybridization probe
Papillomavirus	RNA hybridization probe

Fetal bovine serum is used for growth of mammalian cell cultures in many cases. It is known that many commercially available sera contain bovine viruses. Because of the possible contamination of cell cultures by bovine viruses present in serum, cells which have been cultured on media containing sera should be screened for the presence of contaminating bovine viruses shown in table 4. A filter down to 0.1 micrometer in size can be used for normal filtration. Under these conditions, Infectious bovine rhinotracheitis virus and Parainfluenza 3 virus can be removed, but Bovine viral diarrhea virus cannot be removed. Recently, it's been reported that bovine lentivirus may be related to Human Immunodeficiency virus-1. So, much attention is paid to this virus.

Table 4. Possible Bovine Virus Contaminants
in Fetal Bovine Serum

- Bovine viral diarrhea virus
- Parainfluenza 3
- Infectious bovine rhinotracheitis
- Bovine lentivirus

5.3. Retrovirus Testing

The next group of requirements for mammalian cell derived drugs are assays for retroviruses. The Master Cell Bank and cells from the Master Working Cell Bank beyond the production level must be checked for retroviruses by appropriate methods. There is no single detection method that can detect all retroviruses, so a combination of various assays must be used. The use of inducing agents such as bromodeoxyuridine or iododeoxyuridine are recommended in Japan and some cases by the FDA and EC. Two types of cell culture infectivity assays are used to detect the presence of infectious retroviruses. The XC plaque assay can detect ecotropic murine retroviruses. The Mink S+ L- focus assay can detect xenotropic retroviruses as well as amphotropic retroviruses. In some cases, infectious retrovirus particles do not exist in culture fluids even if the cells are infected by retroviruses. In this case, co-cultivation with a highly sensitive detector cell lines is recommended. Examples of these detector cell lines are mink lung, bat lung, dog thymus, and fetal rhesus lung cells. Host cells are co-cultured for five passages. The cell culture supernatant is harvested at passages one and five and tested in the S+ L- assay and reverse transcriptase assay and the cell are screened by transmission electron microscopy.

Table 5. Detection of Murine Retrovirus

Assay	Virus Type		
	Ecotropic	Xenotropic	Amphotropic
XC Plaque	+	-	-
S+L- Focus	-	+	+
Co-cultivation with non-murine cells	-	+	-

Other Assays
- Reverse Transcriptase assay
- Electron Microscopy

Retroviruses have RNA-dependent DNA polymerase activity, that is reverse transcriptase activity. This enzymatic assay is also useful to detect retroviruses. One problem is that sometimes cellular DNA polymerase reacts nonspecificaly with the RNA template used for reverse transcriptase assay. Elimination of cellular DNA polymerase can often be done by centrifugation, but a double template assay system that controls for DNA polymerase should be used for testing mammalian cells.

The particles of retroviruses may be observed using transmission electron microscopy. In particles of up to 100-nanometer diameter, budding particles can be detected.

5.4. Inactivation and Elimination of Viruses

There are many methods to inactivate or to eliminate viruses. Table 6 shows examples of such methods. An acidic or basic condition will inactivate many viruses. Some viruses are inactivated by lipid solvents such as tri (n-butylyl) phosphate and sodium cholate. These solvents are effective for lipid-enveloped viruses. Heating is also effective for inactivation of some viruses. Filtration with a 0.025 micron filter eliminates the presence of viruses. Besides these methods, commonly used methods of virus removal are chromatography, detergents, ultraviolet light, and polyethylenglycol.

Table 6. Inactivation and Elimination Methods

Virus	Inactivation				Elimination
	pH<4	pH>9.5	Lipid Solvents	Heart >56°C	0.025μm filter
Adenovirus	*	*	-	-	+
Herpesvirus	+	+	+	+	+
Hepatitis B	+	+	+	+	+
Polyomavirus	*	*	-	-	+
SV-40	*	*	-	-	+
Vaccinia	+	+	+	+	+

+: Effective   -: Inefficient   *: Necessity for harsher condition

5. 5. Evaluation of Purification Process

One other element to be checked is the purification procedure of the final product. Process validation is required to confirm elimination or inactivation of retroviruses and other adventitious viruses. Process validation is done on a precisely scale-downed purification system. The starting material, usually culture media is spiked with known quantities of model viruses and purified by a separation method which is the same as that used for the production. Table 7 shows an example of the results. In this case the purification procedure has four steps. Before and after each step, infectivity of the model viruses are assayed. Then, finally, the total figure of the clearance factor is calculated. A factor of logs 15.8 resulted from these example purification steps. Normally a factor of eleven or more logs is acceptable. As for model viruses, some commonly used viruses are Murine xenotropic retrovirus, Polio virus, Bovine viral diarrhea virus, SV40, Parainfluenza 3, and Herpes virus, to name a few.

Table 7. Process Validation

Step	PFU in spike	PFU in product	log reduction
1	3.1×10^7	2.5×10^3	4.1
2	9.3×10^6	$<2.4 \times 10^3$	>3.6
3	1.6×10^7	$<1.6 \times 10^3$	>4.0
4	1.6×10^7	$<1.3 \times 10^3$	>4.1
		Total Clearance	>15.8

5. 6. Final Product Evaluation

For the final product, the following tests are required. In the case of the General Safety Test, the final product is inoculated into guinea pigs and mice and the animals are observed for 1 week. Contamination of

Table 8. Preclinical Testings for Biotechnology-Derived Pharmaceuticals

| Classification by Biochemical Characters | Hormones, Cytokines and Other Regulatory Factors | | | Blood Products | | Monoclonal Antibodies | Vaccines |
| | I | | II, III | I | II, III | | |
	a	b	b	b			
Single Dose Toxicity Study	▲	O	O	O	O	O	O
Repeated Dose Toxicity Study	X	O	O	O	O	O	X
Reproductive and Developmental Toxicity Studies	X	▲	O	O	O	O	▲
Dependence Study	X	▲	▲	X	X	▲	X
Antigenicity Study	X	X	▲	O	O	▲	
Mutagenicity Study	X	X	▲	▲	▲	▲	X
Carcinogenicity Study	X	X	▲	X	X	▲	X
Local Irritation Study	▲	▲	▲	O	O	▲	O
Pyrogen Test	▲	O	O	O	O	O	O
Others							
Immunotoxicity	X	▲	O	X	X	X	X
Crossreaction	X	X	X	X	X	O	X
General Pharmacology	▲	▲	O	O	O	O	▲
Metabolism	O	O	O	O	O	O	▲

O : This study should be applied as a rule. ▲ : This study should be applied if necessary. If there is a proper reason, the study can be omitted. X : Normally, this study is not requested.

Classification by Biochemical Characters; I : Human polypeptides and proteins proved to be identical to native. II : Substances similar to human polypeptides or proteins with the same amino acid sequence as known substances. III : Polypeptide or protein which is differ from human. a : Product will be administered at physiological concentration. b : Product will be administered at unphysiological high concentration.

418

cellular DNA should be checked. For this purpose, DNA hybridization is recommended and various DNA probes are used. FDA guidelines recommend that the amount of contaminating DNA be less than 10 picograms per dose. Sterility, mycoplasma, and pyrogen tests are also suggested for final products.

6. Preclinical Testing for Biotechnology-Derived Drugs

Performing toxicology studies on biotechnology-derived drugs is also difficult because we have little experience with the pharmacological and immunological function of new substances such as cytokines and growth factors. To deal with this problem, an Ad Hoc Committee for the safety assessment on biotechnology-derived products of the Japan Pharmaceutical Manufacturers Associ·.ion has prepared a recommended battery of tests in collaboration with MHW.

Table 8 shows a draft recommendation by the committee. There are some notes regarding this table. When a single-dose toxicity study is conducted, the clinical route of administration should be used. Concerning the Reproductive Toxicity Study and Carcinogenicity Study, animal species should be selected with consideration to low antibody formation . If the study is difficult to perform because of antigenicity, this study may be omitted. The repeated-dose toxicity study is not always required if enough clinical experience with the product has confirmed no toxicity. Reproductive and Developmental Toxicity Studies are required in the case of administration to pregnant female patients or female patients of child-bearing age. Also, the "Segment I Study " and "Segment III Study " can be omitted if there is no applicability. If a pharmacological effect to the central nervous system is found at a low dosage, a Dependence Study should be conducted. In the case of Mutagenicity , a mammalian cell culture study should be used. If clinical administration is expected to be long-term or high-dose, and carcinogenicity is suspected from other study results, a Carcinogenicity Study may be necessary. Local Irritation Study should be done if that route of administration is expected. In the case of Metabolism Studies, identification of metabolites in blood and urine are required. If this is not possible, specific data on the metabolic kinetics and metabolic part using *in vitro* systems is required. Metabolites which have specific interactions need further investigation.

As I said before, this is a draft recommendation and not yet complete. The final report is currently being written and will be submitted for publication soon.

7. Summary

To summarize, many types of safety tests for host cell lines are required for biotechnology-derived drugs. It must be proved that host cell lines are free of adventitious viruses and retroviruses. Also, well considered process validation of viral removal or inactivation assures drug safety Techniques using biotechnology are developing rapidly and are being applied to new drug development. This means safety evaluation should also incorporate new technology to give greater confidence in biopharmaceuticals.

Guidelines for Contamination Testing of Bioproducts

Carol J. Marcus-Sekura, Ph.D., Division of
Virology, and Curtis L. Scribner, M.D., Division
of Biological Investigational New Drugs, Center
for Biologics Evaluation and Research, Food and
Drug Administration, Bethesda, MD 20892

ABSTRACT   Recent advances in biotechnology have led to the
development of both new products and new procedures for
producing currently available products. Since product
development and manufacturing techniques are constantly
changing, the methods to assure products are safe and
consistent must also change. Regulatory agencies have taken
advantage of the newly developed methodologies in evaluating
new products and have begun to devise new standards to
assure product safety and consistency. The Center for
Biologics Evaluation and Research has approached this
challenge by preparing a series of documents, the "Points to
Consider" (PTC), based on its own laboratory expertise and
industry-wide experience and then revising the documents as
needed to keep pace with state-of-the-art science. This
flexibility should enable FDA to provide regulatory
consistency for all products by incorporating experience
gained with recently developed products and manufacturing
approaches, requiring additional testing if needed, and
eliminating unnecessary testing, in an attempt to expedite
product approval. This is a dynamic process based on the
best available science.

INTRODUCTION   Advances in biotechnology have radically
changed many aspects of the process of qualifying newly
developed biological products for initial human clinical
trials as well as licensure for human use. FDA has taken
several approaches toward modifying its regulatory process
to more effectively deal with these new products.

Biotechnology has had its impact in five major areas.
First, it has increased both the number and types of
products. Secondly, it has markedly increased the amount of
material available for clinical use. Thirdly, it has led to
revised criteria for product safety. Fourthly, it has
provided new and more sensitive techniques to characterize

419

R. Sasaki and K. Ikura (eds.), Animal Cell Culture and Production of Biologicals, 419–427.
© 1991 *Kluwer Academic Publishers.*

biological products, and finally, it has resulted in the introduction of many new production methods.

The types of new biological products include recombinant proteins (t-PA, the cytokines and subunit AIDS vaccines), recombinant viruses (vaccinia expressing the HIV envelope glycoprotein and murine amphotropic retroviruses proposed for gene therapy), and monoclonal antibodies for treatment and diagnosis. Since most of these products are produced in continuous cell lines, certain common safety concerns arise. These include contamination of the final product with potentially oncogenic or mutagenic material, adventitious or endogenous infectious agents, nucleic acid, potentially active or immunogenic cellular proteins, and defective product caused by instability of the continuous cell line during production.

Similarly, biotechnology has produced many new techniques to characterize products biochemically. For proteins these include peptide mapping and sequencing, end-terminal sequence analysis, amino acid composition, polyacrylamide gel electrophoresis (PAGE), Western blot, high pressure liquid chromatography (HPLC), and isoelectric focussing. For nucleic acid analysis, hybridization, polymerase chain reaction (PCR), nucleic acid sequencing, and restriction enzyme digestions allow detailed characterization. Additionally, it is now possible to analyze carbohydrate content and modifications, covalently bound lipids, and the distribution and sites of sulfhydryl or disulfide groups.

A variety of production methods are available to manufacture these products. Bacteria and yeast as well as insect, animal and human cells are now being used. Some products, such as monoclonal antibodies, can be made either in vivo (ascites) or in vitro (tissue culture production). Other cells may be propagated in tissue culture for a small number of passages followed by centrifugation or for an extended number of generations using continuous perfusion with fresh media while the spent media is removed and processed to produce purified product. In the future, techniques such as continuous in vitro translation, production in animal milk, and complete synthesis may be feasible.

The purification must be designed not only to give high yield, but to maintain product integrity, function, and purity. It should also be designed to eliminate contaminating viruses, microbes, nucleic acid, antigenic materials, and undesirable chemicals introduced by production procedures.

APPROACHES TAKEN BY CBER In order to deal effectively with these issues, FDA has instituted three basic changes in its approach to regulating biotechnology products. "Pre-IND meetings" which take place several months or more before anticipated IND submission are encouraged. The fostering of early scientific exchange, the availability of up to date "points to consider" documents, and the revision of regulations to permit joint manufacture and testing are all important. In all cases however, the basis for the interactions is good science. Without thoughtful and rigorous science, both product and patient suffer.

In the "pre-IND" meeting and other preliminary interactions, the manufacturer submits preliminary product information and meets with appropriate agency personnel to discuss cell substrate testing, manufacture, biochemical characterization, preclinical testing, rationale and proposed Phase I studies. This bilateral scientific information exchange early in product development can help both the sponsor and the Agency to identify potential problems that can be corrected early. FDA becomes aware of new products that might require policy decisions and early communication can expedite processing of the IND application. The manufacturer on the other hand, with early FDA input, hopefully can avoid unnecessary time delay and expense in product development. This procedure has been found to be extremely effective in expediting the approval of IND submissions.

It should be noted that the information discussed in "pre-IND meetings" is held to be confidential or trade secret as appropriate.

The use of "Points to Consider" (PTC) documents supplements this evaluation process. The purpose of such documents is to provide manufacturers and FDA personnel with a timely consensus of suggestions for the characterization and testing of the new biotechnology products. It is important to be aware that the PTC are merely suggestions and are not requirements, regulations, or guidelines in the legal sense. Rather, the PTC are drafted by CBER committees and sent in draft form to other government agencies and to industry for comment and review.

The applicability of the PTC will vary with the product, the intended patient population, and the "state-of-the-art." Since all are constantly changing, the appropriateness of the PTC will change, and it is our intent to revise them as frequently as necessary to maintain their usefulness based on the best available science. Three of these documents are of general interest to the characterization of cell lines and products: the "cell substrate," "monoclonal antibody," and "recombinant DNA" PTC.

The Cell Substrate PTC outlines suggestions for characterizing cell lines to be used for the production of biologicals. These include an outline of the history of the cell line and the characterization of certain properties of the cell line which are thought to be useful for identity and stability testing during storage and production (karyology, isozyme analysis, tumorigenicity). The development of a cell banking system to insure uniformity of subsequent lots of product and the need to demonstrate freedom from adventitious agents is also discussed.

Of these suggestions, the demonstration of freedom from adventitious agents is the most problematic. Decisions must be made early in product development as to which tests are appropriate for a particular product, at what stages in production (MCB, MWCB, post-production cells, unprocessed bulk, processed bulk, final container) and with what frequency (once, occasional lots, every lot) such tests should be performed. Appropriate tests for bacterial contamination and endotoxin are well established. Testing for mycoplasma, both cultivable and non-cultivable, especially on unprocessed bulk lots of product is reasonably straight forward. Virus testing should be for specific viruses which might be expected to contaminate a given cell line (MAP test). Such viruses would include those which infect the species and tissue from which the cell line is derived, viruses of other species which might infect the cell line, any human viruses which are capable of infecting the cells, and any viruses which might be introduced during propagation of the cells such as bovine viruses from serum or other biological products.

Appropriate virus testing might also include specific tests designed to detect particular viruses (HBsAg for hepatitis B virus, EBNA for EBV, and cocultivation of with indicator cells to detect HIV or CMV), nucleic acid hybridization or virus specific PCR probes. Tests for inapparent viruses are also important and might include cultivation of fluids from the cell cultures with a series of cell lines susceptible to infection with a wide range of viruses to observe for cytopathic effects, hemadsorption and hemagglutination, and electron microscopic changes. It is also advisable to test for infectious retroviruses using the reverse transcriptase (RT) assay with both Mg^{++} and Mn^{++} as the divalent cations. For rodent cells, other appropriate tests might include the XC plaque and the S+L- assay, electron microscopy and nucleic acid hybridization.

Monoclonal antibodies as a group have similar
characteristics and safety concerns even though each
antibody is treated as a completely different product.
There are many overlapping concerns related to the use of
cell substrates in their production, so the monoclonal
antibody PTC should be used in conjunction with the cell
substrate PTC. In the monoclonal antibody PTC, cell
substrate issues appropriate to products made in rodent and
human cell lines are discussed and virus testing appropriate
for rodent cell lines is presented including MAP testing for
common rodent viruses. Endogenous intracisternal A
particles are frequently observed by electron microscopy of
hybridoma cells, but tests for infectious retrovirus may be
negative. If Type C particles are present, their removal
during manufacture should be validated to assure product
safety. If an infectious retrovirus is present,
demonstration of its identity and elimination on a lot to
lot basis is likely to be desirable. If the expected
clinical application involves only minimal benefit to the
recipient, the presence of infectious retroviruses in the
hybridoma line might be considered excessive risk.

Human monoclonal antibodies, since they are produced using
malignant or EBV transformed human cells, require additional
testing. Tests for human viruses are suggested and
presently such testing would be expected to include tests
for EBV (virus genome, EBNA, VCA), CMV, HIV-1, HIV-2,
HTLV-1, HTLV-2, HHV-6, and the hepatitis viruses.

Biochemical and functional characterization of the purified
monoclonal antibody product and preclinical testing are also
discussed in the PTC. Purity and testing requirements for
a particular monoclonal antibody product will depend on its
intended use.

The recombinant DNA "Points to Consider" is designed to give
guidance for the appropriate testing and characterization of
products produced by recombinant DNA technology. This
includes suggestions for characterizing the coding sequence
of the gene and the vector used for its expression,
describing the host cell which will be used to produce the
product and establishing a master cell bank and its
characterization. Suggestions about monitoring production
and manufacture and specific tests to analyze the identity,
potency, and purity of the product are also made.

DISCUSSION OF SPECIFIC ISSUES There are several frequently
asked questions or areas of concern. (1) What are the
implications of the choice of a particular cell line and
cell culture procedures for manufacture? (2) What is the

result of the unintentional use of contaminated sera? (3)
What is the approach to testing when production involves
multiple biological sources? (4) How does one proceed if a
cell substrate contains endogenous retrovirus-like
particles? (5) How can one validate virus and nucleic acid
removal?

In developing a new product, the ideal situation is to
utilize a stable, clearly defined "parent" cell line and to
establish a cell bank which is free from contamination with
adventitious or endogenous agents. This parent cell line is
then modified to derive the actual cell substrate for
production. This situation is not always possible to
achieve, but in all instances, it is advisable to set up a
master cell bank (MCB) of the cell line, and to perform
appropriate testing to qualify the cell bank early in
product development. It is also advisable to keep the
number of generations between the MCB and MWCB small to
decrease the likelihood of both mutations and contamination.
The species of origin of the cell substrate will in part
determine the testing which is appropriate.

The choice of culture conditions, particularly the duration
of time cells are in culture, will also influence the
testing required. When manufacturing employs continuous
perfusion, the number of generations after which continuous
perfusion is terminated should be specified. During
culture, cells should be tested for: sterility, mycoplasma,
stability of the recombinant portion of the genome,
characterization of the product for consistency, consistency
of the output level of product by the cells, and potency of
the product, as well as absence of viral contamination or
changes in endogenous particles in the cells at or beyond
the longest proposed generation time. Cells kept in culture
for extended periods may have to be tested on a frequent
schedule to assure product consistency.

The sources of contamination with foreign materials during
production and purification are legion. Potential
contamination of animal sera is an ever present problem and
in-house quality control procedures to assure freedom from
adventitious agents are needed. Similarly, if antibiotics
or growth promoters are used in cell culture, their absence
from the final product will need to be assured. The use of
other biological products such as monoclonal antibodies or
animal proteins in manufacture will involve appropriate
quality control certification and testing as well as
validation of removal from the final product. For example,
the potential for an antibody to leach from an affinity
column needs to be defined and the amount of antibody
contaminating the final product quantitated.

Many questions are raised about "process validation." The purpose of validation is to ensure that the final product is free from known or unsuspected contaminants which are introduced during the production or purification process. This is usually done by testing the ability of the production process to remove or inactivate contaminants. Frequently, easily detectable virus or radioactively labelled DNA is added or "spiked" into partially purified product at various stages of production. This "spiked" sample is then processed through the next step in the manufacturing process and the eluate tested for the presence of the "spiked" material. The scale of the production process is usually reduced by 100 to 1000 fold to facilitate testing, but it is important to mirror as closely as possible the flow rates, product to bed ratios, types of carrier fluids and eluates and all other production specifications.

When carrying out validation experiments, it is helpful to keep several factors in mind. For example, with retrovirus removal validation testing: 1) It is best to choose high titer virus stock since the higher the titer of the input virus used to measure inactivation/removal, the greater the amount of removal/inactivation which can be determined. 2) Inclusion of a specific viral inactivation step(s) in addition to removal steps will increase confidence that the product is safe. 3) Procedures which include at least one step where substantial removal of virus occurs (>5-6 logs) will increase confidence more than a series of steps of minimal removal capabilities (<2 logs). 4) If a manufacturing step is repeated during a process (e.g., use of the same column/buffer system twice) the second use would not be expected to produce the same amount of removal as the first time. 5) If virus testing is by tissue culture assay, appropriate controls should be included to assure that the bulk material is not directly toxic to the indicator cells being used.

Chromatography columns and resins should be dedicated to one product only. If a chromatography column is to be reused, it should be regenerated by procedures validated to remove residual virus, protein, and nucleic acid. If changes in the manufacturing process are introduced, it may be necessary to reassess validation.

Another area of concern is retrovirus-like particles which are frequently visible by electron microscopy, but are not infectious in the S+L- or XC plaque assays. Several varieties of such endogenous particles have been defined. Intracisternal A Particles (IAP) are frequently found in murine myelomas, but they are apparently defective in their membrane protein and do not hybridize with nucleic acid probes from infectious murine retroviruses. Type A particles are not currently thought to present a major risk.

Retrovirus-like particles are also found in CHO cells. The CHO cell particle is also apparently noninfectious and defective in its membrane protein. The CHO cell particle is present in much smaller numbers than the mouse IAP, has an intracytoplasmic pericentriolar location rather than intracisternal, and exhibits very low levels of inducibility with a variety of chemical and biological agents.

A related particle, termed a Type R particle because of the spoke-like structures radiating from the core, is found in the cisternae of the endoplasmic reticulum of BHK cells. Its genome has been cloned and sequenced and exhibits weak cross-hybridization with nucleic acid from CHO cell retroviruses. No reverse transcriptase activity has been demonstrated and it is also apparently noninfectious. Interestingly, this particle, unlike the CHO cell particle, is inducible with azacytidine.

Based on the level of contamination determined in post-production cells, it should be possible to validate that the purification process can remove and inactivate several logs more retrovirus than is thought to be present in the unpurified bulk.

A murine cell substrate is considered positive for infectious retrovirus if either the MCB, MWCB (manufacturer's working cell bank), or extended (post-production) cell bank tests positive in one or more of the following tests: reverse transcriptase assay, S+ L- assay, or XC plaque assay. Large numbers of extracellular or budding particles evident on EM may also suggest functional retrovirus is present.

Long term experience with other biological products unexpectedly contaminated with animal retroviruses (e.g., inactivated influenza vaccine and live yellow fever vaccine which were contaminated with avian leukosis virus) has suggested that the potential for human problems has been very small to date. The safety records are impressive, and influenza vaccines are still produced using eggs from ordinary chicken flocks. However, the risks of using contaminated cell lines for production needs to be carefully considered and the margin of safety demonstrated during the validation of the purification process readily demonstrated.

When considering the acceptability of a cell line for production, the following information is useful: the species of origin of the cell substrate, identification of the probable source of the contamination, the degree of characterization of the contaminant, the host range of the contaminant (ecotropic or xenotropic; simian or human), the level of viral contamination in post-production cells, the method of cell propagation (e.g., tissue culture vs ascites), the method and validation of product processing for both inactivation and removal steps, the sensitivity of testing planned as process controls, the intended use (patient population, dose, and frequency) and the availability of alternative therapies.

<u>Conclusion</u> The U. S. Food and Drug Administration is continually developing and analyzing new technology and good science to apply to the manufacture of recombinant products, monoclonal antibodies and the cells used for their production. This information is shared with all sponsors through the use of frequently revised "Points to Consider" and the encouragement of "pre-IND" meetings. All of these areas are based on sound, scientifically valid evaluations and discussions. When good science is used in the thoughtful development of these products from the very earliest stages, an expedited process to bring safe and effective biological products to the public should be expected.

Author Index

Subject Index

432